Megacraters on Earth

Introduction to Cosmic Geology

How exploding comets
- *sculpted the surface*
- *killed almost everything and*
- *set up humanity and civilization.*

By

James M. McCampbell

ISBN: 1-4107-7707-3 (e-book)
ISBN: 1-4107-7708-1 (Paperback)

Library of Congress Control Number: 2003096216

This book is printed on acid free paper.

Printed in the United States of America
Bloomington, IN

1stBooks - rev. 02/02/04

To

Noma Jean Kennedy

who loves Jesus, football, gardening,

small animals, and large

ideas.

Previous books by the author

UFOlogy, New insights from science and common sense, Jaymac-Hollmann, 1973. Reprinted as UFOlogy, A major breakthrough in the Scientific Understanding of Unidentified Flying Objects, Celestial Arts, 1976.

UFOlogy II, Some final answers, 3½-inch disk, Digital Books, 1997.

Apollo's Message, Growth of civilization under leadership of the gods, unpublished.

Acknowledgements

Questions raised in research almost always encounter complications far exceeding one's expectations. Whole fields of science are opened beyond the author's kin. The achievements of previous research, sophisticated theories, complex apparatus, and the volume of detail is beyond normal belief. A general practitioner of science can not possibly absorb all the information in the great range of specialties that beckon his attention. The most optimistic hopes are that a) every reader can grasp the essences of obscure subjects, and b) that scientists in every discipline find this exposition acceptable. That challenge has guided the present effort.

Sources consulted in writing this book are too numerous to cite individually. Otherwise, the text would be too cluttered with footnotes or other inserts. Any single source may have provided information used in many places. Or many sources may have contributed to a single point. To circumvent these problems, selected sources have been assembled in the **Bibliography.** While lengthy, that list represents only about 20 per cent of the sources consulting in the research. It is retained as the most comprehensive bibliography for megacraters because there are no others. In a few essential cases, specific references appear in the text as the last name of the principal author followed by the year of publication in parentheses. The **Bibliography** provides details. In addition, sources of some data in tables are acknowledged in footnotes. Apologies are extended to the thousands of authors not mentioned.

The author is especially indebted to Stanford University for its policy of public access to its technical libraries, such as, the Branner Earth Sciences Library. Thanks, also, to the diligent Research Librarians of San Mateo County, California, who were always effective in obtaining volumes from distant collections through the Interlibrary Loan Program.

Personal gratitude is owed to Dr. Robin M. Canup who generously provided reports on massive cratering research at Southwest Research Institute.

Also a friend of several decades, D. William Berte, never failed to help unravel knotty problems with computer software and furnish encouragement along the way.

Bonnie Saligo, Director of Schoolhouse Technology provided much support and issued permission to base certain maps upon those generated in software, Schoolhouse Map Factory.

Many thanks to all contributors to a difficult project.

Preface

The last half of the 20th century saw much excitement in geology that is well-described in standard textbooks. Let it suffice here to introduce some of the more recent concepts along with a few key words that may not be familiar. New ideas spawned vigorous debates while continued research and special projects of international scale provided more data to feed the debates. Eventually, most issues were settled to the satisfaction of a large majority of the experts.

Historical Geology.

After a few billion years of cooling, the primeval earth developed a crust of rock like thin ice on a pond. After much jostling about, all land areas became joined in a single, great, land mass known as Pangea. Great cracks appearing in the land slowly grew wider as they were flooded by the sea. Resulting fragments of Pangea became continents, as we know them today, separated by oceans with the original cracks running down their middles. As the sea floors spread apart, molten rock oozed up through the cracks, solidifying into new floor. Obviously, the seabed close to the mid-oceanic ridges is fresher than that farther away. Thus one speaks about the age of the ocean floors at particular places where the oldest is adjacent to the retreating continents. Immediately offshore, however, thick, alluvial deposits washed down from the land to form continental shelves that have covered the oldest rock. The rate of continental separation is not the same around the globe and varies in time at any given location. More rapid separation produces thin crust whereas very slow motion allows magma to build up mountain ranges along the ridges. Forces driving continental drift must derive from rotation of the earth and not from pressure from the magma as some would say. An instructor in elementary physics once told the author that "…you can't push with a string." Neither can an unrestrained fluid drive continents apart.

When a crack appears, the expected motion may not start immediately. Nothing can move until some geological process provides some place for the ocean floors to go. This situation resembles a logjam on a river: the logs are ready to move but are prevented by some obstruction. The geological process needed to get things moving is the creation of *subduction trenches* where some surplus floor can dive under another, stationary portion. This process is commonly associated with linear or curved island chains and continental margins.

The entire surface of the earth is paved with a dozen or so *tectonic plates* that are associated with continents or oceans. Some smaller fragments exist. These plates jostle against each other or dive below a neighbor with extreme lethargy and unbelievable power. Plates sliding past each other produce earthquakes when temporary blockages let go. Earthquakes are also produced between the top surface of a diving plate as it rubs against the underside of a stationary plate.

Several tens of miles shoreward from a subduction trench, the diving plate reaches a zone of molten rock. The junction of overlapping plates along with fractures in the rock provides pathways for lava to escape to the surface. Continued flows over long periods produce volcanoes. Certain conditions allow the lava to spread out over vast areas of the land creating lava plateaus. These regions are also known as *Large Igneous Provinces, flood lavas,* and *traps* (after the Middle Dutch *trappe,* meaning *stair* as seen on the slopes of eroded canyons).

If a volcano should grow underwater, its lengthy lifetime produces some interesting patterns. Eruption on the sea floor produces a mountain as lava flows upward through a vent before spilling over the rim and running down the slopes. The top may eventually break the surface to form a volcanic island that continues to grow to impressive heights sometimes blanked by snow. Several elements of nature combine to limit the growth or reduce the height. Rain, reduction of lava supply, and sinking of the ocean floor under the island weight will reduce the volcano to sea level. Then coral grows in the shallow water on top of the flattened peak. A ring of coral islands may remain exposed on the edges of a shallow lagoon. A visitor can usually see the neighboring islands but frequently can not see across the lagoons because they are too large. For example, the lagoon within the ring of coral islands at Kwajalein in the South Pacific cover 655 sq mi. Thickness of coral can increase as the mountain continues to sink. But Charles Darwin observed that coral could not grow in the weak sunlight below a certain depth. If the mountain sinks faster than the coral can compensate by growth then the whole ensemble will disappear into the depths.

These stages in the life cycle of underwater volcanoes are illustrated in Fig. P.1. Island Life Cycle, where each stage is identified with their common names of *seamount, volcanic island, coral atoll, and guyot* (pronounced GEE-yoh).

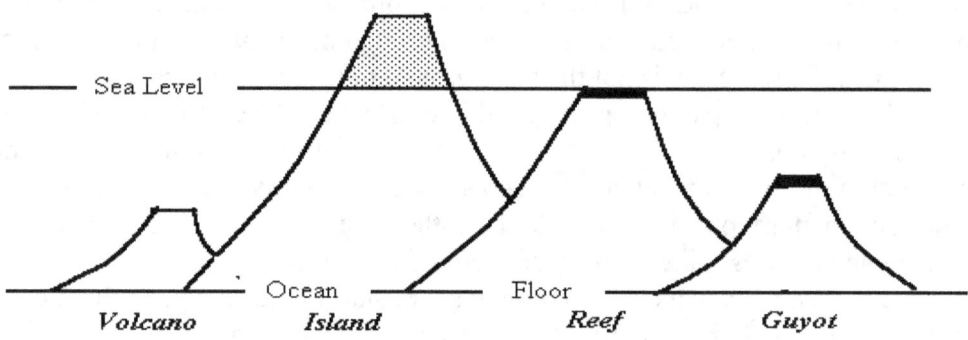

Fig. P.1. Island Life Cycle

Hotspots.

Some disturbance deep within the earth may cause a plume of magma to rise through more viscous material approaching the under side of the *lithosphere,* that is, the crust. It rises on a very narrow stem through which more fluid is pumped by high pressure from the depths. The plume might grow to a diameter of 300 mi. That reservoir of molten rock, known as a hotspot, remains stationary as the crust drifts slowly by overhead. If a pathway from the hotspot through the lithosphere is available, a volcanic island can be created. After a while, however, the island moves too far away so its lava source if cut off. But the flow continues at the original location, building a new island whose ultimate destiny is the same as it predecessor. Thus, a chain of volcanic islands is created as the crust drifts ever so slowly over that hotspot. The Hawaiian islands are the classic example. Kauai, Oahu, Maui, and Hawaii were created in that order by floating over a hotspot. Even today, lava has found a new opening to the southeast of the main island and is building a sea mount named Loihi. In the distant future, the lava supply to Hawaii itself will be shut off as the new island is born, matures, and dies. Native mythology of the Polynesians concerning Pele, the goddess of fire, had it right. She was born on Kauai, moved to the younger islands, and then settled on Hawaii where she continues to put on a fine display.

Cosmic Bombardment.

Earth has shared a common environment for billions of years with the moon, Mars, and Venus so it must have experienced a similar history of cosmic bombardment. Impact craters on the moon and Mars have been perfectly preserved in the near vacuum of their atmospheres. They are not subject to loss by erosion nor burial under miles of sediment. The surface of earth, on the other hand, undergoes vast changes as rain washes away the rims and fills the bowls with sediment. Many must remain undiscovered at the bottom of the oceans and some have been lost forever in the process of subduction.

It has been well established that the pattern of cratering on earth, moon, Mars, and Venus are essentially identical. Smaller craters are progressively more numerous that their big brothers. Naturally, little craters on earth are most easily erased and, as expected, are less numerous than on the other bodies.

On the high end of the scale, the moon displays vast, circular structures outlined by mountain ranges and filled with dark lava. Most astronomers believe that they were produced billions of years ago when the barrage of massive projectiles ceased but the indelible record on the surface speaks otherwise. The density of small craters on the lava plains is much less than beyond their boundaries so the lava flows must have covered the old craters, presented a new slate upon solidifying, then recorded impacts of more recent vintage.

Craters On Earth.

The very large circles, known as multi-ring structures, have been studied intensely but strangely neglected. It has been claimed that no such scars were ever created on earth and never will be. Also, they have been neglected in statistical analyses perhaps because they are too rare to be meaningful.

Large impact craters known on earth now number about 150 with the largest being near the tip of the Yucatan Peninsula in Mexico. It is named *Chicxulub* after a nearby town and, at a diameter of 200 km, is widely accepted as the dinosaur killer.

The present work, in contrast, is based upon 200 multi-ring structures larger than Chicxulub ranging in size up to 3640 mi in diameter! Take note. These babies reach global proportions. A few of them had been suspected by previous authors who had noted distinctive, circular forms of some geographical features. For the present purpose, it became necessary to assign names to the new discoveries that would provide clues to their locations. Such designations are used throughout the text and appear in bold-italics.

Esoteric Language.

As in all professions, a special vocabulary has been developed by geologist that is most useful to them but incomprehensible to outsiders. Fossils had shown the history of earth falling naturally into distinct periods of time. In some zones, the remains of certain creatures were prevalent whereas, in higher levels, they had thinned out markedly or disappeared altogether. The new layers also disclosed new classes of creatures. So discussions in geology focused upon ages of invertebrates, insects, fish, reptiles, dinosaurs, and mammals. Corresponding changes in plant life were also observed. Major geological eras were dubbed *Tertiary, Createous, Jurassic, Triassic, Permian, Carboniferous, Devonian, Silurian, Ordovician,* and *Cambrian.* These periods, in turn, were broken into dozens of subdivisions with equally strange names. It is unlikely that any of the above will be widely recognized except the one made famous by a movie. For members of the club, these names indicate specific times in earth's history but they must be ignored and avoided here. Instead of using these technical terms, a specific time in the remote past will be designated as "umpteen million years ago" where *umpteen* is some number and million years ago is abbreviated to My with ago being implied. For example, everyone knows that the dinosaurs vanished about 65 My.

Organization.

This book is laid out to provide a condensed version of the whole story at the very beginning. With comic-book ease, the reader can absorb a few paragraphs of explanation while scanning an illustration without even turning a page. Such a format pushes details toward the back where they may be studied at leisure. Curious readers will find them helpful in understanding the locations of craters and the extent of their damage. Specialists in geology will encounter a smorgasbord of new information that will probably challenge many prevailing opinions.

Behind these details, one will follow an exploration into related questions such as, emplacement of minerals, the tilt of earth's axis, earth's position in orbit around the sun, and orbits of comets.

While the demise of dinosaurs accompanied the most recent, major cataclysm on earth, many other episodes have been recorded in the fossil record. A major section is devoted to the question if all extinctions were caused by cosmic collisions and attempts to identify which ones might have been responsible.

The database for this study is then searched for clues to distribution of crater sizes, comparison with other planets and satellites, distribution by latitude, arrangement on a time scale, and collisions by cometary chains.

Finally, new concepts requiring some knowledge of physics, astronomy, and mathematical analyses are treated in appendices. These may well be skipped by many readers but deserve the attention of scientists. Everyone is invited to probe into this presentation as far as individually possible and find exhilaration at every step.

Contents

Chapter 1. Sculpting The Earth.

> *"Man's mind, stretched to a new idea never goes back to its original dimension."*
> - Oliver Wendell Holmes

This chapter will lead the reader, step-by-step, into a broad appreciation of an incredibly complex subject. Because the discussion encompasses the entire globe, hundreds of place names must be used. Most will not be recognized. A gazetteer could be assembled at the back of the book but it would be unwieldy. Standard gazetteers can be found in libraries but they are not always available for checkout. The most convenient way to identify strange mountains, rivers, deserts, oceans, seas, islands, volcanoes, deep trenches, etc is by consulting an atlas. Most homes have one while others may have lain hidden for years in the family encyclopedia. In any event, the most rewarding and convincing study of the descriptions will be achieved by personally verifying the geography that is described. Clues that reveal the impact structures and their affiliated damage to earth's crust will then become clear.

The term, *crater*, will be applied loosely to cosmic scars even if they are actually multi-ring structures. The truth is that nearly all such scars were produced by explosions at high altitude. Only a few craters in the present collection were produced by objects colliding with the surface and digging massive holes in the ground. Even they are substantially larger than a famous example on the Yucatan Peninsula, ***Chicxulub***, that is universally blamed for destruction of the dinosaurs 65 million years ago. Another doubtful belief is that ***Chicxulub*** is the largest crater on earth whereas, in the present study, it is almost the smallest. The original intent was to consider none smaller but some showed up with such significant roles that they had to be included. Many small ones were too numerous to allow treatment in a limited book. They had to be ignored. Also discarded were several craters somewhat larger than ***Chicxulub***. They have been noted but not formally identified, cataloged, described, or included in statistical analyses. These omissions should have negligible effect upon the results. Much work is still required to develop a comprehensive picture of earthly bombardment.

Since the complete subject at hand can not be understood in a single glance, some tough choices had to be made. A linear and logical pathway through the mass of data would have been convenient but none could be found. Some starting point on the geological time-scale had to be selected. The breakup of Pangea, when all land mass on earth was contiguous, seemed to be an appropriate beginning although some major events occurred 100 million years earlier. The breakup could be followed continent by continent where, in some cases, more than a single map was required. Also, portions of the world's oceans had to be included to account for evidence of impacts that are now flooded with salt water or nearly so.

The following pages contain 22 snapshots of earth as impacts changed the landscape on a massive scale. Descriptions are summarized by text on the left page while the subject is illustrated by a map on the right page. This plan presents an overview of earth's bombardment by cosmic missiles that can be grasped in a single reading. Explorers can don their pith helmets for an exciting safari beginning of the next page.

Fracturing Pangea

Long ago when Pangea alone ruled the earth, a cosmic body of some sort was attracted by gravity toward the sun. Before plunging into its surface or whizzing off into an open or closed orbit, it felt the urging of another gravitational field. Earth beckoned it away from its original path into a collision course. Its explosion at high altitude near the present Timbuktu in North Africa created a giant, multi-ring structure 1930 mi in diameter. The outlying, circular fault from that event carved the great bulge seen today protruding into the Atlantic Ocean as the western Sahara Desert. Radiating from ground zero was a linear, rift system that fractured Pangea along the straight line of Africa's west coast.

Three, previous explosions participated in this drama. As the rift lengthened, it encountered the edge of a crater whose evidence at this point is the concave, coastline of Angola. Rather than rigidly adhering to a straight course, it proceeded along a curved path following a trail that had already been weakened by a crater rim that remains. No portion of this crater remained on the continent. But the author has surveyed the huge, nearly vertical escarpment behind the coastal plain of Angola from a small airplane.

Next, an ancient crater named *Angola* was circumvented. It had been associated with a massive, lava flow known as Parana-Etendeka Trap. Most of the latter was sheared off with the Parana portion moving westward and only a small fraction, Etendeka, remaining in Africa.

Finally, a crater 1180 mi in diameter, called *Kalahari*, had carved the southeastern corner of Africa that extended westward beyond the rift as shown in Fig. I.1. Beginning The Breakup. That portion was simply sliced away to join other remnants that eventually drifted about 4300 mi to the present location of South America.

Compounding damage to the crust from three rift systems *(Timbuktu, Angola, and Kalahari)* probably triggered the eruption of Parana-Etendeka. Starting 133 My and lasting about 1 million years, the lava spread out over 1.25 million sq mi (Renne 1992). Opening of the Atlantic Ocean was delayed until nature could provide for disposal of surplus crust. The traditional cause of Pangea's fragmentation has been continental movement over the Tristan de Cunha hotspot.

A legitimate question may be raised about the reason, if any, that the rift system from *Timbuktu* elected to choose its particular course instead of any number of other possibilities. Such systems are, indeed, common features of the largest craters but their directions vary widely. A value of the angle can be established by noting the angle from true north as an azimuth to the right or a negative value to the left. In this case, the angle is - 20.6°. Some noteworthy points arise from this examination of a major impact. In a single stroke it sculpted the west coast of Africa with the round bulge in the north and the linear coast in the south. The time of the event can be established from radioactive decay of frozen lava and core samples from the ocean floor. Interactions with earlier events are marked by snaking around some older craters or slicing through others. Earth's crust is most severely damaged when effects of more than one explosion are compounded. Finally, some intelligence may be deciphered from the geometry of linear rifts. This question is probed thoroughly in a later chapter.

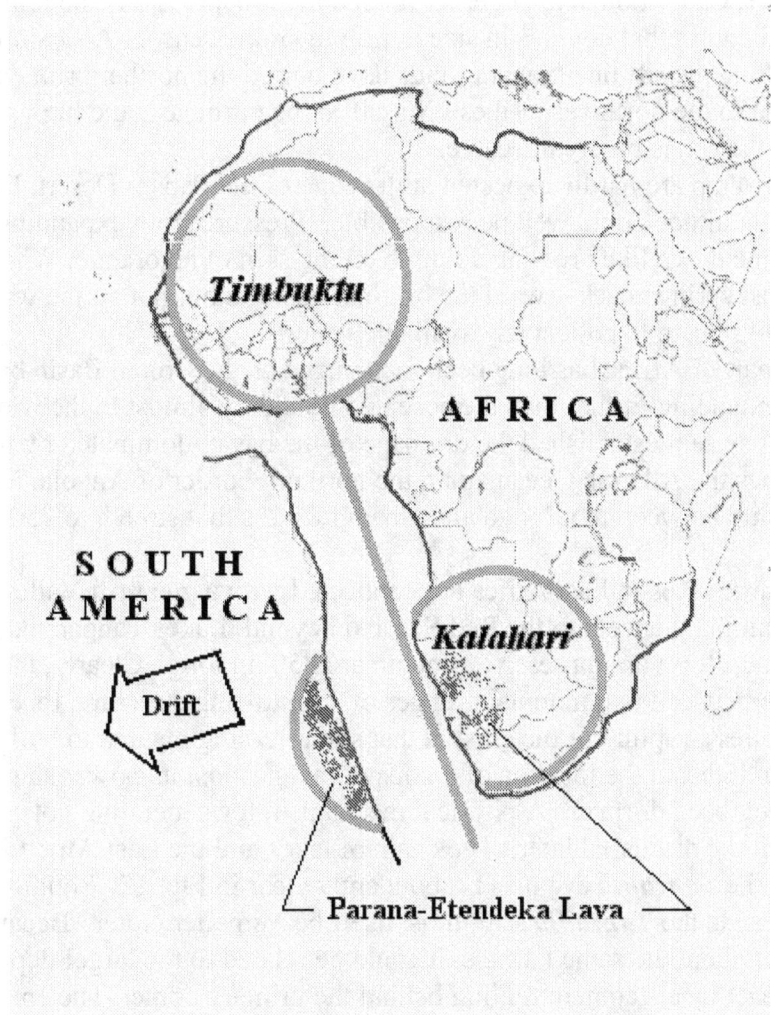

Fig. 1.1. Beginning The Breakup.

Congo Pals

A straight chain of islands at the "arm pit" of Africa named Bioko, Sao Tome and Principe, and Annobon, project to the shore at Mount Cameroon, the highest peak in Cameroon. Such geography suggests a linear, rift system radiating from an impact center that opened up channels for the release of magma. If so, the center must be on that line along with substantiating evidence. Suitable conditions are found at Lake Chad whose shores are shared with Niger, Cameroon, and Nigeria. It occupies a depression containing a shallow lake about 60 mi across that doubles in size during the rainy season. A swamp extends another 60 mi along the rift line. Sahara sands have buried the northern rim except for Massif de Tarazit to the northwest, Tibesti Massif to the northeast, and high plateaus to the east. Elsewhere the rim is barely traceable.

A lake and swamp are hardly expected at the edge of the Sahara Desert. But a powerful explosion at high altitude could well be responsible. Pressure from expanding gases would push the landscape down like pressing a thumb on the skin of an orange. When the force is releaved, the crust springs back toward the original position but not all the way. A shallow depression would remain to collect any available water.

The central part of Africa has long been recognized as the Congo Basin but satellite photos show it extending farther than the accepted boundary almost to the width of the continent. Its rift system established the drainage of the basin, dominated by the Congo River, that reaches the Atlantic Ocean along the northern border of Angola. Near its center lies another swamp between the Congo and Ubangl Rivers that stretch to 250 mi parallel to the rivers.

A chain of large lakes in East Africa are connected with a great rift Valley that runs northward to Ethiopia thence into the Red Sea and beyond. Lakes Tanganyika, Kivu, Edward, and Albert form an impressive, smooth arc 850 mi long. Clearly, they lie in a circular rift zone between two mountain ranges of a multi-ring structure. Its center at Mt. Kilimanjaro implies a rupture of the crust at that site releasing magma to build that famous peak. It may also indicate the location of a hotspot that has lost its power since the volcanic mountain has long been dormant. A physical mechanism for generating hotspots below impact points will be discussed later. The chain of lakes and the East Africa Rift Valley, were created by the *Tanzania* event as is abundantly clear in Fig. 1.2. South-Central Africa. Lake Victoria, inside the *Tanzania* scar, must itself be a smaller crater. Because it lies west of Kilimanjaro at about the same latitude, it could be related to the larger depression, that is, a subsidiary impact by a fragment trailing behind the primary comet. The epicenter of *Victoria* lies 1.85° north of *Tanzania* but alignments along latitudes need not be perfect. If a real daughter, then the difference in longitude of 4.42° means that it must have struck about 18 minutes after the primary and, traveling in space behind the primary at 40,000 mph, it would have been lagging behind about 12,000 mi.

Another rift system along the coast of the Indian Ocean forms the eastern borders of Tanzania and Mozambique. Running almost due north to south, it gently bent to the west around a small crater that has been named *Mozambique 1*. Farther south a similar detour around *Mozambique 2* completed the surgery that excised the large island of Madagascar off the African continent.

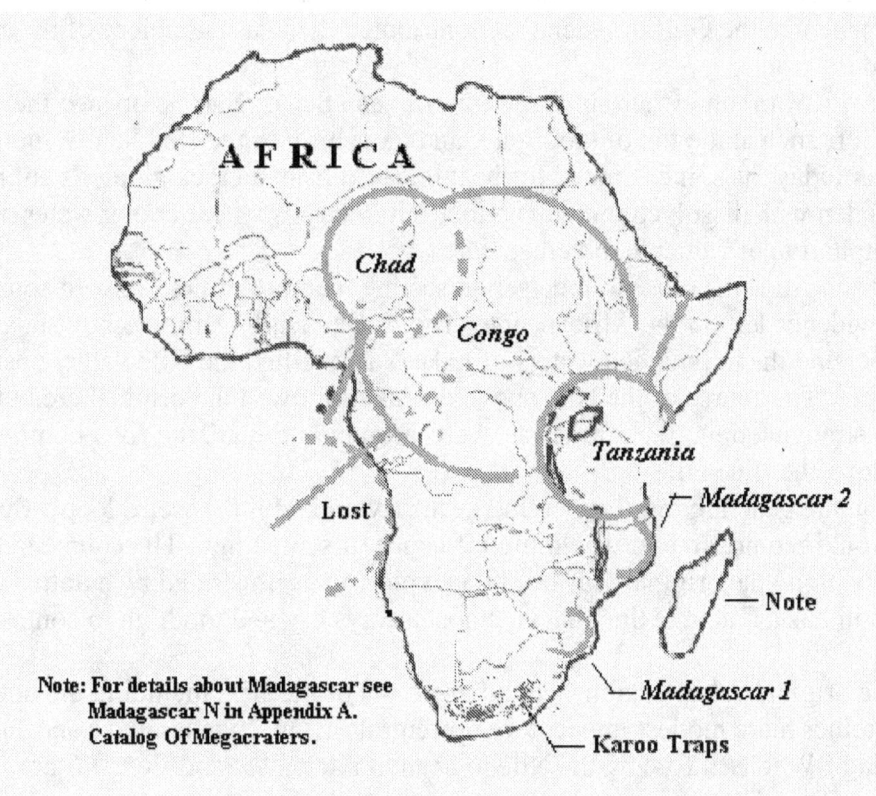

Note: For details about Madagascar see
Madagascar N in Appendix A.
Catalog Of Megacraters.

Fig. 1.2. South-Central Africa.

Splitting Africa - Arabia.

Two cosmic events dominated the region when Africa and Arabia were joined. One was centered a short distance west of Addis Ababa, the modern capital of Ethiopia, while the other was very near Jiddah, Saudi Arabia. Both left ample evidence of their local destruction but, being about 400 mi in diameter, are easily overlooked. In addition, two major rift systems sliced through the region creating the Red Sea and Gulf of Aden.

Speckled areas in Fig. 1.3. Afar And Environs, designate a lava plateau covering 200,000 sq mi up to 7000 ft thick. Ruptures supplying this magma were giant volcanoes in Simen Mountains National Park just west of Addis Ababa. All that remains today are their bare cores soaring high above the landscape with the mountains removed by erosion. A linear rift opened up the Gulf of Aden that contributed to isolating a piece of the crater and its lava field in Arabia.

Another rift from somewhere in the Mediterranean Sea or Europe opened the Gulf of Suez, the left branch at the top of the figure, and gave birth to the Red Sea. While still getting wider today, its spine is noted for heat released from a crack along its entire length. Hot water laden with dissolved minerals bubbles into basins where cooler water of the sea causes precipitation of valuable mineral salts.

This second rift came close to precisely bisecting the northern crater with some surprising evidence left on the African side. The Aswan High Dam was built in a gorge left by the impact and the impounded waters of Lake Nasser filled the Nile valley upstream for nearly 350 mi. The course of the Nile above the lake followed the original arc until executing a strange loop to the northeast. At Khartoum in Sudan, the river continues upstream along the Blue Nile.

Why would a generally northerly trend of the river exhibit the weird loop? Such a large deviation would require an overwhelming influence of some kind. The course in that region was probably along the original scar from the explosion as illustrated by a dotted curve. That alignment still has a drainage line that is almost always dry and too high to connect with the river.

After the original valley established the *great arc* of the Nile including the dotted segment, another more modest impact tore up central Sudan. It generated a short mountain range running SW to NE, forcing the Nile to abandon its valley and flow NE around the end of the range before finding its way back to the undisturbed channel.

As in many other instances, Khartoum was founded at the confluence of two rivers the better to profit from commerce. The White Nile arising far to the south and the Blue Nile converge there. Two additional craters of modest size forced the White Nile to snake around them before continuing to Khartoum. These details are too fine to illustrate but will be fully described later.

The Afar Triangle, as it is sometimes called, is a great, inhospitable desert that must have been formed after the Red Sea and Gulf of Aden had reached some maturity. It is situated at the north end of the East Africa Rift Valley that in Ethiopia hosts a chain of lakes. Overflow and linking of this chain into a large river would be inevitable whenever sufficient water were available. Then water would have gushed down the East Africa Rift *River* to form an expansive delta. One of the branches of the ancient river still flows underground as the only source of water for Djibouti, capital of the small country of Djibouti, at the end of the little bay in Afar shown in the figure.

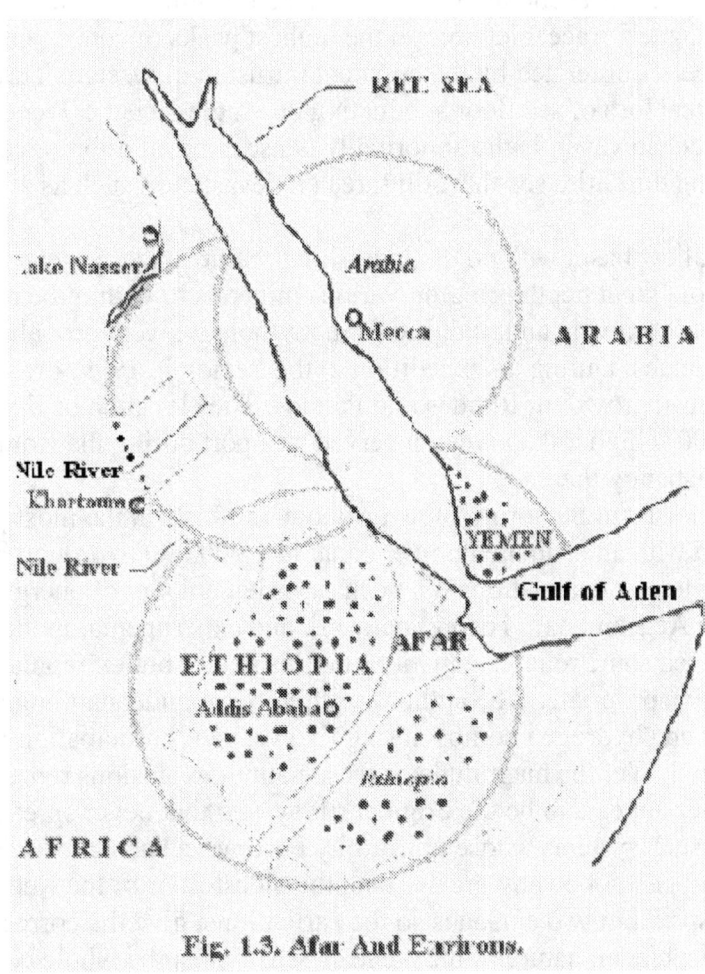

Fig. 1.3. Afar And Environs.

Surprise In Europe.

At the lower right in Fig. I.4. Europe And Middle East, the Gulf of Suez points to the culprit as a finger of accusation in a courtroom. The Aegean Sea is revealed as the point of an explosion that separated Africa and Asia by opening up the Red Sea. Could the Aegean Sea really be the remnant of a cosmic impact? Supporting evidence abounds. The most obvious clue is the arc of major islands along the southern border, Kithra, Crete, Karpathos, and Rhodes. Careful examination of the peninsulas on the Peloponnesos together with seamounts and underwater ridges clearly show the elements of a multi-ring structure. Other evidence points toward a true contact with the surface resulting in excavation of a huge crater. The rim is vaguely traceable through the highest peaks of central Greece. The northeast rim has been obliterated by the apparent intrusion of western Turkey about 70 mi into the sea or the real loss of sea floor subducting down the Helenic Trench. The most compelling evidence, however, is the abnormally dense concentration of earthquake epicenters encircling the entire sea that still threaten devastation such as at Izmir, Turkey in 1999.

Instead of a shallow basin with no distinguishing features, the Aegean contains many islands that rose from great depth bringing various minerals to within the reach of humanity. These islands are still growing and, judging from mythology, were probably observed at birth by modern humans. During an expedition to this region in 1999, the author noted local traditions of continued growth of Rhodes and Patmos. The elevation of Ephesus was estimated to be 1000 ft higher than when it served as a port during the Roman times when Paul preached Christianity there.

A famous Greek island, Santorini, blew up about 1500 BC in the most powerful volcanic explosion of record with an estimated power equal to 4,000,000 hydrogen bombs. Yet it still stands as a popular tourist destination. Who can say what amount of energy was required to excavate the whole Aegean Sea? Theoretical work and experimental models in the post-nuclear age developed some reliable equations on this matter but extrapolation to sizes being considered seems inappropriate. The author acquired some understanding of the damage inflicted by atomic and hydrogen bombs during five years of participating in their testing in Nevada and Eniwetok. Yet the magnitude of these natural explosions remains inconceivable.

The Gulf of Suez appears to be the original rift system that never spread. Its width of 30 mi is typical of all such systems whose uniformity becomes a subject of scientific inquiry. The age of the event can not be any greater than the oldest floor of the Red Sea, about 30 My. The Red Sea spread in two episodes so the earlier must give the correct value.

In contrast to the hidden nature of the Aegean Sea, almost the whole country of Hungary lies in the well-known Hungarian Basin. It must also be treated as an impact scar.

A somewhat larger event separated the British Isles from Europe proper by creating the North Sea where oil resources are indicated by black blobs. Oil is also associated with the conjunction of two other craters that formed the Caspian Sea. *Caspian N* as a shallow basin is most likely a multi-ring structure whereas *Caspian S* is a deep, true crater. The geography of Scandinavia strongly suggests an impact whose rift formed the Gulf of Bosnia that separated Norway and Sweden from Finland.

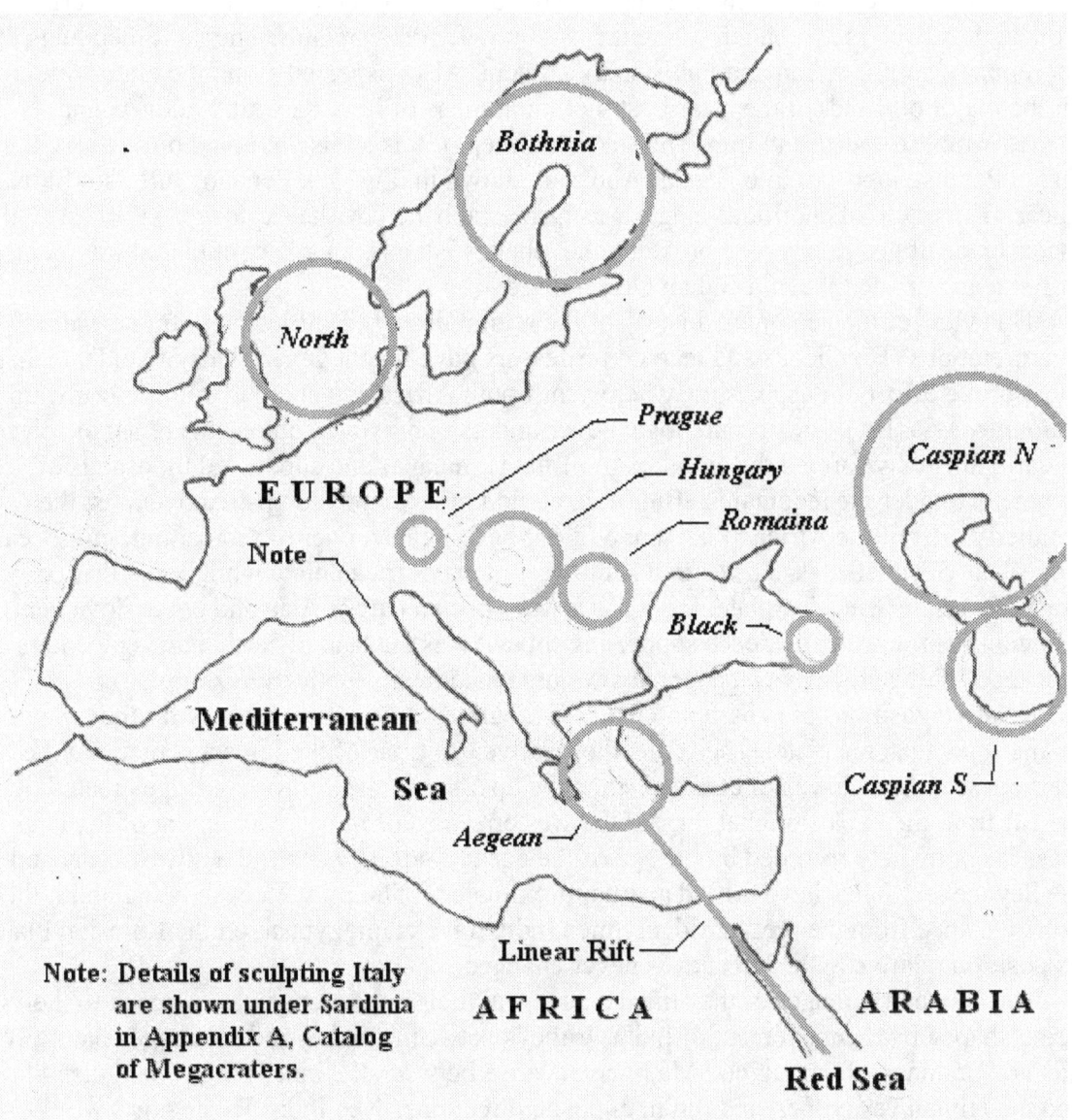

Fig. 1.4. Europe And Middle East.

Note: Details of sculpting Italy
are shown under Sardinia
in Appendix A, Catalog
of Megacraters.

9

Southern Asia.

A major explosion in the distant past had a profound influence upon the landscape from Turkey to India by opening the Persian Gulf. Striking about 250 mi northeast of Mashhad, Iran, the missile created an arc so perfectly circular that radii to various points along it differ by less than the accuracy of measurement. The fault controlled the course of the Tigris River from its headwaters to Abadan where it flows into the Persian Gulf. The fault then hugs the shoreline of southern Iran and Pakistan to Karachi. Also dispersed along the crater rim are all the major oil fields of the Near East including those of Iraq, Kuwait, Saudi Arabia, United Arab Emirates, and Iran. The smooth sweep of this arc is disrupted only where the Oman Peninsula pushes into Bandar Abbas as shown in. Fig. 1.5. Persian Gulf And India. A linear, rift system of enormous length was also established by the explosion. It sculpted a major piece of geography as it broke the Seychelles Islands, a large continental fragment, off of India and created the mid-Indian Ocean ridge.

This view of the geological history of the region directly conflicts with traditional interpretations. Text books and learned professors have taught generations of students that India broke off from Pangea somewhere near South Africa and charged boldly northward to slam into Asia. By some obscure logic never understood by the author, this renegade piece of earth allegedly formed the circular arc of the Himalayas and subducted thousands of kilometers under the mountains. But India could not have moved upstream against the southerly drift of the African Plate nor could it have jumped over the spreading, mid-ocean ridge. The Seychelles Islands drifted southwest on the African plate while India drifted northeast on the Eurasian plate. India has always been a part of Asia and never dove beneath that continent. Further evidence supporting this view is the lack of earthquake epicenters and volcanoes in the Himalayas that are always associated with subduction zones.

Relative positions of geographic features are not altered no matter how far they may ride on the *same* tectonic plate. Therefore, the relative positions of the impact center and the west coast of India have not changed since the opening of the mid-ocean ridge created by the rift from the explosion that formed the western coast of India. Movement of the land mass is completely recorded by the age of the ocean floor between India's west coast and the Seychelles. They have drifted apart approximately 1500 mi with each being about the same distance from the present ridge. Since India traveled piggyback on the Eurasian Plate, its position relative to the Himalayas never changed.

The concentric ranges of the Himalayas reveal their impact origin as opposed to their being shaped by the movement of India. Valleys between and beyond the ranges naturally dictated drainage of the region. Major rivers arise between the inner range, Gangdise Shan, and the Himalayas proper, and run in opposite directions. The Indus River skirts westward around the end of the chain near Islamabad then flows directly to the sea near Karachi. The Brahmaputra River flows east in the mountain valley about 750 mi before turning back to the west around the eastern end of the Himalayas in Arunachal Pradesh. Then it meets the Ganges River arcing in front of the Himalayas. From a confluence in Bangladesh, the waters reach the Bay of Bengal through a delta nearly 300 mi wide.

One of the largest lava flows in the world, known as the Deccan Traps, found in central India, consists of 500,000 cubic miles of lava.

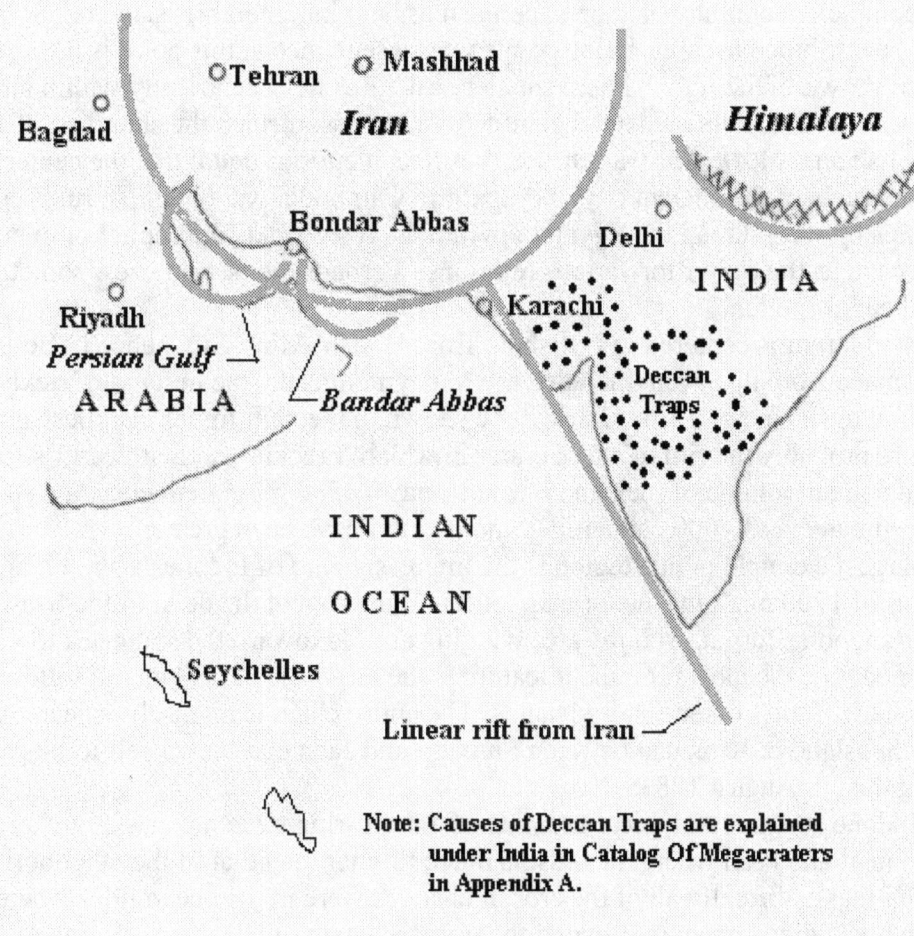

Fig. 1. 5. Persian Gulf And India.

11

Southeast Asia.

Sometime after a circular rift set the stage for evolution of the Himalayas, another crisis struck southern China that obliterated all evidence along the eastern sector of its predecessor. Fig. 1.6. Vietnam Cluster illustrates this point and shows how the resulting structure in China, 1820 mi in diameter, determined the east coast from the vicinity of Beijing in the north to Hong Kong in the south, a distance of about 1400 mi. Any clear evidence of the structure in western China has been erased by the growth of high and distorted mountain ranges. However, a central depression, as expected, is marked by Lake Dongting Hu. A chain of many lakes toward the northeast and the gorge of the Yantze river record the radial rift to the sea near Shanghai.

The complex geography of shores, peninsulas, and innumerable islands in Southeast Asia can be attributed to three major events whose sequence at this point is uncertain. This trio is also shown in the figure where smaller structures lie eccentrically within larger ones.

The baby, still an unprecedented giant 570 mi across, formed the shoreline of South Vietnam. Its name of *Angkor* was chosen to reflect a curious detail that the center lies quite close to the ruins of the ancient city of Angkor in Cambodia where temple ruins contain more stones by weight than the Egyptian pyramids. It served as the capital of many powerful kingdoms and as the center for various religions. Perhaps the people knew something that has long been forgotten.

The next larger is centered just offshore from Ho Chi Minh City hence its descriptive name. Damage from this event included the Malay Peninsula, the mountain backbone of Borneo, and the island of Taiwan. It will be shown in the next figure that these geographical features are probably the rim of a true crater in which direct impact with earth's surface excavated a great volume of rock and created a basin 1290 mi in diameter. The spoil can be roughly estimated as 30,000 cubic miles and it could have been greater.

The largest example in this region is the impressive arc of Indonesia stretching 3800 mi on a radius of 1700 mi. On a better map, one sees the root at the delta of the Irrawaddy River at Rangoon, Burma, with the arc swinging around toward the southeast to Timor island. Of course, the most prominent feature is the large island of Sumatra followed by Java then the chain of the Lesser Sundra Islands. The entire chain is intensely volcanic and includes the island of Krakatao between Sumatra and Java that blew itself to bits and globe-encircling dust in August 1883.

The Indonesian arc is a typical example of a multi-ring structure caused by an aerial explosion near an ocean where the crust is much thinner than that of the neighboring continental mass. Virtually all of the crustal changes were in the ocean floor whereas little or no evidence remained onshore. Beyond the primary chain of islands lies a ring of smaller islands that are the exposed ridges and peaks of a second range of mountains. These include the Adaman Islands in the north followed by the Nicobar Islands, many large islands off shore from Sumatra, and some stragglers at the eastern end. The entire structure not only fractured the ocean crust but thrust its rim upward and outward over the outlying region. The weight of the overthrusted portion of crust forced the underlying portion downward into the molten rock below. The final configuration is the newly created, deep ocean trench.

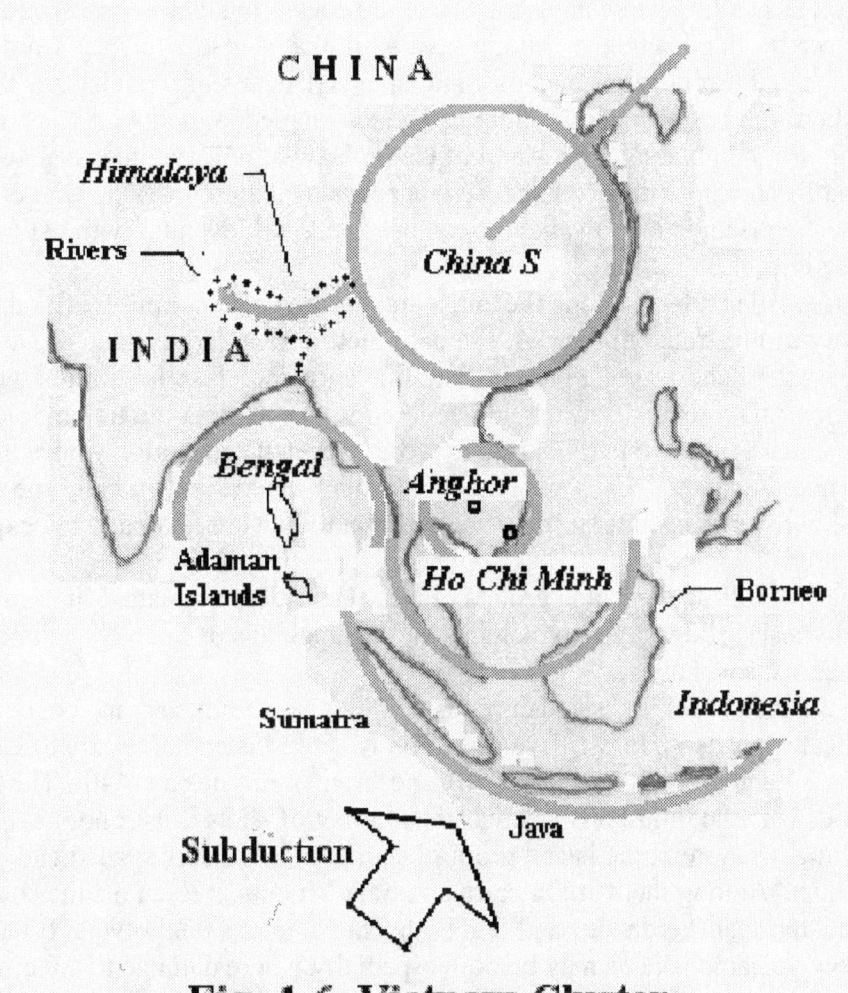

Fig. 1.6. Vietnam Cluster.

South Pacific Islands

Creation of the Indonesian and Ho Chi Minh arcs was accompanied by massive ruptures of earth's crust that separated the Pacific Plate from the Indo-Australian Plate then continued toward the southeast for a total of 6000 mi. Lava oozing out of numerous fissures spread out across the ocean floor covering an area almost the size of Alaska to a depth of 6000 ft. Known as the Ontong Java igneous province, it is the largest in the world. Lava spilled over into adjacent basins that may themselves be impact scars. Geologists have established the time scale for this region with major surges of lava about 125 My and 95 My. Sometime in the more distant past, a huge volcano grew out of the ocean that, upon loss by erosion of the portion above sea level, became the Ontong Java Atoll that is also known as Lord Howe Island, that is located just north of the Solomon Islands. Dozens of coral islands there define the flanks of the former volcano surrounding a banana-shaped lagoon with an area of more than 400 sq mi. For comparison, the island of Oahu, Hawaii will eventually be reduced to a ring of coral atolls corresponding to its present beaches and encompassing a lagoon of only 230 sq mi. The "big island" of Hawaii, on the other hand, will eventually transform into an atoll of 1600 sq mi.

Much later, existing islands along the entire fracture zone were built by flood lavas and volcanic activity as illustrated in Fig. 1.7. Global Fracture. The large island of New Guinea dominates the western end where a present fault line corresponds to its straight, northern shore. The primary fault is under water and covered with sediment out to the Solomon Islands. These consist of a two parallel ridges at the expected orientation whose higher elevations are above sea level. Only the island of Santa Isabel reveals a fault line along its axis. The parallel rows of islands strongly suggest that they were related to the explosion at Ho Chi Minh.

Then to the east follow the island groups of New Hebrides, Vanuata, Fiji, Samoa, Cook, and French Polynesia. The last group includes the archipelagos of Society, Tuamotu, Gamier, Marquesas, and Tubuai.

The total land area of all the islands is considerable. New Guinea alone covers 309,000 sq mi with volcanic peaks up to Mt. Jaya at 16,503 ft. Other island areas are the Solomon Islands (10,954), Vanuatu (4,707), Fiji (7056), and French Polynesia (1440). The number of inhabited islands is in the thousands whereas the number of islets is astronomical. At the eastern end of the rift system, the island groups become more widely spaced and their numbers diminish. Among them are Pitcairn and Sala Y Gomez. Even off the coast of Santiago, Chile, the Juan-Ferdandez and San Felix islands may be involved. It will be shown later how volcanic islands may be born especially where damage to the crust is caused by more than a single explosion or where the ring and rift of a single explosion intersect.

For an estimated width of this complex fracture zone of 500 mi, the total area involved in this grand rupture must be somewhere near 5 million sq mi! ***Indonesia*** and ***Ho Chi Minh*** thus set the stage for development of the largesst lava plateau and largest atoll on earth and essentially all the islands of the South Pacific from Asia to South America.

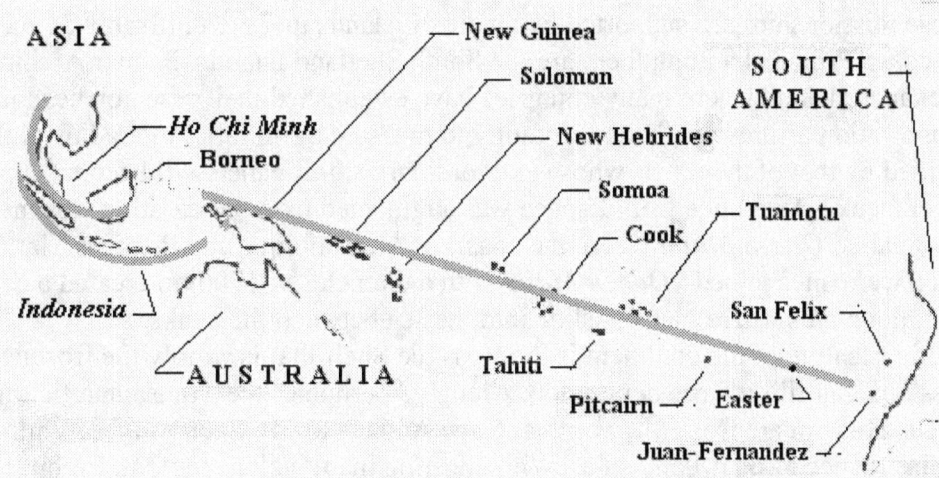

SOUTH PACIFIC OCEAN

ASIA

New Guinea

Solomon

SOUTH
AMERICA

Ho Chi Minh

Borneo

New Hebrides

Somoa

Tuamotu

Cook

Indonesia

San Felix

Tahiti

AUSTRALIA

Pitcairn

Easter

Juan-Fernandez

FIG. 1.7. GLOBAL FRACTURE.

15

Antarctica.

A large chunk of Pangea was blasted away by the Kalahari event in South Africa. Once set free, is drifted 2000 mi to the south and rotated about 75° counter clockwise where it now sits covered with ice at the South Pole. Fig. 1.8. The Seventh Continent shows the general configuration of Antarctica along with several mammoth craters that carved holes in it and trimmed its borders.

Before separation from Pangea, the Antarctic Peninsula shown at the bottom in the figure lay along the southwestern coast of Africa between that land mass and another fragment that became South America. The coast of South Africa east of Cape Town follows the prominent Drakensberg Mountains to Maputa. The western end terminates abruptly where the severed chunk had been. The lost portion of that range is found as the backbone of the Antarctic Peninsula whose smooth arc implies a cosmic impact. It has been identified as *Weddell* corresponding to the sea which it borders. Naturally, the diameters of *Weddell* and *Kalahari* should be the same and they are; 1140 mi versus 1180 mi respectively by independent and unsuspected measurements.

These relationships are supported by an arc of islands that are uniformly 52 mi off the Pacific coast. Principal among them are the South Shetland Islands, Palmer Archipelago, and Alexander Island where many countries have established their research headquarters. This observation affords another opportunity to measure a rim ratio, the radius of the outer rim divided by that of the inner, whose extraordinary significance will become apparent.

Severance of Antarctica from Pangea was augmented by a pre-existing weakness from the event called *Queen Maud* where the coasts of Mozambique and Princess Martha in Antartica were once joined. *Queen Maud* with a diameter of 1700 mi created a coastal range of mountains where glaciers slide into the sea between the peaks.

Another feature of the continent is the Ross Ice Shelf that is simply the frozen surface of the Ross Sea that fills a large depression. Along its perimeter the Transantarctic Mountains describe a 2140-mi arc from Mt. Minto at Cape Andane to Mt. Seelig in the Whitmore Mountains thence to Bear Peninsula. While hosting many volcanoes, Mt. Erebus (12,448 ft) is the most active. However, at the midpoint of the arc at Mt. Kirkpatrick is a world-class lava flow known as Kirkpatrick Basalt. The arc was clearly formed by an impact within the present Ross Sea with a diameter of 1680 mi called *Ross*.

Throughout the displacement of Antarctica, it carried all its geological features along on a single plate. Thus, running the movie backward to the initial position aligns the tip of South America with the Ross Sea. Accounting for the southward drift of Antarctica also nestles the Chilean Alps closely within the Transantarctica Mountains on the western shore.

Near the end of the Antarctic Peninsula the radius of curvature diminishes toward the east. This distortion resulted from another impact that created the South Sandwich Islands. Adjacent on their convex side, a trench was opened into which a relatively narrow ribbon of the Atlantic bottom disappears. The resulting Scotia Plate to the west of those islands is virtually static between the inexorable, westward drift of the Antarctic Plate and the South American Plate. More accurately, both continents rotate around a pivot near the South Pole. Resistance along the inter-plate fault along the edge of *Scotia* bent the Peninsula progressively near its tip.

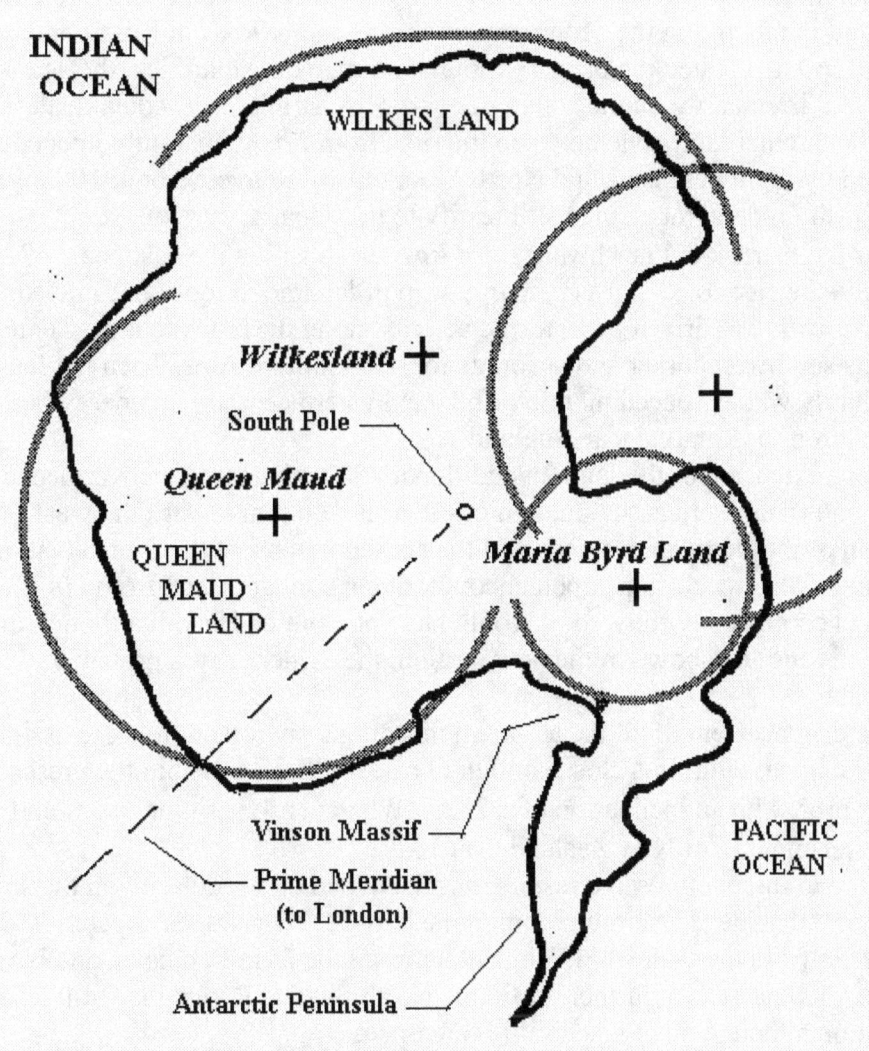

INDIAN OCEAN

WILKES LAND

Wilkesland +

South Pole

Queen Maud +

QUEEN MAUD LAND

Maria Byrd Land +

Vinson Massif

Prime Meridian (to London)

Antarctic Peninsula

PACIFIC OCEAN

Fig. 1.8. Seventh Continent.

A New Continent

A much delayed impact on the runaway continent broke off the northern third that became itself a new continent as shown in Fig. 1.9. Birth Of Australia. *Wilkes*, so named for Wilkes Land with its center marked by +, raised a giant rim 2180 mi in diameter. That fracture created the Great Bight of southern Australia and evolved into the Southeast Indian Ridge through the South Indian Ocean. Ever so slowly the two parts inched away from each other to the present spacing of 2000 mi including Tasmania.

A second, massive explosion in the same neighborhood played its own role in sculpting Antarctica and in modifying the geography of a vast region. *Ross* in Ross Land with a diameter of 1680 mi created the primary mountain range on Antarctica, the Transantarctica Mountains. An irregularity on the southern coast of Australia near Adelaide deserves attention. The circular fault zone between the rims from *Ross* apparently opened up Spencer Gulf between the Eyre Peninsula and North Mount Lofty Ranges. About 100 mi farther north, the North Finders Range may still testify to this local disruption.

A linear fault struck out northward from *Ross* establishing the east coast of Australia and, probably, the east coast of Tasmania. A strip of land pried loose became New Zealand. Subsequent growth was driven by volcanic activity along the crack that now enters New Zealand at its southwest corner and accounts for the Southern Alps. Then the fault changes direction slightly while proceeding to North Island where geothermal energy is still released in boiling caldrons at Rotarua near Aukland.

Northward extension of the fault an additional 2500 mi created the Kermedec-Tonga Trench that ends near Samoa. Subduction of the ocean bottom toward the west lifted the straight chain of the Kermedec Islands and the similar pattern of the Tonga Nation. At its end near Samoa, the trench stops upon encountering a similar trench from *Ho Chi Minh*. If the trench had continued northward, it would have pinpointed Hawaii although direct evidence is missing. As shown in the next section, the evidence was probably obliterated by another impact.

Tectonic displacement of all the actors in this drama are too complex to assign a specific azimuth to the linear fault. It is close enough to true north to be within the present regime of earth's solar orbit with an inclination of ± 23.5°. Whatever the original azimuth from *Ross*, the straight trench north of New Zealand is at 18.2°.

In addition to these impressive results must be added an irregularity on the south coast of Australia near Adelaide. The circular fault zone between twin ridges apparently opened up the Spencer Gulf and the Gulf of St. Vincent between the Eyre Peninsula and North Mount Lofty Ranges. About 100 mi farther north, the North Finders Range may still testify to this local disruption although other possibilities are open.

The oldest ocean bottom along the coasts of Australia and Antarctica, 22 My, indicates the beginning of the separation, the birthday of Australia. This date is reinforced by recent research with radioactive decay showing a value of 75 My. As to *Ross*, it must have followed rather closely behind before Australia had moved out of range.

A third crater in this region, *Byrd* after Maria Byrd Land, with a diameter of only 680 mi must be considered the little brother of the other two. Besides carving the coastline between West Longitudes of 50° and 150°, no evidence has been found of other damage on a global scale. Nor is the relative timing of this event known.

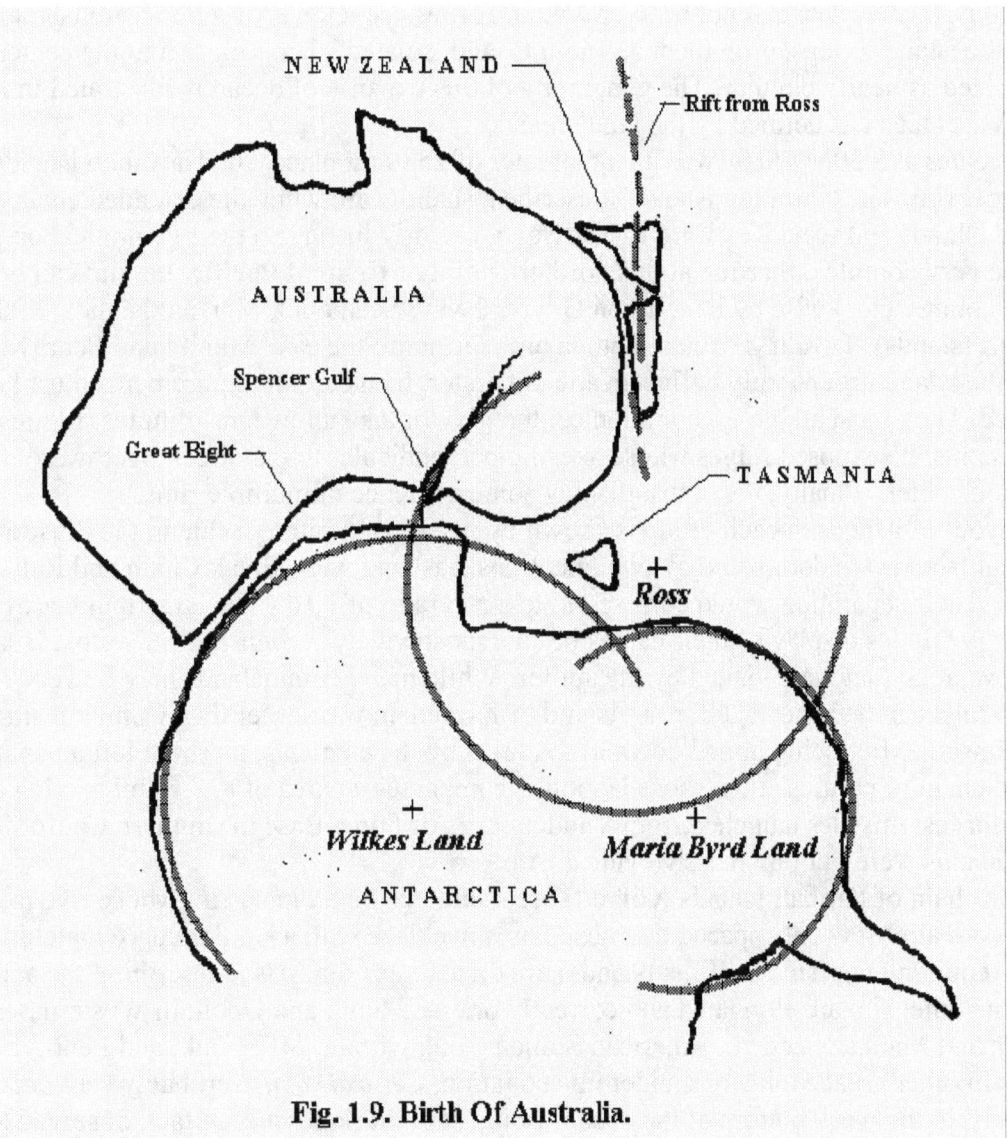

Fig. 1.9. Birth Of Australia.

Central Pacific.

The idea that the Kermedec-Tonga Trench may have extended northward beyond Samoa is reinforced by the existence of the North Tokelau Trough, a long, narrow but non-subducting depression with the correct alignment. Beyond the cluster of Phoenix Islands lies a vast, open ocean to Hawaii known as the Central Pacific Basin. Only tiny Johnston Island breaks the surface and furnishes an extremely remote site for testing nuclear weapons and destroying biological and chemical warfare materials. The prospect of the Basin being an enormous, cosmic scar can not be overlooked. A raw suspicion grows into astonishment when the island groups, prominent seamounts, and isolated islands closest to the center are recognized as nearly circular. The geography of this expanse of ocean is illustrated in Fig. 1.10. Vast Ocean as defined by the land areas.

A center at 8.30N 170.90W with a diameter of 2580 mi places the Hawaiian islands on the upper rim. The "vacation islands" describe a shallow arc while the extended chain of related islands and reefs westward to Midway are nearly linear. So their geological origins may be very complex. Starting at this northern point of **Central Pacific,** the rim can be traced counter clockwise by Resolution Guyot, Ewing Seamount, Marshal Islands, Kiribati (Gilbert Islands), Tuvalu, Tokelau, Samoa and, farther to the east, Maniki and Penrhyn.

If the island groups truly define an ancient crater, then certain requirements must be fulfilled. They must all have a common center, give or take a few tens of miles. Elongated islands must be exposed ridges whose axes lie perpendicular to the directions toward the common center. Finally, they must display some evidence of multiple rims.

Layout of islands in each group is shown by inserts of arbitrary scale in the illustration. Marshal Islands are dominated by two lineations, east and west, Ratak Chain and Ralik Chain. Their separation would correspond to a rim ratio of 1.16. The large number of islands precludes display of their names but the most widely recognized Eniwetok, Bikini, and Kwajalien that were visited by the author. While many coral islands may be very small, entire atolls can be huge. The former island of Kwajalein was larger than Oahu but smaller than Hawaii. After being honed down to sea level by the elements, its shore left a banana-shaped chain of coral islands whose lagoon encompasses an area of 839 sq mi serving as a target for test missiles launched from Vandenburg Air Force Base in southern California. Such islands were built by massive vulcanism.

The chain of Kiribati Islands (Gilbert) are seen to be on an inner rim whose five major islands are almost evenly spaced like beads on a necklace with a similar curve matching one of the requirements. Tuvalu (Ellis Islands) are on an outer rim. Also, describing the smooth arc of an outer rim are elongated and correctly oriented Savii and Upolu in Western Samoa, Tutuila and Manua Islands in American Somoa along with far off Maniki and Penrhyn.

As in other great explosions under investigation, *CentralPacific* probably sent out a radial rift in the weak bottom of the ocean. Long, straight ridges are, in fact, observed to the southeast from the general interior of the crater although offset from its center. The northern end of Christmas Ridge is marked by Palmyra whereas the southern extremity of the Tuamotu Ridge is marked by Pitcairn at a distance of 3050 mi. The azimuth from Palmyra to Pitcairn of 136.9° may be taken as the rift angle from this crater if only tentatively.

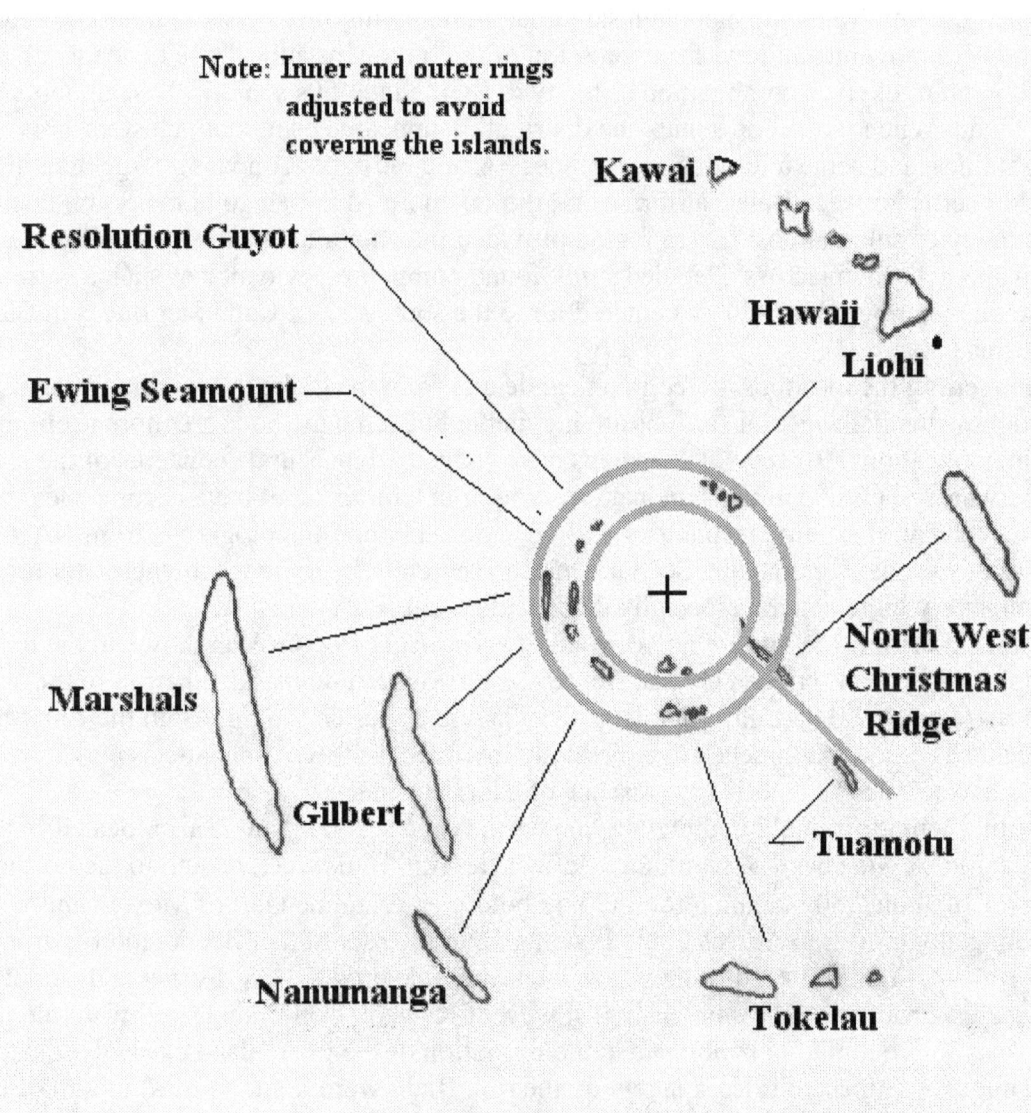

Note: Inner and outer rings
adjusted to avoid
covering the islands.

Kawai

Resolution Guyot

Hawaii

Liohi

Ewing Seamount

North West
Christmas
Ridge

Marshals

Gilbert

Tuamotu

Nanumanga

Tokelau

Fig. 1.10. South Pacific Islands.

East-Pacific

In the early days of modern oceanography, seamounts captured the attention of many scientists. Numerous expeditions mapped the bottom topography of the Pacific Ocean discovering many seamounts and successfully dredging material from the tops of some. Recovering coral samples from flat tops revealed sea mounts that had once stood as islands above the waves then sank to their present depths. The latest age of fossilized creatures indicated the time when the mountain sank that, in many instances, was found to be sudden. This relative movement allows for depression of the bottom or a local rise of sea level. Or, possibly but unlikely, a combination of the two. By the late '50s some 1400 seamounts had been found. When plotted on a map, the distribution appeared as random clusters, arcs, straight lines, and some individual occurences. A notable exception was a large area in the eastern Pacific between Baja California and the Tuamotu Archipelago in the South Pacific. A distinctly circular, almost barren region provided the author a clue to a possible crater that had to about 3000 mi across. Detailed study found compelling evidence of such a crater whose diameter became 3430 mi centered far to the southwest of Cabo San Lucas in Baja California.

Converting the locations of geographical details from the globe to two-dimensional maps is fraught with challenge. Distortions are inevitable but some methods are more useful in certain applications. However, these maps have sponsored inaccurate concepts of the earth that pervade the normal mind. True patterns with much more detail have become clear only with the advent of satellite technology and wide distribution of photographs from NASA programs, such as, LandSat and SeaSat. One can search ordinary maps in vain without detecting very large craters, especially those underwater.

Some interesting details are noted around the rim *East-Pacific S* as shown in the lower example of Fig. 1.11. Hidden Giants. Dense clusters of seamounts skirt the rim to the southeast. South of Baja California, Islas Revillagigedo lies on the rim. Both the Christmas Ridge and Tuamotu Archipelago are distinctly involved, however, the latter's curvature fails to match exactly and suggests the presence of a larger crater.

With a center offset about 600 mi to the northeast, a second giant is also located in the East Pacific. A wide band of seamounts defines the rim from directly south to the northeast on an arc of about 150 ° running for 4500 mi before entering the Gulf of Mexico and deflecting along the San Adreas Fault. It swings into the root of the Mendocino Escarpment off the coast of northern California where the lower side dropped. A distinct and smooth arc then passes through the Hawaiin chain just west of Kawaii. Continuing southward along Christmas Ridge implies a major role in the formation of the Line Islands. Finally, the axis of Tuamotu Archipelago rides squarely on the rim. Radii were found to be a) mouth of the Colorado River in the Sea of Cortez - 1960 mi, b) Kawaii - 2020 mi, c) Isle do Desappointement in Tuamotu -1930 mi, and d) Pitcairn in Tuamotu -1980 mi. The average diameter would thus be 3940 mi.

As this latter impact struck a thin, oceanic crust, evidence of a linear fault radiating from its center should be easy to find. The Hawaiian Chain of coral atolls, reefs, and seamounts aim directly at the heart of *East-Pacific N* with uncanny accuracy. They extend west of Kawaii for 900 mi to a sharp bend at Lisianski where they link into the Emperor Seamounts trending northwest to Kamchatka, Siberia. These details raise disturbing questions about the local geology, tectonic drift, and the role of hot spots.

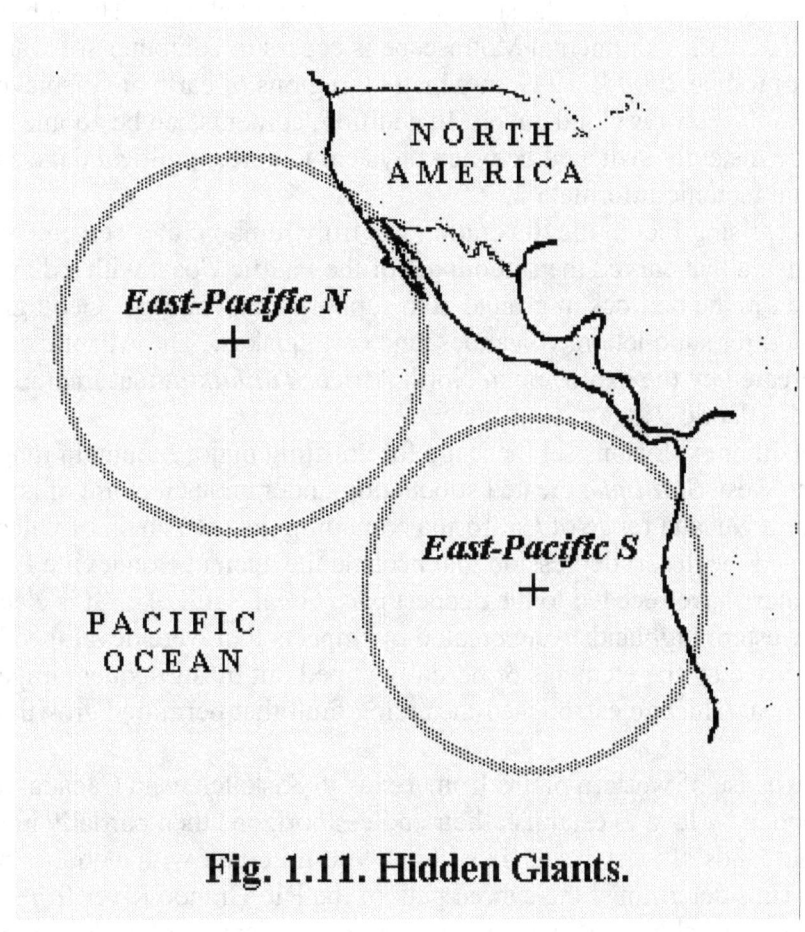

Fig. 1.11. Hidden Giants.

North America.

A glance at the accompanying map reveals a pattern of craters in North America that would be expected from the cosmic records on the Moon, Mars, Venus, outer planets, and their moons. The continent is paved with craters! These arcs and circles are not idly drawn to highlight features on the surface but are included because they provide further and, in most cases, compelling evidence of cosmic origin.

The essential problem in recognizing these craters is that they are *too big to see.* Vantage points from the ground are out of the question. Views from aircraft are not broad enough or are blocked by haze. Even photographs from space craft usually fail for two reasons; a) area coverage is much too small, or b) the surface is obscured by clouds. The only hope for understanding Fig. 1.12. Continental Moonscape is access to computer software such as Microsoft Encarta Suite 2000 © 1999 in which all regions of earth are displayed in photographic montages *sans* cloud cover. In addition, contents can be zoomed over a wide range and can be instantly switched between physical features, political data, satellite photographs, and tectonic information.

The most surprising fact in the illustration is a truly jumbo crater centered on the Michigan Peninsula that carved the smooth arc of the Pacific Coast with a diameter of 4000 mi. A large basin in the bedrock at ground zero supports that concept. Geologically, the entire arc is noted for subduction, volcanoes, and earthquakes. The Atlantic coast, it will be recalled, was created by the explosion in North Africa, *Timbuktu*, that fractured Pangea and opened the Mid-Atlantic Ridge.

Faults from other explosions set the stage for uplifting major mountain ranges on the continent. In the west, *Salt Lake* created subduction under an ancient arc of islands that has become the Sierra Nevada range of California containing its own chain of volcanoes. In addition, it formed the desert depression that became the ancient Bonneville Lake of 50,000 sq mi whose waters have receded to the deepest part, Great Salt Lake. It is clear from the figure that the western highlands were created by impacts that were developed by tectonic forces then modified by the elements or partially wiped out by subsequent impacts. In the east, an impact near Chicago established the arcing fault that permitted growth of the Appalachians.

The great expanse of western plains from Texas to Saskatchewan Canada, must have been mares flooded by lava to establish their endless horizons then partially buried by miles of sediment. Flat lands of west Texas lie within a 900-mi crater with a double rim. The valley between rims determined the curved path of the Rio Grande River from southern Colorado through central New Mexico and along the Mexican border. Big Bend National Park houses a chain of extinct volcanoes that are significantly inside a continent. A right-angle turn of the river around Emory Peak traces a switchover from the outer channel to an inner one. Additional clues to *Lubbock* are easily found in elevation changes of Balcones Escarpment from San Antonio to Ft. Worth.

The map also shows how missing craters in the central states have simply been covered up by sedimentary rock. Some of the craters on dry land reached into the oceans where rims are partially exposed as island arcs. *Alberta* formed the large islands off British Columbia and the popular, inland passage to Alaska. *Zacateca* raised the Islas Tres Marias off Mazatlan. And *Guatemala* created a long chain of islands from the Bay of Honduras to a prominent peninsula in Quintaneroo.

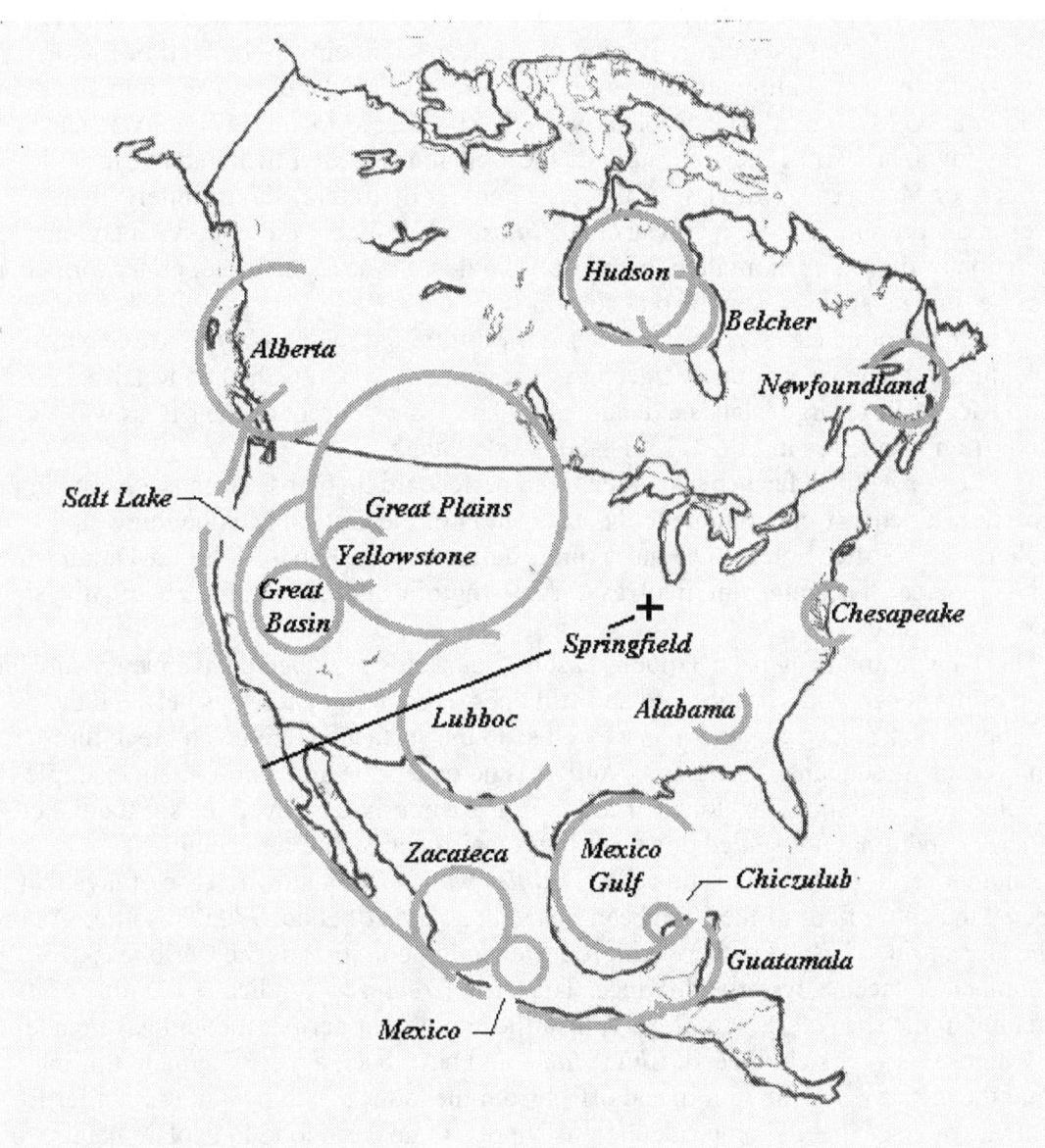

Fig. 1.12. Continental Moonscape.

Caribbean.

Everyone who has followed the scientific news and entertainment media in the last decade is well aware that dinosaurs died off about 65 My. Worldwide appearance of abnormally high concentrations of iridium in the proper geological stratum convinced most scientists that the destruction of dinosaurs could be attributed to an asteroid impact. An especially large crater for that time was discovered with its southern half buried in the Yucatan peninsula and the other half lying under deep sediments in the Gulf of Mexico. No one has actually seen it although it is assuredly there. Additional clues were found that dated the crater at 65 My. Voila! The problem was solved. *Chiczulub*, as the crater was named for a nearby town, had killed the dinosaurs. Its location and size are illustrated in Fig. 1. 12. Continent or Moonscape. In view of the large number of craters with infinitely greater power in the present catalog, it appears that *Chiczulub* probably did not give the global population of dinosaurs more than, perhaps, bad colds. Some much larger crater formed at the same time must have been the real culprit.

As in many other cases, a linear rift system was emitted by the *Gulf of Mexico* toward the southeast. It can be accurately traced to the eastern tip of Cuba then on to Hispanola and Puerto Rico. These large islands extending 1700 mi from the root of the rift grew by lava issuing from fissures as in Fig. 1. 13. Islands And Craters.

The Lesser Antilles farther east is seen as a perfect arc that arose from a separate impact. A missile with energy comparable to the excavation of the Gulf of Mexico threw up a double rim as so often noted. The inner rim is defined by St. Kitts- Monserat-Dominica-St. Lucia- Grenada. The outer rim involves Virgin-Anguille and much farther south, the single outcrop of Barbados.

The Tobago and Brownson Troughs east of the islands provided an avenue down which a strip of the ocean floor disappeared and still does. By finding pathways between the overlapping plates, lava can make its way close to the surface where it can break through to form volcanic mountains. The Lesser Antilles is no exception and a prominent example is Mt. Pelee on Martinique. Its violent eruption in 1902 utterly destroyed the substantial city of St. Pierre. And it has not settled down yet but remains active and threatening.

Another result of the explosion called *Antilles* was the very slow release of lava that carpeted the entire floor of the Caribbean Sea with an area of about 970,000 sq mi. As one of the largest flood basalts in the world, it is known to geologists as the Caribbean-Colombian Cretaceous Igneous Province. The term, *Cretaceous*, indicates the time of its formation in the span of 38 to 68 My. If near the end of that period, it could have made many dinosaurs very sick, indeed. *Chiczulub* could have been a mere fragment trailing behind the primary by 1 hr 38 min and offset from the latter's path by 480 mi, not far for cosmic travelers. (These simple calculations happen to take too long to explain here.)

While a strip of ocean floor dove under the Lesser Antilles, the floor to the north continued its inexorable march. Drill cores have shown the ages in the Atlantic steadily increasing toward the west. Thus bottom ages range from creation of the Mid-Atlantic Ridge to the present. A missile could have hit anywhere in the ocean at any time so dating of the event is not obvious. It can, however, be estimated by careful study of the ocean floor where relative ages are known.

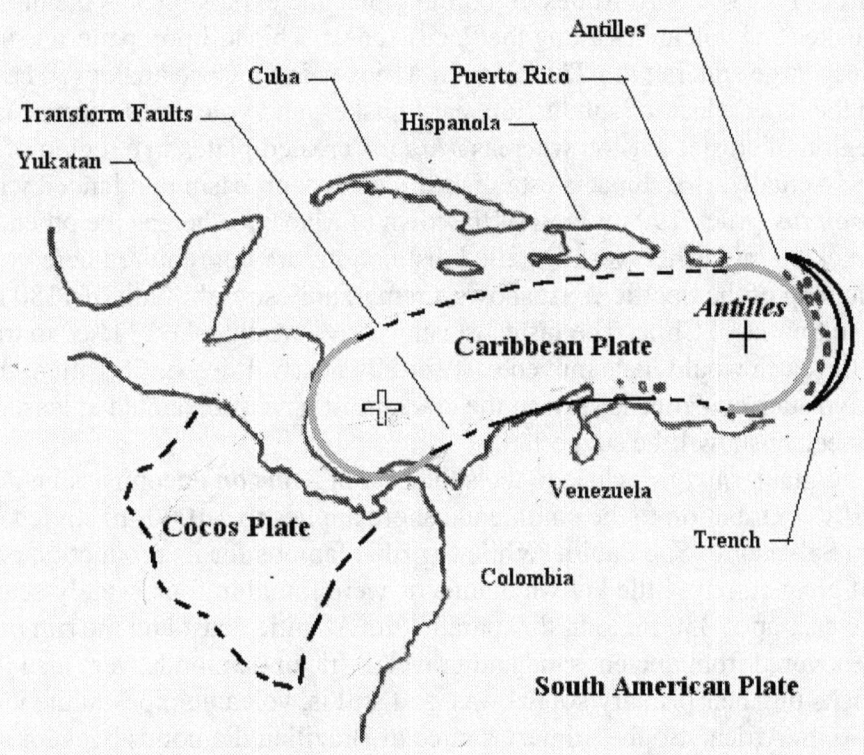

Fig. 1.13. Islands And Craters.

South America N.

Two giant, multi-ring structures shaped the northern half of the continent and controlled the topography from the Andes Mountains to the high plains of Brazil.

In the west, the rounded coast was formed by an explosion called *Manaus* running from Guyana on the Atlantic to the southern tip of Peru on the Pacific. The typical, double rim is easily discerned on this 2010-mi crater along a) the Cordillera do Merito from Caracas, Venezuela to Bogota, Colombia as the inner and b) the mountain range of Medellin, Colombia to Quito, Equador as the outer. Those details are supported by the concentric courses of three rivers. The explosion created a perfect, circular fault with the South American side overthrusting the western side thus initiating subduction and growth of the northern Andes. In addition, a linear fault was directed southward with the same results leading to birth of the southern Andes. In both regions, the explosion was the ultimate cause of all earthquakes and volcanoes along the Pacific coast. This lollipop pattern is universal among the very large craters with *Timbuktu* in Africa being a close analogy. The only difference is the latter's lack of subduction, earthquakes, and volcanoes because it created plates in a region of crustal tension whereas *Manaus* created plates in a region of crustal compression. Actually, two impacts established the Amazon basin that landed within a span of 23 min. *Manaus* struck 190 mi west of the town of Manaus whereas the other, called **Amazon**, struck 80 mi farther west. Detailed distinctions are not required here.

A seismic map of the Pacific coast shows a remarkably straight fault of 1180 mi from Antofagasta to Santiago, Chile. The fastest seismic wave would take 211 sec to travel so far. Meanwhile, the earth would have moved eastward by nearly 1 deg or 58.1 mi at the latitude of Santiago. No such evidence is seen so the crack must have propagated at least 30,000 mi/hr and a mechanism will be suggested.

The second giant, snuggles close to the southeast of *Amazon* encompassing nearly all of Brazil. *Brazil's* contribution to the continental shore applies to a 1000-mi stretch on the Atlantic from Salvador to Sao Paulo. While Brazil is famous for its production of gemstones of great variety, little known diamonds were found in four, widely-scattered regions in the east up to 350 mi long that parallel the Atlantic coast and the rim of the crater. They were recovered from ancient sand and gravel. African diamonds were also found in alluvial deposits but their primary source was, and still is, volcanic pipes in the vicinity of Kimberley, South Africa. So the primary source of Brazilian diamonds begs for an explanation.

Overlapping both of the above, a smaller crater called **Rain Forest** is shown in Fig. 1.14. *Amazon And Andes* because the only clear indication of its rim is the boundary between rain forest in the Amazon basin and the plains of western Brazil.

Three smaller craters must be pointed out with all too brief comments. *Venezuela* in the north controlled the arcing gorge of the Orinoco River and released lava that formed the Venezuela Igneous Province. *Maracaibo* created a famous lake of that name surrounded by a distorted circle of mountain ranges. It is a major source of oil in South America and nearly all of the producing fields on that continent lie along the coastal arc created by *Amazon* and *Manaus*

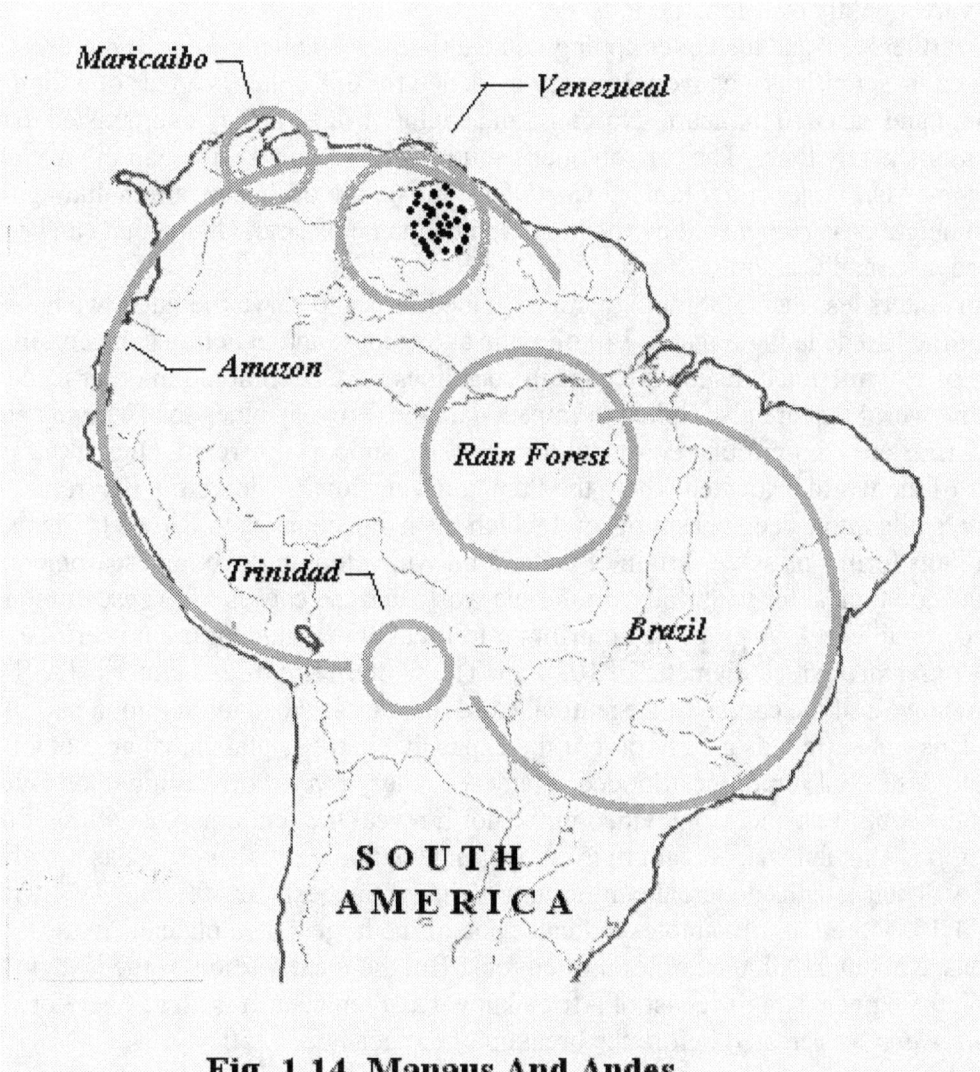

Fig. 1.14. Manaus And Andes.

South America S.

.

Campo Grande struck near that city's location 670 mi southwest of the capital, Brazilia. With an area of about 200,000 sq mi, it is somewhat inferior to *Brazil* that it overlaps about 50 per cent. It carved its signature much later than *Brazil* upon creating a basin drained by three rivers whose confluence followed the Parana River along a linear rift. Waters from intermittent swamps in the plains of Moto Grosso do Sul eventually escape by that route. Lakes, swamps, and deserts are formed in impact basins depending upon the rainfall that usually varies wildly over long periods.

Still farther west and also overlapping its neighbor is a smaller *Chaco*, a region recognized in Spanish as a canyon. Primary evidence for this crater is a pair of arcing ranges to the west and east 410 mi apart. Northern and southern rims, if they ever existed, have been demolished by time. The canyon opens southward through a wide gap in what appears to be a somewhat larger crater labeled *Cordoba.* The nature of this terrain including the northern neighbor is recognized by the geographical name of Gran Chaco that can be translated as Grand Canyon.

Twin craters less than 500 mi in diameter joined forces to carve the huge, wedge-shaped mouth of the Rio de la Plata that is 140 mi wide at the coast and penetrates the continent a distance of 200 mi. Such distinctive gaps in coastlines have been noted in other places around the world that are also linked to impact damage. Primary clues for *Uruguay* and *Buenos Aires* are their circular coasts with impressive support upstream. One of the scenic wonders of the world is located where the Parana River, flowing down the rift from *Cordoba*, spills into a deep chasm beyond which the name changes to Rio de la Plata. The Iguassu Falls form a massive curtain of water 2 mi wide plunging over an escarpment. These conditions are typical for waterfalls around the world and the causes of the escarpments can usually be established. A probable contributor to Iquassu is a much more powerful event called *Montevideo* with a diameter of 1020 mi. The western rim of this crater is the Sierra del Nevada west of the capital of Argentina, Cordoba, that is close to, but not a part of, the Andes. This rim surrounds a large part of the expansive Pampas plains that are notably dry tending to desert but sometimes flooded. Many secondary rivers converging upon Buenos Aires in the southwest and Montevideo in the north reveal the fundamental nature of this entire region. This fan-shaped pattern is compelling testimony of a shallow basin with axial symmetry. Their lengths dictate the minimum radius of the basin as 400 mi.

Fig. 1.15. More Latin Examples indicates one of the largest lava plateaus in the world that released about 240,000 cu mi of molten rock. But the local portion of the lava, known as Parana, left fragments on the coast of Africa known as Etendeka. It is clear, therefore, that the *Montevideo* was created before the breakup of Pangea about 150 My.

Catamarc State in northwestern Argentina received a modest hit so long ago that no surface clues remain. Only a perfect circle of earthquake epicenters delineates this 500 mi crater that are well east of those associated with subduction along the coast. A minor detail at the center of *Catamarc* is an apparent sinkhole 100 mi across.

The southern half of the continent clearly experienced a history of bombardment in its interior equal to the northern half.

Fig. 1.15. More Latin Examples.

Three Oceans

When ***Timbuktu*** fractured Pangea, it will be recalled, its linear rift split the southern portions of both Africa and South America along a straight line. Thus the east coast of South America is seen to be essentially straight except for a notable bulge from ***Brazil*** and an indentation at Rio de la Plata. Still another irregularity in the coast is hardly noticed. Immediately south of Rio de Janeiro, the coast swings inland in a smooth arc toward Sao Paulo and continues for about 500 mi as shown in Fig. 1.16. Continental Trimming. About 870 mi off the coast is a large region known as Rio Grande Rise where water depths are substantially less than the surrounding ocean floor. Satellite research shows ***Rio Grande*** is a nearly perfect circle of mountain ranges and valleys 450 mi in diameter and its rim fits the bay at San Paulo. The Rise became anchored on the ocean bottom by an impact while South America continued to drift to the northwest. This motion is dictated by the location of ***Rio Grande*** relative to the bay it created and is confirmed by faults in the Atlantic. An approximate time of the impact can be established by comparing its location with the total distance that South America has receded from the Mid-Atlantic ridge. A ratio of one third indicates an age of about 50 My.

South of La Paz, Bolivia, the west coast is generally straight because it was formed by the rift from ***Amazon***. The coastal rift from Santiago to Antofagasto, both in Chile, projects to the epicenter of that event. To the north in the arm pit, a major event pushed the shore toward land with its center on the Nazca Ridge and a diameter of 1720 mi.

A second rift system east of the coast that defined the southern Andes projects to the center of ***Manaus*** just 80 mi east of ***Amazon***. That would correspond to a time delay of 23 min for a string of comets. Respective azimuths of 185° for ***Amazon*** versus 188° for ***Manaus*** support that idea. Bending of South America's tip was previously ascribed to resistance along the boundary between the South American and Scotia plates. But this matter is more complex than that.

Consider the missing half of ***Montevideo***. Since portions of its lava were left in Africa, one would expect to find a mountain arc of the same radius concave to the west. While Pangea remained intact, the present Antarctic Peninsula fit between the southern tip of South America and the point now occupied by Cape Town, South Africa. Thus the missing half of ***Montevideo*** should be found there. The eastern bend of the outer portion of the peninsula was mentioned before in the context of Antarctica's shape. That outer portion has been dragged to the east by the same forces that distorted the tip of South America. However, the primary mountain range on the peninsula from the root to the midpoint curves the other way. It is an arc that fits the requirements for the lost half of ***Montevideo***! That range as a whole has no known name but includes such peaks as Mount Andrew Jackson at 13,737 ft. A thick blanket of snow obscures most of the range but ice-free ridges and peaks provide an accurate reference to determine that the radius is compatible with ***Montevideo***. A small problem remains. The missing half from South America would have an internal angle of 180° while the arc in question is only 119°. The geographic feature that would fill this void is none other than the outer half of the Peninsula that has been turned inside out by the Scotia drag. A dotted arc in the illustration shows its original position that completes the missing half of ***Montevideo***. It had taken the long way home.

Fig. 1.16. Continental Trimming.

33

Down Under.

On the opposite side of the globe from Montevideo at the same latitude an entire continent has been left hanging. Deserts in the outback with severely limited rainfall and minimal erosion have favored preservation of large craters. The southern coast, known as the Great Australian Bight was created by *Wilkes* in Antarctica when those continents separated. Within Australia itself, twelve craters have been identified with diameters up to 1020 mi. In addition, faint and dubious evidence requires that three more with diameters up to 590 mi be relegated to a separate section entitled Candidates. Fig. 1.17. Same Old Thing shows the location, size, and name of each crater but space will permit discussion of only those of special significance.

Although tiny among the Australian giants, *Bass* helps explain the history of Tasmania. It rode the Australian plate away from Antarctica but, at some unknown time, an explosion depressed Bass Strait that became flooded by the sea leaving the present island. Primary clues are the arcing north coast, King Island and Flinders Islands. Also, the short mountain range behind Melbourne. This explosion was too weak to leave any tectonic signature.

At the other end of the scale, *Artesia* formed the Great Artesian Basin covering the eastern third of the continent whose outlying faults defined the East Coast from Townsville to Melbourne. The Great Escarpment in Queensland records the resulting change in elevation and hosts the two highest waterfalls in Australia. Similar escarpments associated with craters are found behind the coastal plain of Angola and the Pinnacles along the New Jersey side of the Hudson River.

The central part of the continent was dominated by *Alice* whose basin, 930 mi across, includes three major deserts, at least one of which was formed by another impact. MacDonnell Ranges including Mt. Ziel at 4955 ft constitute a central uplift that is associated with major lava flows. Similar conditions apply to the Musgrave Ranges 200 mi to the south. Ayers rock along with other outcroppings in the Uluru National Park lie within this crater but appear to be more closely linked to another event farther south. East-West alignment of the outcroppings must be compatible with a crater rim and the vertical plane of rock strata must be explained.

A dense cluster of earthquake epicenters located north of Alice possibly indicates an impact site that is surrounded to the southwest by a few others on a common arc. Because mid-continent earthquakes are rare, these epicenters could represent a large and very old site whose surface clues have vanished.

Topography of the south-central region is noted for the Great Victorian Desert inland and the Nullarbor Plain along the coast. This 840-mi crater is called *Small Bight* in contrast to the Great Australian Bight. A portion was apparently left undetected in Antarctica under a thick blanket of ice and snow.

On a more modest scale are two craters in the east, *Victoria* and *Brisbane*, that were formed in that order The latter created a linear rift penetrating the formers rim and setting a course for the Darling River as shown by a dashed line in the figure.

Australia is famous for all kinds of gemstones although not especially noted for production of diamonds. However, a world-class source was discovered in 1970s within the extremely remote domain of *Kimberley*, aptly named as it turns out. Primary indications of a crater are the abundance of intrusive rocks and suggestive bays.

34

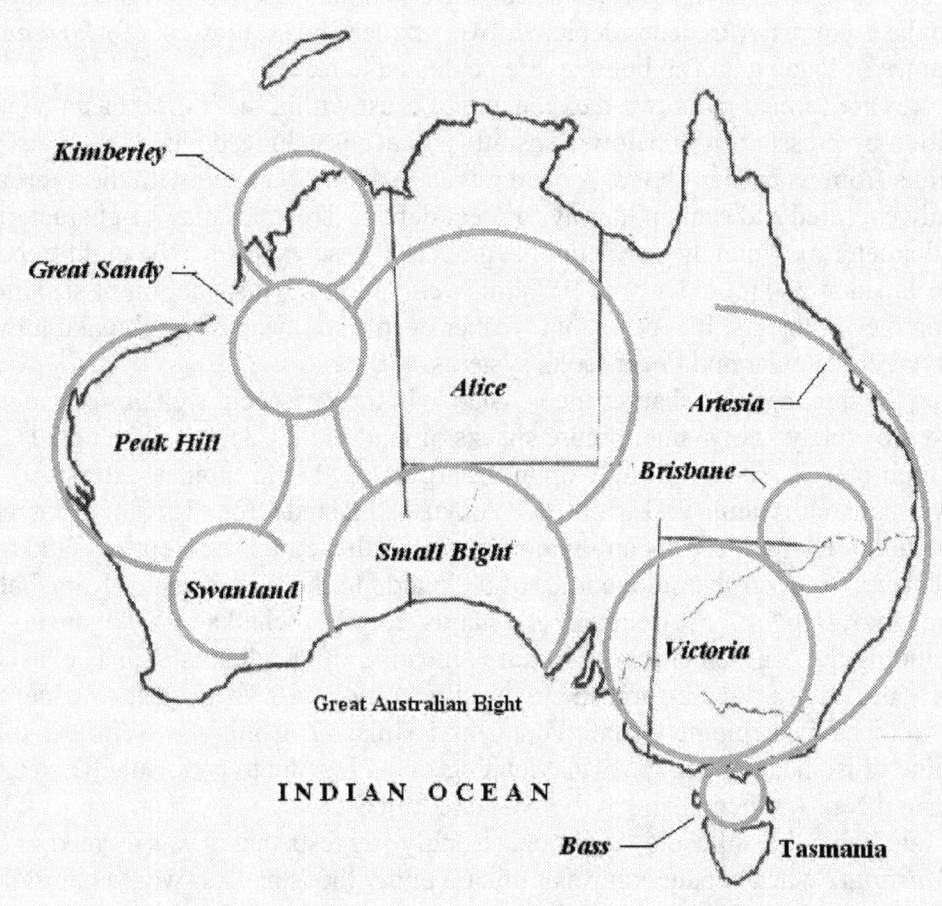

Fig. 1.17. Same Old Thing.

Subduction Trenches.

As the Pacific Plate moves ponderously toward the northwest, it encounters a defensive wall protecting the Far East. The only alternative for the ocean floor is to plunge under the Asian Plate along a 5000-mi front running from the southern tip of the Philippine Islands to about Anchorage, Alaska. Fig. 1.18. Disappearing Ocean Floor shows the line of deep trenches deflecting the aggressive mass that eventually melts in earth's mantle from whence it originally crystallized. Island chains of volcanic origin are found west of all the trenches. Only massive explosions from asteroids or comets could have created this geography. In addition to the eight major trenches depicted, two smaller ones southwest of *Mariana* have familiar names of Palau and Yap but they are neglected here.

Island arcs are formed by outward expansion of crust within the impact basin coupled with elevation by pressure from below. Isostatic balance is achieved when the underlying seafloor sinks from its burden above. A deep trench is thereby created with the approaching plate already captured and sent on its way to great depths. Trenches may be characterized by the crater diameter of which they are minor segments. These values for the eight trenches range from Japan at 710 mi to Kuril at 1980 mi, well within the distribution of sizes so prevalent on the continents. It appears that weaker events are incapable of breaking through the crust to create circular and linear faults systems.

To bring the trenches into sharper focus, islands in their respective chains can identify them. Most are widely known but obscure specks of land in the ocean will be noted whenever their presence bears strongly upon the argument. 1) Philippines - all islands from southern Mindanao to Manila on Luzon. 2) Ryukyu - all islands from Taiwan to Japan's Kyushu including Okinawa. 3) Japan - major islands of that country excepting Hokkaido. 4) Kuril - all islands from northeastern corner of Hokkaido to the East Coast of Kamchatka where is ties into *Aleutian*. 5) Aleutian - all islands from Kamchatka to Yukutat Bay in Alaska including the Alaska Peninsula, Kenai Peninsula, and Kodiak Island. The last two items clearly illustrate a double-rimmed crater with a ratio of 10.8%, a typical value. 6) Marianna - all islands including Guam, Saipan, and Tinian. 7) Bonin - Iwo Jima, and 8) Izu - a straight line of islands including Miyake leading to the Izu Hanto peninsula between Yokohama and Nagoya then inland to the volcanic Fuji San.

Benin-Izu, being the only straight trench, strongly suggests that is was created as a linear rift from *Mariana.* When subduction was initiated under the islands, it would naturally proceed northward along Bonin-Izu. While it was instantaneous under the arc, progress to the north may have been very slow. If, however, compression of the Pacific Plate was substantial at that time, it could have sprung west under Bonin-Izu briskly.

Naturally, prominent chains of earthquakes and volcanoes back up all the trenches. Relative movement of the overlapping plates creep past each other in jerks that are most often minor but, when major, generate devastating earthquakes. The fracture zone between plates provides a pathway for lava to find its way to the surface where it can build seamounts and islands. Lava simply oozes out of fissures and caldera most of the time but occasionally energy is released in cataclysmic eruptions of volcanoes. Examples in this part of the world are the ongoing activity of Mt. Mayon on Mindanoa, island building by Myosin-syo in Izu, colossal explosion of Mt. Bezymianny in Kamchatka, and Mt. Katmai near Anchorage. Earthquakes in Japan have reached 8.9 Richter.

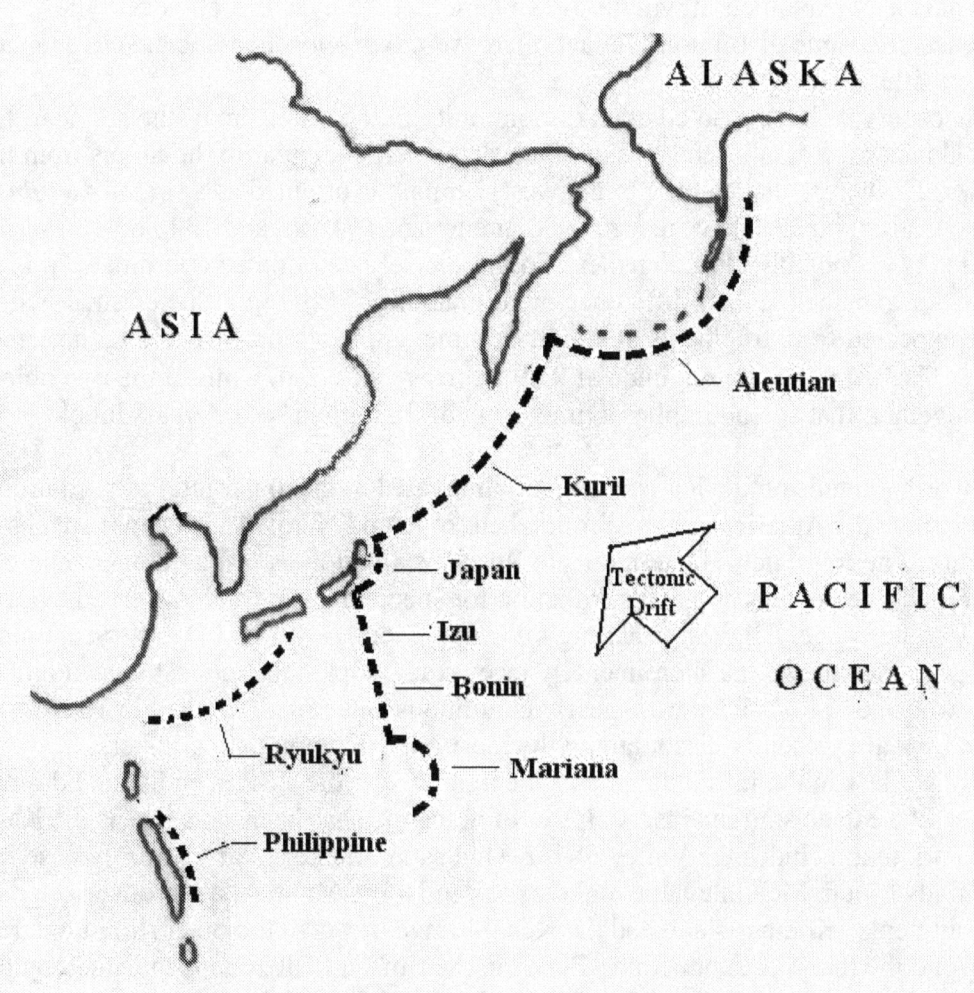

Fig. 1.18. Disappearing Ocean Floor.

Aleutian Islands.

Almost everyone has marveled at the beautiful rainbow of the Aleutian Islands illustrated at the top of Fig. 1.19. Bering Twins. Some authors have suspected that they were left as a scar from an asteroid or comet impact of unprecedented proportions. The majority of geologists, on the other hand, have rejected that idea. They staunchly defend the notion that the arc was created by tectonic forces related to subduction of the Pacific Plate down the Aleutian Trench. But that theory carries little weight, is vacuous, and never has been satisfactorily explained. It will be shown here that the islands are, indeed, the consequence of cosmic disruption. In fact, there were two, closely associated events hence the figure's title.

Astute observers have pointed out a discontinuity in the curvature of the arc near the island of Unalaska. Actually, the axis of Unalaska is perpendicular to the radius from the site of *Aleutian E* whereas its neighbor to the west, Umnak, is oriented as a rim of *Aleutian W*. These two distinct impacts have respective diameters of 1140 mi and 1340 mi.

Epicenters are notably close together. From precisely determined coordinates it is a simple matter to calculate their difference in latitude of only 109 mi. Such a small value would be expected from fragments of a mother comet approaching earth along the same trajectory. The difference in longitude of 9.69 ° corresponds to 671 mi. If the two objects traveled together, that spread implies a time delay of 38 min 46 sec. A real, double whammy!

Noteworthy details on the Russian end are illustrated at the lower left in the figure. Two islands continue the American chain, Probrazhenskoye and Nikol'skoye whose shapes and orientation are appropriate. The terminus in Russia is an oddly configured peninsula. An outer rim in this region has avoided subduction for special reasons and preserved evidence as an underwater ridge. Highest peaks are known as Meiji Guyot and Detroit Seamount. Again, a peninsula marks the shoreline. Klyuchevskaya Sopka, the highest volcano in Asia, is nearby whose eruptions have been nearly continuous since first recorded in 1697. Sprays of molten rock are presently discouraging the most intrepid, mountain climbers.

The long Alaskan Peninsula dominates the inner rim at the eastern end where dramatic evidence of the cataclysm remains. A cluster of volcanoes has been set aside as the Katmai National Park that includes the Valley of Ten Thousand Smokes. Farther north close to this rim, one finds Mount McKinley, the highest peak in North America, a curious juxtaposition on two continents. Kodiak Island and the Kenai Range preserve the outer rim in this region being beyond the threat of subduction. Between the rims, a fault zone is indicated by the Shelikof Strait separating Kodiac from the mainland and Cook Inlet leading to Anchorage. A large chunk of that city's business district slid into the bay in response to an earthquake on March 27, 1964.

Whether the two impacts are related remains somewhat open without a detailed examination of the Aleutian Trench. Movement of vast areas of earth's crust must be very lethargic so the first strike established the geography just described. But the second strike hit too soon for subduction to be initiated in Cook Inlet. The second one from the more powerful of the two opened the trench 75 mi south of Kodiak leading to the Gulf of Alaska. Subduction down that gap pre-empted any such phenomenon at Cook Inlet.

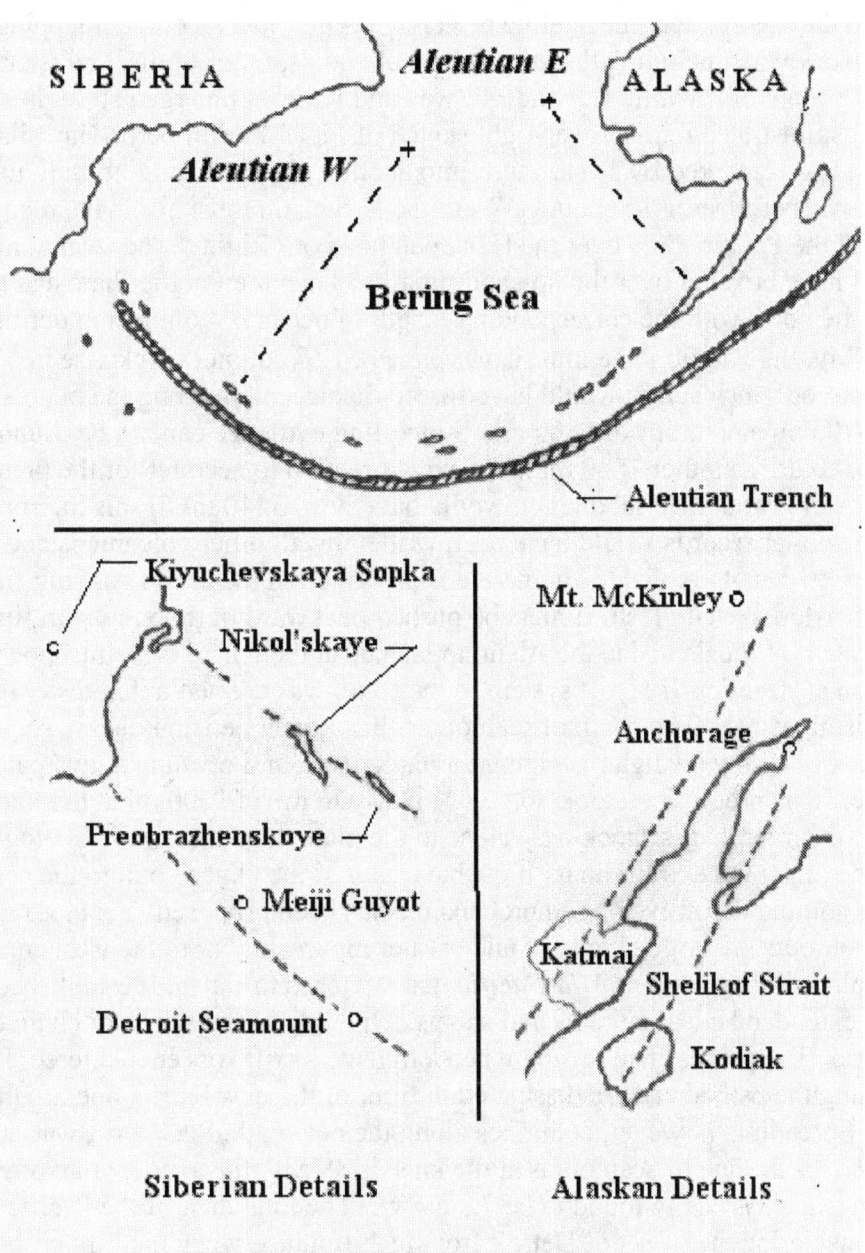

Fig. 1.19. Bering Twins.

Emperor-Hawaian Seamounts.

The concept of drifting plates has provided modern geology with an explanation for chains of volcanic islands. A large chamber of molten lava rising from great depth comes to rest not far below the crust. Referred to as a *hotspot*, it furnishes lava to build mountains underwater that slowly reach the surface. As a plate moves over the hotspot, the pathway for lava flowing to a mature island is cut off while another one opens closer to the hotspot, according to the theory. Recent satellite research has provided accurate mapping of seamounts worldwide including the entire chain of shoals, reefs, atolls, and islands that constitute the State of Hawaii. Beyond Midway and Kure, submerged structures run a total of 1840 mi as illustrated in the lower-right sketch of Fig. 1.20. Emperor-Hawaiian History. The western end is marked by a sharp turn into another chain running north to the Aleutian Islands, known as Emperor Seamounts. Well, the N-S chain must also have recorded movement of the Pacific Plate over the Hawaiian hotspot. Right? The seamount Detroit at 70 My must have hovered over the hotspot initially. Then the Pacific Plate had to move 1600 mi to the north with the corresponding length of ocean bottom lost under the Bering Sea. At 38 My, the Pacific Plate must have rotated 61.6° counter clockwise in a brief geological period. Such action would have forced displacement along the plate's periphery of at least 6400 mi at a radius of 6000 mi. Supporting evidence can not be found. Finally, the plate had to drift another 1840 mi in the new direction to account for the present reality. The total loss of seafloor by subduction would have been 3440 mi. If this interpretation were correct then similar records would have been written by all other volcanic islands in the Pacific. They were not. A viable alternative is presented in the accompanying figure starting with the upper-left sketch. Detroit must be pushed backward in time to its original position that is indicated by shoals inside the Aleutian Islands at the island of Semisopochnoi. Upon the explosion at *Aleutian W*, a rift system to the south was created much faster than the speed of seismic waves. The chain of seamounts then developed and died in sequence as each one sank of its own weight, wedging the fault apart, and opening a lava path for raising the next one. This progression took 70 - 38 = 32 My to travel 1600 mi at the rate of 50 mi/My or 8.05 cm/yr that is shockingly close to the plate drift at Hawaii. Geologists have realized that something extraordinary must have caused the "knee" but, to the author's knowledge, nothing definitive or comprehensible has been proposed. Yet clues are provided by bottom contours. A large cluster of underwater mountains far to the west constitute the rebound peaks of a huge crater, *Mid-Pacific* that stretches to Japan. It established the weak zone of the extended Hawaiian reefs and shoals. Opening of the Emperor chain could only proceed so far. The process had to stop when a transverse rift was encountered. Lateral forces opening the path ahead are dissipated in front of the new barrier and are thus ineffective. Spreading, however, continues along the new path. Such a rift was created by *Mid-Pacific.* As the age of seamounts at the knee is 38 My, the average rate of progress to the vacation islands is easily found to be 7.8 cm/yr. The total drift of the Pacific Plate is found by considering the offset of Detroit from its birthplace when the Emperor Chain was aligned with the epicenter of *Aleutian W*, amounting to 850 mi instead of 3440 mi. The Emperor-Hawaiian chain can be explained from the geological evidence of impact craters but not by plate tectonics.

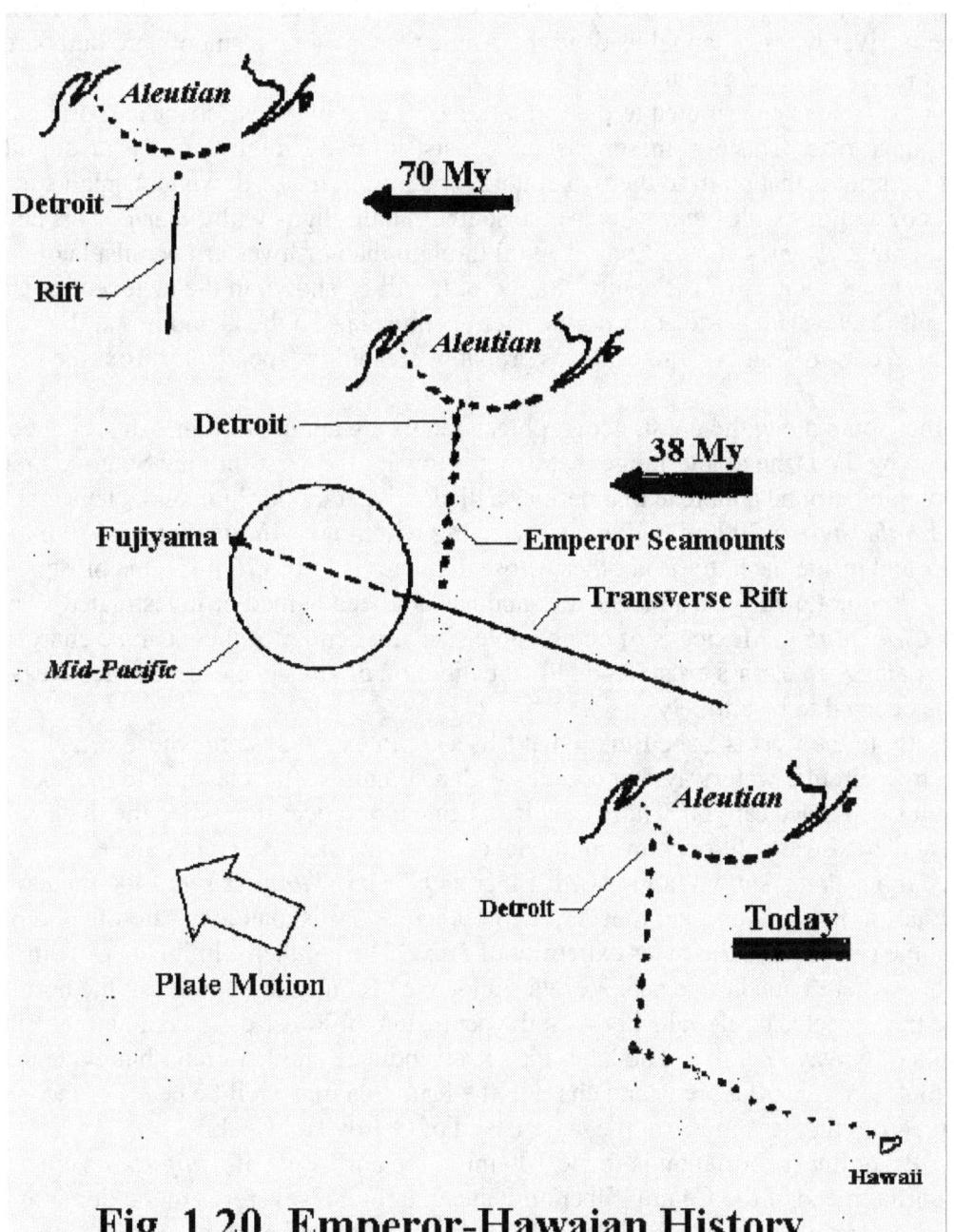

Fig. 1.20. Emperor-Hawaian History

Siberia.

Two impacts in eastern Siberia, *Siberia* and *Lena* each about 500 mi in diameter, overlap about 30 percent. The inner rim of *Siberia* is indicated primarily by the arcing Khrebet Gyddan range opposite Kamchatka Peninsula that is concentric with the curved coast. An outer rim is suggested by oddly shaped peninsulas in the Sea of Okhotsk near Gizhiga and Yamak although most of it has been lost. The Siamese twin, *Lena*, is named for a major river whose channel leads to the Arctic Ocean along a smooth arc outboard from the Verkhoyansky Khrebet range.

Both explosions contributed to the formation of Sakhalin whose shape and orientation suggest linear rifts. This 600-mi long island consists of two straight sections on slightly different azimuths that point to the two craters. The northern half down to a small spur on the east coast aligns with *Lena* whereas the southern half aligns with *Siberia*. This latter mountain range extends an additional 190 mi through the northwestern peninsula of Hokkaido to the vicinity of the capital, Sapporo. It will be shown in the next section that the Kuril fault reaches Hokkaido at its northeastern peninsula. So the island lies at the intersection of two fault systems and was probably created independently from the rest of Japan.

Farther south along the coast, geographical details are again familiar. An arcing coast is backed up by the Dzhugdzhur range that extends toward the west into the Stanovoy range, both wrapping around a plateau that defines a distinctive, circular highland. This crater named *Khabarovsk* is 790 mi in diameter. Nesting within it is *Aldan Plateau* 330 mi in diameter and inside the latter is another whose diameter is only 170 mi. This pipsqueak certainly does not qualify as a megacrater and has not been named or investigated. The famous *Chicxulub* in Mexico is of comparable size and neither of them can be charged with massive damage to earth's crust. And global extinction of any species from such weak events is deemed to be unlikely.

Directly to the west is Lake Baikal that fills a complex rift system whose center is well defined by a circular waterway that couples with a straight section leading to the lake through a steep-sided canyon. These details indicate a rare occasion when the thick, continental crust was split open by an impact of moderate size, 320 mi diameter.

The largest crater shown in Fig. 1.21. Siberian Cluster, *Tunguska*, is 1400 mi across. While clues to its rim are rather meager, its presence is unmistakable in satellite photos. In the east, the rim touches the outer extremity of *Lena*. The southern limit is near Tomsk while the western boundary is near Voruta. Evidence becomes ambiguous in the northwest because the expected circle falls between the peninsula of Polustroy Yamal and the curved peninsula of Novaya Zemlya. The latter may itself indicate another crater but evidence is too flimsy to accept. An offshore island chain in the Kara Sea may well be peaks of the submerged rim that terminates at the large island of Ostrov Bolshenk.

An extraordinary formation is found 100 mi to the northwest of *Tunguska*'s epicenter. At the northern end of the Central Siberian Plateau, a round, severely eroded dome of intrusive rock provides for a radial drainage pattern. This large mound with a diameter of 330 mi rises to an elevation of 5600 ft extending over an area of 86,000 sq mi. It must surely mark the place where a powerful explosion depressed the bedrock and opened a channel for the escape of lava.

Two larger craters in Siberia are described in the following section because intricate overlapping in Fig. 1.21 would have otherwise become unintelligible.

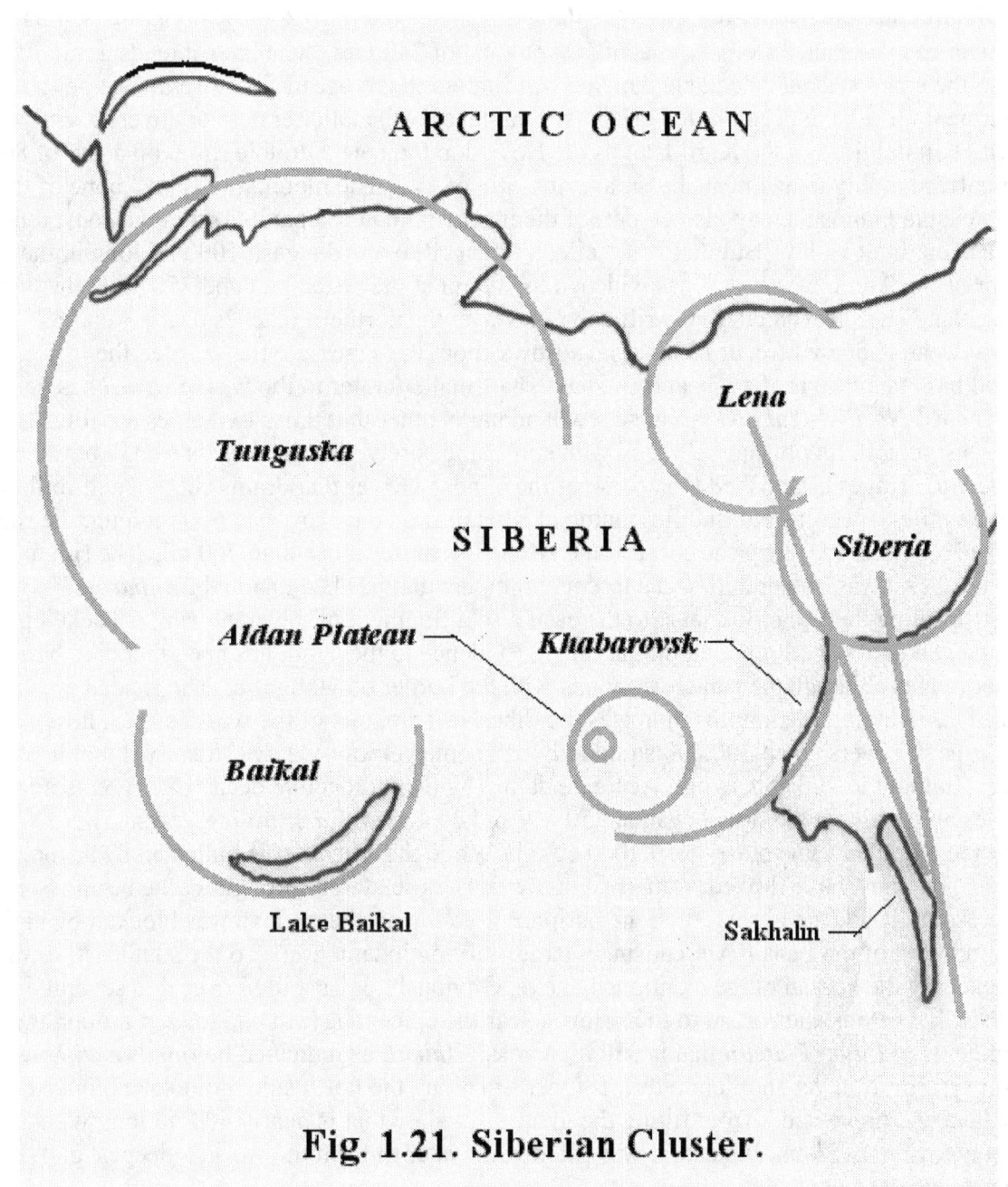

Fig. 1.21. Siberian Cluster.

Terrestrial Mare.

Logging in at 1650 mi, a brutal event overthrusted the Pacific Ocean floor forming the Kuril Trench. The Kuril Islands were raised from the northeastern peninsula of Hokkaido, Japan, to the southern tip of Kamchatka, Russia, then along the coast past an old friend, Klyuchevskaya volcano. The coastline was formed by the explosion up to its root above 60° north latitude. This entire record in the Pacific Ocean stretches 1200 mi along the islands then an equal distance along Kamchatka for a total of 2400 mi. The arc subtends a full 80° from the epicenter but detectable damage was limited to the ocean floor. Evidence on the continent must have been blotted out by several layers of smaller craters or covered with lava. Parallel lines of the Kuril Islands off Hokkaido indicate a double rim with a gap of 80 mi corresponding to an unusually small, rim ratio of 4%. The mountainous backbone of the Kamchatka Peninsula may also be part of the inner rim that is separated from the coast range by a deep, fault valley. But that is not all. A linear rift opened a gash 1000 mi long in the floor of the Arctic Ocean that has widened to 200 mi at the Asian end and 350 mi at the Greenland end. Beyond that, the rift coincides with the northern reaches of the Atlantic Mid-Ocean Ridge where, at Iceland, lava flows about the same rate that created the major, flood basalts millions of years ago. A somewhat smaller crater to the west of *Kuril* has been christened *Noril'sk-Yakutsk* because few landmarks other that those two cities are found in this vast region. Its rim can be traced with surprising surety. To the west, a trough between concentric ridges established the course of the Yinesey River that drains about a half million square miles of central Russia. Beginning at Khatanga on a narrow bay 600 mi long, the rim carved a perfect arc along the coast of the Arctic Ocean for more than 700 mi. The rim to the northeast was obliterated by a subsequent impact that has been named *Khabarovsk*. The coast fronting the Sea of Okhotsk to the east, also a fragment of the crater rim, is backed up by the Dzkugdzhur Khrebet mountain range. Evidence to the south has been muddied by other ranges although the rim corresponds with the border of Mongolia. The shaded area in Fig. 1.22. Russian Mammoths indicates the Siberian Trap, one of the world's great lava plateaus. It covers about 300,000 sq mi although some geologists have suggested that it might amount to 700,000 sq mi. Accurate dating by the radioactive decay of Argon isotopes has established that the flow began at 250 My and continued for 1 million years. The volume extruded came close to 500,000 cu mi. While the epicenter is indicated by a small cross, lava must have flowed from great many fissures scattered throughout the basin. Some interesting details are observed on the periphery. To the northeast, lava was blocked by high ground west of the Lena River causing a large, curving indentation. To the southeast, it was blocked by the coastal range mentioned above. Obviously, lava spilled over into several of the smaller basins and spread to their rims. Near the epicenter, lava had to skirt around the highlands of *Paton Plateau* that is 330 mi across. *Baikal*, as indicated before, is a complex, fault system from yet another crater. Its inner rim along the north side of the lake blocked the lava and prevented it from filling the great, rift valley that is nearly 400 mi long with an average width of 30 mi. Instead, it was filled with water to form the world's deepest and oldest lake.

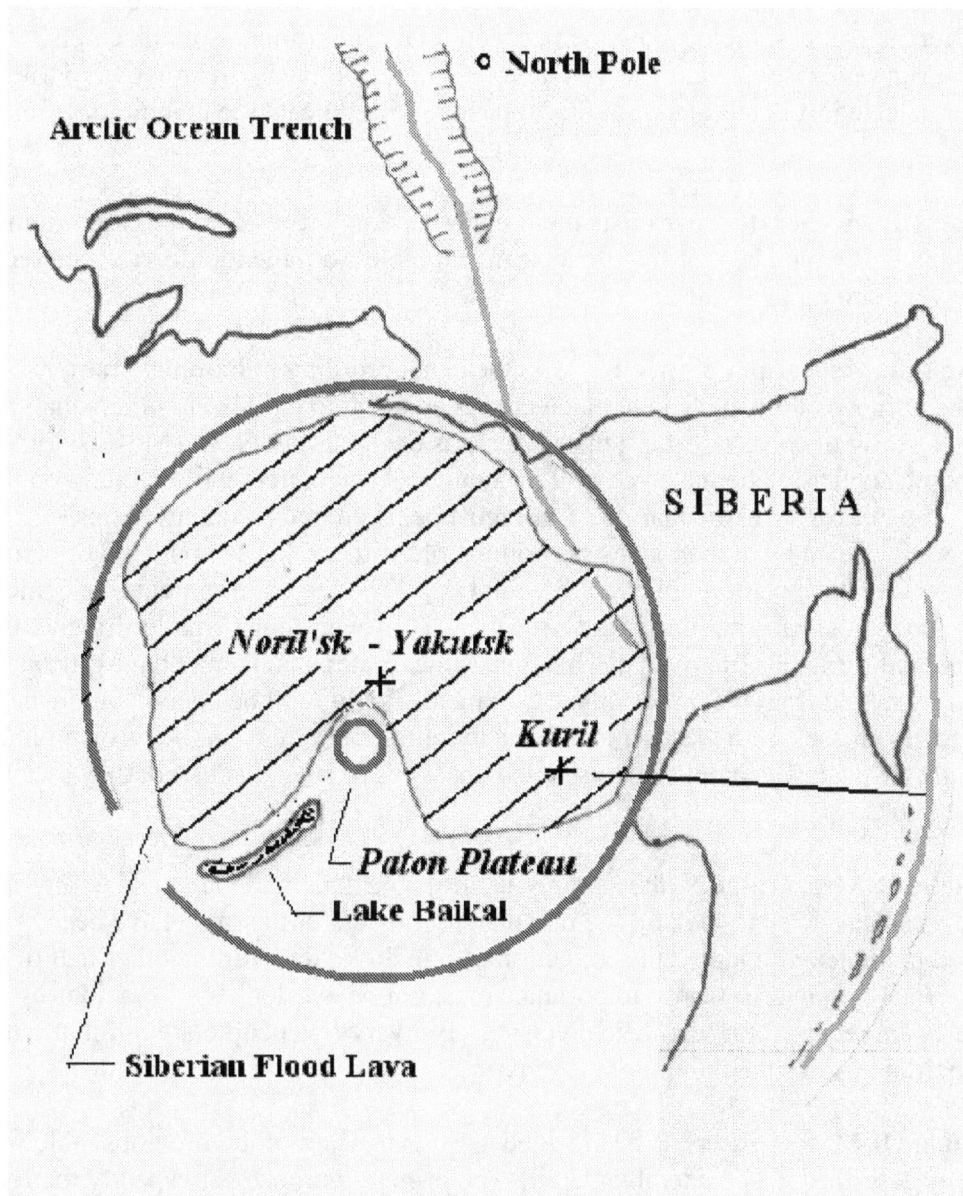

Fig. 1.22. Russian Mammoths.

Database.

Exploration to this point has involved 103 craters selected primarily for size although smaller ones were included if they appeared to be especially significant. This number amounts to only half of the total 208 events that have been discovered, named, traced, measured, and analyzed in terms of geological effects. Data for all of them are tabulated in Appendix A where the information is presented in several ways to facilitate investigations along different avenues. Included are names, continents, coordinates of epicenters, diameters, rim ratios whenever observed, and the angles of linear rifts relative to true north whenever detected. A quick glance at this appendix will indicate the magnitude of the database.

Table A.1. Megacrater Sizes lists them in descending order of their diameters in miles. They are huge. It would be worthwhile to scan that table from the top down to appreciate the number over 1000 mi in diameters.

Table A.2. Megacrater Latitudes sorts the data according to latitudes from 90°N to 90°S. The purpose of this table is to place craters adjacent to or close to others that might belong to a string of comets striking in sequence as the earth turned west-to-east under the aiming point, such as, Shoemaker-Levy 9 on Jupiter. Some craters will be found on the same latitude, give or take a narrow margin. Care must be taken that candidate series of craters have the same age and that their sizes are compatible with the known patterns of comet strings that have been photographed in the night sky. The largest fragment of a comet is expected to lead the pack but that is not necessary. However, much smaller fragments will trail far behind the major fragments. This means that crater chains may have much smaller craters far removed from the initial impact. Some difficulty will be encountered when the trail of craters crosses from western to eastern longitudes but they can be resolved in each case. While only a few examples of suspected chains have been discussed, they appear to be rather common.

Table A.3. Megacrater Names simply presents the data in alphabetical order. It is provided as an easy way to look up the raw data for any crater that may have been mentioned in the text or maps. This collection also includes data for the other half of the craters, not yet introduced, whose importance must not be overlooked. Locations by coordinates are required for scientific precision. But very few people are familiar with the global coordinates so something more is required.

Catalog Of Megacraters has been added as an expansion of the previous table where each one is described in considerable detail. The primary purpose is to specify their locations relative to major landmarks such as cities and geographical features. Most instances try to delineate the crater by specific clues.

Members of the safari can now return to camp, take off their pith helmets, and relax over a spot of gin. They should have a lot to talk about.

Chapter 2. Life Extinctions.

*"Discovery consists of looking at the same thing
as everyone else and thinking something
different."*

- Albert Szent-Gyorgyi

Despite an earlier promise to avoid esoteric words of science, a few must be called upon to make sense out a complicated subject. Early studies of geology recognized distinct layers in rock that were thought to be consistent around the world. Twelve major zones were identified starting about 3.4 billion years ago when earth was very young. The major zones, called epochs, were separated by thin layers of a different character than the adjacent deposits. Also, each of the epochs was broken down into several subdivisions with a total of 36. All of the periods were given names to facilitate discourse among geologists. Emphasis was upon the layers rather than their interfaces. So a reference to the Upper Permian and Lower Triassic would convey a date of about 325 My give or take 10 million years.

One thing was obvious. Fossils of formerly living creatures changed in character at the boundaries between geological periods. A much simplified sequence ran through a) bacteria and algae, b) jelly fish and worms, c) tribolites resembling horseshoe crabs, d) sponges, e) fish and amphibians, f) insects and sharks, g) mammal-like reptiles, h) reptiles, i) dinosaurs, and j) mammals and birds. Something was surely destroying life forms at the boundaries. Recovery was accompanied by new species of ever-increasing complexity. Life extinctions at the interfaces were assigned new names, such as, Cambrian and Ordovian. As common in most scientific disciplines, the technical jargon was incomprehensible to outsiders. Communication across boundaries of neighboring sciences was at least hampered.

The concept of slow changes in life forms over billions of years, from the simplest to the most complex, originally ran afoul a belief that God made all the creatures within the astonishingly short period of one week. But Charles Darwin himself realized that the starting clock for evolution had been reset several times. Most living creatures were exterminated at the end of each period while some survived as seed stock for the next evolutionary development. This "punctuated evolution" continued until the recent appearance of *homo sapiens* that is probably not at the end of road.

Great debates on these matters have reverberated through the halls of academia, justice, public forums, and literature. Even the raw concept of evolution generated vociferous objections. Some of the opposition even claimed that the devil himself designed all the fossils and hid them in solid rock around the world to fool scientists. While still being debated, that hassle generally gave way to a similar contest. The new issue was what caused the disappearances of so many life forms. One camp favors earthly causes of life extinctions (volcanism) versus those who argue for cosmic intervention. Volcanic activity, goes one argument, was accompanied by massive release of toxic gases and carbon dioxide that upset a delicate balance in earth's atmosphere. Impactists, on the other hand, believed that explosions of asteroids or comets near the surface created clouds of dust that spread around the globe and lingered for an indefinite period. Extinctions have been attributed to blast waves, vaporizing, burning, poisoning, starvation, suffocation, death of embryos, and hurricane winds of unprecedented velocity. Some creatures far removed from the catastrophe would freeze or starve from the lack of sunlight so vital for plant growth.

Scientists have not reached consensus on the exact causes of death that probably involved combinations of the several means.

Further insights may be found by concentrating attention upon the kinds of creatures that *survived* each crisis. Why, for example, did creatures in fluids live through the catastrophes? Crocodiles, turtles, sharks, and coelacanths were apparently protected by the water in which they swam. The airborne dinosaurs, the family of pterodactyls, survived to seed the evolution of birds. Both groups were immersed in fluids. They were able to live through intense shock waves, earthquakes, and air pollution. Recent experiments have shown that people are not harmed by extraordinary acceleration when encapsulated in a water-filled container. Perhaps the present concepts of mega-events will shed some light on this subject.

Chicxulub

A suggestion by Nobel Prize winner, Louis Alvarez at the University of California, provided a necessary connection between asteroids and craters opposing the idea that large holes in the ground were always of volcanic origin. While the element, iridium, is rare on earth it is more common in asteroids. A great explosion of an asteroid must have spread its iridium around the world that should be found at the interface between geological epochs corresponding with demise of the dinosaurs. Continuous hammering by the media has brought into the public awareness that dinosaurs were destroyed about 65 My. That geological record is the Createseus - Tertiary boundary, often abbreviated as K-T. Samples of that layer from around the world revealed an abnormal concentration of iridium.

The final step in this research was to locate a crater of the right age that was large enough to have caused the observed effects. About that time a monster crater was discovered in Yucatan, Mexico near the town of Chicxulub. It lies deep under ground with little evidence on the surface. Half is suggested by an arc of water-filled sinkholes in sandstone. The other half under the Gulf of Mexico was accidentally discovered by research vessels surveying for oil. This crater, called **Chicxulub,** fit the bill and generated an enormous amount of publicity. The diameter seemed to grow in the press but stabilized at about 200 mi. In comparison with other asteroid craters, it was prodigious.

This geological formation was confidently established as 65 My. The search was over. Explosions of a huge asteroid near Chicxulub, Yucatan scoured the giant basin and killed the dinosaurs. Chicxulub became the largest crater known on earth and still retains that reputation. However, several authors suspected that other evidence pointed to much more devastating events based upon circular features of the global landscape. Appendix B. Previous Research summarizes those suggestions although others may have been overlooked.

In the present catalog of 208 megacraters, **Chicxulub** is among the smallest. Could it alone have been more destructive that its numerous, big brothers? Was there a much more powerful explosion close to the same time? One thing is clear. Dinosaurs were not wiped out by **Chicxulub** because it is too small.

The Sample

Correlating specific megacraters with each of the extinctions is an overwhelming task The full range of crater sizes is too great to handle. Nor can the time scale back to the earliest examples be encompassed. And a critical deficiency exists in the knowledge of crater ages. What can be done?

Extinctions, it is assumed, should be associated with the largest craters. So candidates should be found in Table A.1 of Appendix A near the top of the list. But most of the Age Column in the table is blank. Only those with assigned ages can be used now. Ten examples with ages are noted down to a diameter of 1000 mi as follows:

1.	*Noril'sk - Yakutsk*	6.	*Tanzania*
2.	*Arctic*	7.	*Iran*
3.	*Weddell*	8.	*Kuril*
4.	*Kalahari*	9.	*Ethiopia*
5.	*Himalaya*	10.	*Great Plains*

Uncertainties in their ages may be large but are probably small enough to preserve their correct sequence. A very serious weakness is related to ignoring the 53 other craters larger than 1000 mi. They are the very ones most likely to be linked to extinctions. Any investigation along these lines is similar to glimpsing a three-ring circus through a tiny peephole. Gross errors of interpretation are likely. However, the present method may be valuable.

Lava Plateaus

Vast outpourings of molten rock have covered hundreds of thousands of square miles all over the world. The volume of material extruded to the surface has approached a million cubic miles in a few cases. Some of them were indicated in the previous chapter as

Fig. 1.1	Etendeka in southern Africa that was torn apart with the major portion, called Parana, riding away on South America.
Fig. 1.2	Karoo in southern Africa.
Fig. 1.3	Ethiopia in east Africa.
Fig. 1.7	Ontong Java in the western Pacific.
Fig. 1.12	Columbia River.
Fig. 1.13	Caribbean Sea.
Fig. 1.22	Siberian Traps.

There are many more. A collection of scientific papers on large igneous provinces, edited by Mahoney (1997), provided an up-to-date review in great detail. A world-wide, color map prepared by co-author Coffin (1997) showed the provinces on the front and back covers of the book. The provinces vary in size so much and are so numerous that they are hard to count. It would be fair to say that the major examples number in the dozens.

Dates for the onset and termination of lava flows can be determined in a laboratory by radioactive decay of Argon. The method is accurate to within 50,000 yr for lava that is hundreds of millions of years old. An important thesis of the present work is that major lava flows were initiated by cosmic blows strong enough to fracture the crust of earth, start the flow, leave multi-ring structures, and their linear, rift systems. In other words, ages of great lava plateaus tell when the missiles struck, created the damage, and turned on the lava faucets.

The age of Siberian Traps has been accurately established at 250 My that is, of course, the same as the Great Extinction. That point has been recognized by some scientists who attribute extinctions to the attendant emission of noxious gases from the lava. This plateau is critically important because of the close fit to the boundaries of ***Noril'sk-Yakutsk***. In addition, some details support the relationship. The lava flowed around ***Paton Plateau*** and stopped at ***Baikal*** and the outer rim of ***Lena***. This pattern in Fig. 1.22. Russian Mammoths should be reviewed.

Lava has no independent option to break through the surface somewhere to create volcanoes, islands, or plateaus. A powerful force is required to rupture the bedrock and provide escape routes for molten material from great depths. The only thing in nature capable of that feat is the explosion of cosmic missiles. That is, indeed, a basic thesis of this book. Explosions rupture the surface and release lava inside the craters they generate or on their boundaries. Thus, four things are intimately related: explosion of comets, formation of megacraters, massive lava flows, and extinctions.

No simple creature can anticipate the onset of disaster and communicate that knowledge around the world. "Uh-oh! A comet is approaching. We should all start dying." No way! The lava flow must be triggered by massive explosions that strike without warning. Then death follows. Lava slowly fills the basin over a few million years long after a global, dust canopy has settled out.

Noril's'k-Yakutsk thus becomes a convenient anchor point for the time scale to be investigated. At 250 My, it predates the breakup of Pangea at least 100 mil yr. And it displays the basic requirements of a) a great crater, b) the lava plateau of Siberian Traps, and c) the most intensive destruction of life on record called Great Extinction.

Such events must have been common from the time that cosmic material agglomerated into the new-born earth. However, no record of impacts would have been left in the molten rock and it is unlikely that any form of life existed at that time. Only after the earth had cooled enough to gain a veneer of solid rock would impact craters be preserved. If impacts really initiated lava floods then a correlation must be attempted to identify which impact was responsible for each one. Data in Table 2.1. Impacts And Lava Flows constitute a key to unlocking this door back to 250 My.

Table 2.1. Impacts And Lava Flows.

Time (My)	Lava Age (My)	Lava Flow Name	End of Era [a]	End of Epoch [a]	Probable Crater	Diam (mi)
0						
—	18	Columbia		Miocene	*Great Plains*	1060
—	30	Ethiopia-Yemen		Oligocene	*Ethiopia* [b]	650
—	65	Deccan Traps	Cretaceous	Gulfin	*Iran*	1370
—	90	Madagascar		Cenomanian	*Tanzania*	1100
100						
—	117	Rajmaha		Aptian	*Himalaya*	1640
—	135	Parana-Etendeka	Jurrasic	Upper Jurrasic	*Kalahari*	1120
150						
—	190	Antarctic-Africa		Plienbachian	*Weddell*	1140
200	198	NE America	Triassic	Upper Triasic	*Arctic*	2400
250	250	Siberian Traps	Upper Permian		*Noril'sk-*	1410

					Yakutsk	

a. From Courtillot, V., Evolutionary Catastrophes, Cambridge, 1999.
b. May have been larger.

The first conclusion to be drawn from the table is that each of the great lava flows and their associated extinctions corresponds to a specific multi-ring structure. And the final, logical step is inescapable. The designated cosmic explosions killed most of the life forms living at their respective times.

The second column shows the time of the lava flows listed in the third column. Columns 4 and 5 indicate the ends of geological periods. The last two columns deal with the craters for which age has been assigned. There appears to be an acceptable correlation between lava flows and craters. It should be stressed, however, that many other craters could have been involved. With Siberian Traps as a base line, note the dates of 200 My for *Arctic* vs 198 My for *NE America* plateau. Also notice the 65 My for Deccan Traps in India that is the same age as *Chicxulub* and a more likely candidate for the dinosaur problem. The 135 My for Parana-Etendeka provides a good age for *Kalahari*. The lava plateau must have started about that time before being split apart by *Timbuktu*. The Rajmaha flow at 117 My may reveal the age of *Himalaya*. It may be significant that a majority of craters lie in a narrow range of 1060 mi to 1640 mi. That detail might indicate ease of dating for those sizes or, perhaps, a slight preference for that size in nature.

This tabulation, it will be recalled, considers only 10 of the 36 known extinctions. Also, the confidence level is not uniform across the board. Many uncertainties remain because:

a) knowledge of the fossil record varies around the globe from intensely studied areas to regions like the Congo, Amazon, Siberia, and Antarctic that have barely been scratched,
b) three billion years of cosmic bombardment weakened the integrity of earth's crust unevenly, and
c) rims and rifts from several craters that are still extant combine randomly at different locations

Nevertheless, this approach should point the way toward a reliable reconstruction of earth's cosmic history.

A Map

The story of this bombardment can best be told with a map showing the locations, sizes, and sequence of events as in Fig. 2.1. Some Lethal Craters where enclosed numbers designate their sequence. Each impact reset the evolutionary clock so that new creatures of astonishing variety could develop during the ensuing eras. These repeated starts in the long run achieved the highest level of life to which the author and reader belong. So due credit should be given to the celestial object that snuffed out the dinosaurs and opened the age of mammals. Otherwise, we would not be here.

51

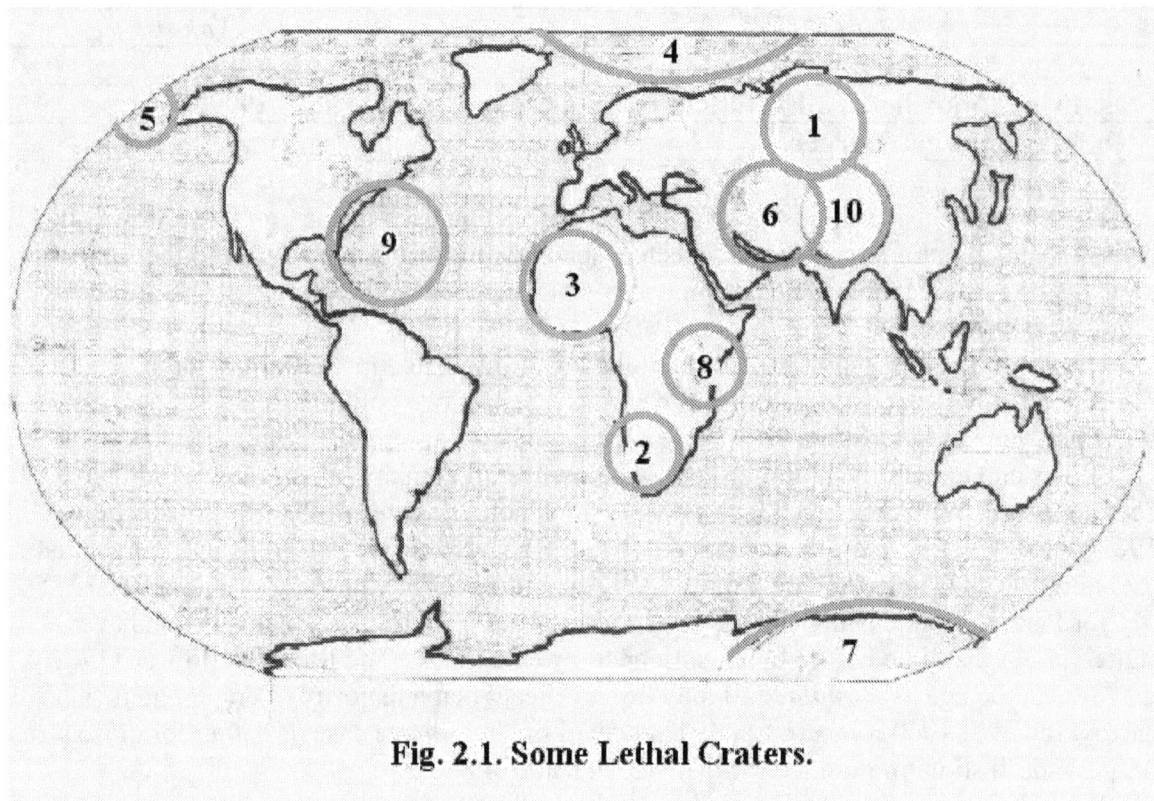

Fig. 2.1. Some Lethal Craters.

Bam! Bam! Bam! Something like this series of explosions has pestered earth from its birth. Future research may eventually allow a reliable and complete reconstruction of the history.

Brief reviews from Chapter 1 should help to bring this picture into focus.

1) *Noril'sk - Yakutsk*. An untold number of craters from earlier times were undoubtedly buried under the Siberian Traps. Only younger ones could be shown in Fig. 1.21.
2) *Arctic.* Almost nothing is known about any craters that might be on the bottom of the Arctic Ocean. The U.S. Navy, and presumably others, has developed detailed maps of underwater topography that show a profusion of ridges and faults. Some knowledge of cratering might be developed from that data.
3) *Antarctica.* Except for the four major events shown in Fig. 1.8. the impact history of this continent is a complete blank. The central depression now filled with ice may have been started by an impact.
4) *Kalahari.* This event preceded the separation of South America from Africa.
5) *Himalaya.* A rather isolated event.
6) *Tanzania.* Fig. 1.2. shows that previous damage of extraordinary proportions may have been inflicted by Congo. That could explain the intense volcanism in the region and lava layers below level grasslands.
7) *Iran.* Few details in Fig. 1.5. are available but this one was capable of the observed results without softening of the target area by earlier explosions.
8) *Kuril.* Another example of an extremely powerful event that needed no help.
9) *Ethiopia.* Smallest in the group but influenced by nearby rifts.

10) **Great Plains**. While the sequence of regional events in Fig. 1.2. has not been established, overlapping of at least four rims focused upon the Columbia River Plateau. Release of lava could have been triggered by any of them or in any combination.

Comparison of data in Table 2.1 and Appendix A reinforces a concept of long standing. Experts have proclaimed that cosmic missiles of greater size and energy struck earth in the earliest times. The average diameter of the largest seven in Appendix A, 3150 mi, can be compared with the corresponding value from Table 2.1, namely, 1400 mi. In other words, craters prior to 250 My appear on average to be 2.3 times larger than those after that date. Experts have also proclaimed that impacts of extraordinary size did not continue past about 4,000 My. The present study fails to support that idea by contrary examples.

Super Giants

Table A.1 in Appendix A reveals many craters larger than those above. In fact, seven of them have diameters from 2400 mi to 3640 mi. If the weaker events were capable of extinctions back to 250 My, then all of these super-giants must have been more frightful. Any one of them could have triggered an extinction after 250 My. They might also have been driving forces for major extinctions at earlier times. The geological record has been reliably dated for the end of periods as follows:

Devonian	345 - 355 My
Silurian	395 - 410
Ordovian	425 - 430
Cambrian	475 - 500
Eocambrian	570 - 600
Precambrian	600 - 700

All but the last one have several subdivisions. The super-giants above may well have been responsible. It has surprised the author to find evidence of a crater as old as 250 My.

The remaining task, not yet addressed, is to expand the dates of lava plateaus older than 250 My and attempt to correlate their geographical locations with the known coordinates of impacts shown in Appendix A. Additional help can come from overlapping craters whose sequences are obvious.

Extinctions

A very broad and intensive study of marine organisms was published by Raup (1984) and colleagues who traced the record back to a little more than 500 My. They found a pattern of extinctions resembling a jagged, saw-tooth curve. In each instance, the percentage of extinctions rose, crested at some maximum value, then fell steeply back to normal. This process typically lasted millions of years. Extinctions were attributed to "...environmental disturbances." The percentage of extinctions suddenly began to decrease and eventually improved when the environmental disturbance was resolved. The population then began its journey back to normal. Strictly speaking, these episodes were not extinctions because later generations of many species survived. Other, more advanced animals in the record did not. The trigger for these disasters must have coincided with the beginning of environmental changes, not at the time of minimum populations. This observation seriously modifies the traditional timeline for extinctions.

Raup's cyclic nature of extinctions was picked up by Muller (1988) at the University of California who smoothed the data into a wave-like curve with a period of about 26 mil yr. None of the known processes in the solar system provided an explanation of the facts. Something entirely new must have been the cause. A promising hypothesis was the gravitational influence of an unknown, companion star to the sun. Calculations required it to travel in an elliptical orbit reaching a maximum of 176,000 times the radius of earth's orbit. In other words, the maximum separation would place the star 2.8 light years away. That value would be about half the distance to the closest known star, Alpha of Centarus. The hypothetical star has not been found although it could be seen with binoculars if it could be identified against a background of millions of others.

Another explanation with more appeal relates to movement of the solar system. It oscillates perpendicular to galactic plane with a round trip of 64 mil yr. Thus it passes through that plane every 31 to 33 mil yr.

Present ideas about extinction seem to be too simple. Rather than a steady periodicity of 26 mil yr, intervals are not uniform. The average period calculated by the author is 32.9 mil yr with variations of + 66% to - 52%. That performance is most unbecoming to any star in a stable orbit. Also, the average period between extinction fits nicely within the uncertainty of oscillation of the solar system.

For the Maestrichtin Extinction at the Cretaceous-Triassic boundary (read dinosaurs), calibration of the time scale and judicious use of dividers yield new data. The onset of their threat began at 78.7 My. That means the trigger for loss of dinosaurs was 13.7 My before the traditional date of 65 My. Therefore, the cataclysm that killed them struck much too soon to be the result of *Chicxulub*. A great many dinosaurs perished swiftly after the disaster or within a few years when food became scarce. Successive generations must have survived for 13.7 mil yr. But the large animals could not adapt to the environmental changes. Smaller species, requiring little nourishment, survived and adapted to the new conditions. Populations grew and diversified.

Looking back along the saw-tooth record, one can measure the beginning of the earlier, eleven extinctions. Rounded to the nearest 5 mil yr, they are 95, 150, 180, 210, 240, 275, 305, 350, 380, 460, and 500 My. Consider the Great Extinction at the nominal 250 My. The numbers above show an event at 240 My, 10 My earlier.

These ages are the expected times when all major disasters struck and set the stage for extinctions to follow. They are the best indicators for the times when the craters were formed. These interpretations will be put to the test of science as more data becomes available. Undiscovered megacraters are, of course, absent from the catalog but any found in the future should also have one of the above ages.

Prospects

The future looks bleak. Since destruction of the dinosaurs and most other creatures, 78 mil yr have passed. In view of the periodicity of extinctions, earth is overdue for another shellacking. On the other hand, earth has been spared for 125 mil yr between calamities in the distant past. During that interval, however, two lesser but very traumatic periods were experienced. The U.S. government and the scientific community are vigorously preparing to guard earth from asteroids whose damage would stretch the imagination. A promising idea is to send a spacecraft out to the approaching bullet and deflect its lethal path with rockets or nuclear explosions. Substantial protection may be in sight. Prospects regarding comets are less cheery. How well would the system work against a dirty snowball 30 mi in diameter? Fracturing it into innumerable pieces might help but most of them would still hit the target. No present ideas offer hope of protection from comets. But mankind can take a bit of comfort that earth has been relatively safe for 78 mil yr. The vast majority of comets, all but an infinitesimal fraction, simply loops around the sun and departs into space with no influence upon earth. But humanity is playing cosmic craps: 7/11 it survives, 12 or snake-eyes it doesn't.

Chapter 3. Statistical Considerations.

> *"A compete assessment of the role of large body impacts in terrestrial evolution will no doubt require new minds and fresh handling on the many problems.*
>
> — Leon T. Silver

The total number of megacraters is large enough to justify some statistical analyses. What is the meaning of the size distribution? Do craters tend to cluster about any particular size? How many such clusters exist? At what sizes? How do the present discoveries compare with terrestrial craters from asteroids? And how do they compare with impacts on other bodies of the solar system? Are the craters concentrated on particular latitudes? How complete is the present tally? Such questions are addressed here.

Summary Of Sizes

Any discussion of statistics must involve some concepts and details that the general reader might find boring or incomprehensible. Anyone bogging down in this chapter should simply scan through the text to get a feel for the subject. Or the chapter can even be skipped altogether without great loss. Perhaps, later review will be more meaningful.

General trends can be seen by scanning down the column of crater diameters in Table A.1 of Appendix A. Very large ones are rare whereas the smaller ones are progressively more numerous. But the overall meaning of the data is not clear. It must be organized and displayed in some illustration to reveal the significant trends.

Diameters are found throughout the range from 3640 mi down to 90 mi. *One hundred ninety five (195)* are larger that **Chicxulub** to which the loss of dinosaurs has been attributed. The author's original intention was to ignore craters smaller than **Chicxulub**. Many smaller ones were found only to be acknowledged then passed over. For special reasons, 13 lesser examples were logged into the overall tally.

For simplicity and graphical convenience, every 20th crater from the whole list was selected instead of dealing with each one individually. This approximation has little effect upon the findings to be presented. A smooth curve of crater diameters is shown in Fig. 3.1. Crater Diameters along with numerical data on the left. The largest craters are enormous. While their magnitudes are somewhat shocking, they are not out of line with major craters on other bodies of the solar system. The largest known to contemporary science is South-Pole Aiken on the moon with a diameter of 1300 mi. It falls within the mid-to-lower ranges of those on earth. Earth's largest, Joplin, is 2.8 times bigger than the lunar giant. The location of South-Pole Aiken raises questions about how the comet got there.

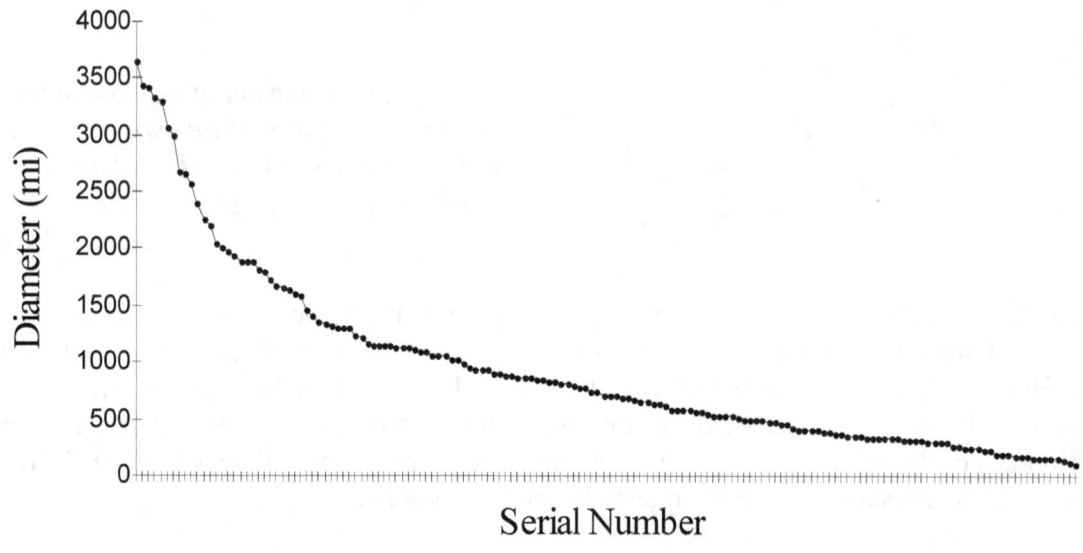

Fig 3.1. Crater Diameter.

Clustering

Two-thirds of the curve on the right is very smooth as indicated by separate, detailed examination for all craters. A minor platform or stair step in the curve at about 1000-mi diameter suggests a tendency for clustering about that size. Ten events lie in the narrow range of 1100 mi to 1170 mi. Even though similar results are reported later in a different context, this real but minor cluster is not considered to be significant. In other words, the population of comets in the solar system spans the full range represented by the megacraters. Their size distribution follows an exponential law without significant deviations.

Some Math

Further insights arise from employing a mathematical trick. When examining numbers over a wide range of values, it is convenient to deal with their logarithms. While a mystery to many people, a logarithm is simply the power to which 10 must be raised to designate the true value in question. Everyone knows that 10 squared equals 100. In the shorthand of mathematics, that relationship is expressed as $10^2 = 100$. Similarly, $10^{2.11} = 128.82$. The largest crater of 3640 mi would have a logarithm of 3.5611. The advantage of dealing with these exponents will become apparent.

Logarithms for each of the 20th craters are plotted in Fig. 3.2. Log Distribution. Actual values are indicated by black triangles with the horizontal axis being the diameters of the mid-range figure divided by 100. Crater diameters, increasing toward the right, are obviously rare relative to the smaller ones to the left. Notice that the black triangles lie close to a straight line. Gray squares in the chart represent the best fit of a straight line to the data based upon the least squares method. Values for the crater groups themselves and values for the trend line are both listed to the left. The chart shows the smallest group to be more numerous than would be expected. Perhaps because they are easier to see. By contrast, a deficiency is noted for the group of craters around 2000 mi diameter. Not many of that size have been found so this result could be caused by only a few that remain hidden.

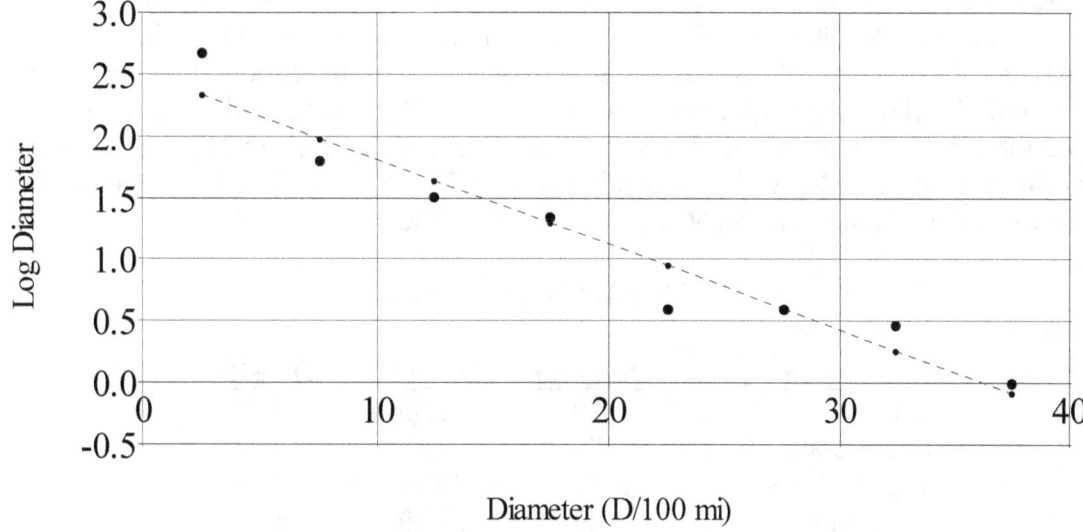

Fig. 3.2. Log Distribution.

The straight line of logarithms for these megacraters on earth is very similar to the distributions of *asteroid* craters on earth and throughout the solar system. Engaging in a detailed analysis of these points would be a divergent path off the main trail being explored. But it should be followed for several reasons.

Straight lines in graphs are defined by their slopes and intercepts with the vertical axis. Slopes have higher values proportional to their steepness and are negative when rising to the left. A specific formula can be established from the data. The slope can be calculated as 2.328333 - 0.265476 / 2.5 - 32.5 = -0.002095. The intercept on the y-axis can be determined graphically as 2.5. So the distribution of megacraters on earth can be expressed as

$$\text{Log } N = 2.5 - 2.095 \times 10^{-5} D$$

Where D is the diameter in miles and N is the number within a size range of D ± 250 mi. Of course, reduction of the bandwidth will generate different numbers. If it is too narrow, the analysis will crumble.

A particular meaning is implied if the slope for megacraters was found to be equal to the well known value for asteroids. Then a simple and common population must exist. Megacraters could be understood as merely an extension of the asteroid curve representing a single population. Differences in slope, on the other hand, would indicate two separate and distinct populations. One must suspect the latter applying to comets versus asteroids. That concept is reinforced by studying the different mechanics of crater formation.

Global Distribution

Comets racing toward earth can not slow down to wait for North America to rotate into their paths. Consequently, craters can be assumed as evenly distributed around the globe at all longitudes. Circumstances are not so simple for distribution at different latitudes from the North Pole to the South Pole. Raw data is presented in Appendix A. Table A.2. Megacrater

59

Latitudes. But again, the complexity begs for some clarifying illustration. This need is best satisfied by summing the number of craters in each Climate Zone as shown in Table 3.1. Climate Zones. Boundaries of these zones are determined by the angle of earth's inclination, that is, the tilt of the spin axis relative to the orbital plane. With minor fluctuations, the nominal value is 23.5 °. So, on earth from north to south, they are defined as the Arctic Circle, Tropic of Cancer, Equator, Tropic of Capricorn, and Antarctic Circle. The northern boundary of the Tropics is the parallel of latitude of 23.5 ° N. Consequently, the northern limit of the North Temperate is 90.0 - 23.5 = 66.5 ° N. The Temperate Zone lies between these latitudes being 66.5 - 23.5 = 19.4 ° wide.

Table 3.1. Climate Zones.

Zone	Number	Area [a]	Density [b]
Arctic	10	31.35	0.32
N Temperate	80	27.89	2.87
Tropical (N+S)	41	39.29	1.04
S Temperate	27	27.89	0.97
Antarctic	6	31.35	0.19

a. Millions of square miles.
b. Craters per million square miles.

Earth's curvature is clearly responsible for the zones having different surface areas. Table 3.1. Climate Zones takes these details into account by using the number of megacraters and the areas in each zone to calculate the crater densities. The last column provides a suitable basis for comparing the distribution along latitudes broken down into climate zones. For example, 80 megacraters were found in the North Temperate Zone, yielding a density of 2.87 craters per million square miles.

Considerable variation in the zonal densities is noted in the last column. Very low values in both arctic regions are easily explained. The Antarctic Zone extends far beyond the Antarctic continent so that the division between land and ocean is about 50/50. While the U.S. Navy has developed contour maps of ocean floors in great detail, it is doubtful that much interest was taken in the southern oceans. At any rate, that data is not available. A major fraction of sea-floor craters are probably lost forever under miles of sedimentary rock and mud. Four, very large examples were found to govern the shape of Antarctica. They are *Ross* (1680), *Queen Maud* (1680), *Wilkesland* (1670), and *Weddell* (1140). Two lesser events contributed their influence along the coasts, namely, *Byrd* (680) and *Bellhausen* (570). Undoubtedly, the entire continent was bombarded by a host of comets whose record remains hidden under miles of ice.

The Arctic Zone is dominated by the North Sea but it includes portions of Siberia, Northern Europe, and Canada. Ten impacts in this zone were essentially on land.

Comparison of North and South Temperate Zones holds no surprises. Land areas in the southern hemisphere are limited to Argentina and its small neighbors, Australia, and the southern third of Africa. East-West breakdown is obtained by determining the longitudes of these land areas. The extent of oceans is about 4 times the extent of land. Densities for these regions have a ratio of about 2.87 / 0.97 = 3, not too far from the expected number of 4.

Continental Record

Considerably more information into global distribution of megacraters is found by examining the number of craters found on each continent. That would leave out the oceans. So, all the salt water in the present geography is lumped together in a category called *Oceanus* in Table 3.2. Continental Distribution.

Table 3.2. Continental Distribution.

Continent	Number	Area [a]	Density [b]	Final No.
Africa	20	11.667	1.714	
Antarctica	6	5.500	1.091	
Asia	41	17.035	2.407	
Australia	11	2.975	3.697	
Europe	11	3.859	2.857	
North America	50	9.435	5.299	
Oceanus	49	140.110	0.350	
South America	19	6.860	2.770	
Total	207	197.43		714 [c]
Average			1.048	

a. Million square miles. *WorldAtlas.Com*, June 10, 2001.
b. Number per million square miles.
c. Can be pro-rated for each continent.

The second and third columns indicate the number of craters in each subdivision along with their corresponding areas. Their densities are easily obtained as in the fourth column. Note the units being employed are the number of craters per million square miles.

Oceanus has by far the lowest because most of its craters are well hidden and the expanse of all the oceans dominates the globe. They account for 70.9% of the surface compared to 29.1% for dry land. Both the Arctic and Antarctic have low densities because of the ice cap and inclusion of ocean areas. Mid-range values are found for Africa, Asia, Europe, and South America. The strong showing of Australia at 3.697 per mil sq mi relates to the barren landscape and lack of eroding rain. The most densely pummeled continent is North America at 5.299 per mil sq mi, possibly biased by the author's nationality.

At any rate, that figure is not unique and should apply to all other parts of the world because comets do not recognize shorelines. Exceptions are the arctic regions. Their shortage is basically understood as failure of discovery. However, a more subtle factor is involved whose influence will be discussed later. Omission of the arctic regions leaves both Temperate and Tropical Zones to share exposure to cosmic assault equal to North America. The value for North America can then be applied to these zones. Tabulated data indicate their combined areas to be 197.43 - 2(31.35) = 134.73 mil sq mi. Based upon the density of North America, the expected total would become 5.299 x 134.73 = 714 events. The 207 craters that have been recognized, therefore, account for roughly 27% of the projected total. That data also indicates that detectable but lost megacraters could be around 714 - 207 = 507 or, say 500. How would the undetected craters be distributed throughout the world? Uniformly impacting oceans and continents would spread them out according to their relative areas. The oceans should contain 0.709 x 500 = 355 whereas the continents should contain 0.291 x 500 = 146. Because of the thick layer of mud on the ocean floors, it is most unlikely that anywhere near the expected number will ever be found. Naturally, all of these calculations must be treated as rough approximations but the results are still revealing

Asteroid Craters

A survey of asteroid craters in 1982 tabulated a total of 103 examples, most of which have been thoroughly investigated. More have been discovered since then and new ones continue to be found. For the present purpose, it will be adequate to examine statistics of those that were known in 1982. Compared to megacraters they are quite small. Some are miniscule. The largest have diameters of

Chiczulub	Yucatan	200 mi
Hanburu	Australia	93
Vredefort	South Africa	89
Sudbury	Canada	88

Smaller ones cover the complete range down to the size of a football field at 300 ft. Untold numbers have undoubtedly been obscured by subsequent impacts. Those that should be detectable can easily be hidden under salt water or forest canopies in remote areas. Examples are the tundra of eastern Siberia and basins of the Congo and Amazon Rivers. Also, permanent ice has blocked discovery in Antarctica and Greenland.

While of little interest in the present context, a few points on asteroid craters deserve attention. Any disruption of the target geology is limited to the immediate area of the crater. Sudbury, for example, lifted invaluable ores to the surface during rebound of the original crater. This treasure is profound but the event did not modify the geology on a regional scale, let alone on a global scale.

A question arises concerning the distribution of asteroid craters on the globe and how that compares with the same data for megacraters. To the author's knowledge, that subject has not been explored. In parallel with megacraters, asteroid craters may be assumed to be uniform around the globe. Among the 103 craters, 84% were found in the northern hemisphere with the remaining 16% in the southern hemisphere. So the northern group was 84 / 15 = 5.24 times greater than the southern group. Something like that ratio is expected because of the greater land areas in the north. Details on this point can be approximated from the longitudes enclosing the continents. Results are

Northern Hemisphere:

Asia	140E - 30E	110
Europe	30E-0 + 10W - 0	40
North America	130W - 70W	60
	Subtotal	210

Southern Hemisphere:

Africa	40E - 10E	30
South America	70W - 40W	30
Australia	155E - 115E	40
	Subtotal	100

The target area of the northern hemisphere is thus seen to be about 210 / 100 = 2.1 times that of the southern hemisphere. But a discrepancy is noted. The north has, in fact, about 5.25 times the number of the southern craters. So roughly 2.1 / 5.25 = 40% of the N/S ratio can be attributed to the greater target area. But that leaves about 60% unaccounted for. Some other parameter must be afoot.

The northern hemisphere must be more prone to asteroid bombardment than the southern. Details must be neglected because of uncertainties in the data, complexity of results, and borderline relationship to megacraters. In summary, the mean latitude for southern craters is 20.4S, well within the expected value for the present inclination of earth. The most southerly crater is at 27.7 S, not badly out of line of the mean inclination. The northern mean latitude of craters, by contrast, is an astonishing 48.1N that raises disturbing questions. Even worse, the most northerly crater is Haughton in Canada's North West Territories at 75.37N. It has a respectable diameter of 12.3 mi. If asteroids fall straight down as suggested by circular scars, the orientation of earth had to be different in the distant past. In this instance at a doubtful 15 My, the axis of earth had to be 75.37° from the perpendicular to the orbital plane. Similar results were found for megacraters. Conditions favoring an excess exposure to asteroids would be the winter months in the northern hemisphere. It is then tipped in the direction of travel by the sun.

Comparing crater latitudes and their ages in the present analysis led to supporting evidence. With only 5 exceptions, northern latitudes were progressively larger than 23.5° back to 700 My. Also, the present inclination was probably not reached until 130 My. All except three are less than 800 My. Exceptions range in ages between 1685 My and 1970 My. They are Teague (1685 My), Sudbury (1840 My), and Vredefort (1970 My). Their large diameters from 17 mi to 87 mi clearly contributed to their preservation.

Several other trends became clear. There is a dearth of craters within the 23.5° band north and south of the equator spanning the temporal range from 600 My to 1600 My. None of any size has been found in the tropics that might have been formed during that period. The vast majority are less than 750 My and they are strongly concentrated in the Northern Hemisphere. An intense cluster lies between 40° N and 60 ° N.

Something fishy is indicated by this data on asteroid craters. Earth's inclination during the winter months in the northern hemisphere exposes that hemisphere to collisions in the direction of travel by the sun. During the other six months, the Sun leads the way protecting earth from impacts by its strong gravity. Asteroids have little chance of passing through that intense field without being captured by the Sun. The preponderance of impacts in abnormally high latitudes suggests much larger inclination of Earth's axis than at present.

The Moon

Indelible records of asteroid impacts on the moon date back billions of years. Once formed, they never change except when damaged by successsors, pelted by debris from later ones nearby, or coated with a thin layer of dust as discovered in the "great leap for mankind." A large number can be seen directly but binoculars or a small telescope reveal exquisite details. In addition, close-ups are available in various atlases and NASA publications.

The moon rotates on its own axis in sync with its monthly orbit around earth. So one side always faces earth and the backside is always hidden. Earlier studies tabulated 100,000 craters on the near side. Nothing was known about the back side until it was photographed from a Russian spacecraft. It found 1,000,000 craters greater than 1.5 mi in diameter. Apparently, smaller craters are too numerous to count. Statistical studies of lunar craters have shown a linear relationship between the logarithms of the diameters versus the logarithms of their numbers. This pattern appearing as a straight line on log-log, graph paper, prevails for all the inner planets.

Mares

Prominent dark patches on the moon gave rise to the mythical "man on the moon." With some optical assistance, they are found to be circular with concentric, mountain ranges closely spaced around their edges. They were originally called *Mares* because astronomers long ago thought them to be seas. The linguistic derivation is from the Latin *mare* from which comes the English *maritime*. In one sense, the early astronomers were right. Instead of basins full of salt water, they are outsized craters that became flooded with lava. Mares are simply lakes of lava that have been frozen solid. Shortly after formation, their surfaces were perfectly smooth. Later impacts created craters in the surface of the "lake." The density of such craters is very low compared with other areas on the moon. Mares are, therefore, young features on the lunar surface. Their ages can be approximated by comparing crater densities inside the craters with densities outside. All too often, the literature speaks of lunar Mares being billions of years old when massive bombardment allegedly ceased. That concept is fatally flawed.

Previously published analyses of crater sizes on the Moon that are known to the author, completely ignored the mares. Perhaps they are too few to have statisitical meaning. Or they may have been considered of volcanic origin and outside the population being studied at that time. Mares, on the other hand, are of fundamental interest here because they are exact counterparts of terrestrial megacraters. How do their sizes compare with those on earth? Relevant data are shown in Table 3.3. Lunar Megacraters.

Table 3.3. Lunar Megacraters.

Name	Diam (mi)	Name	Diam (mi)
Southpole-Aiken	1300	Nubrium	680
Oceanus Procelluc	900	Tranquillitatis	380
Foleunditatis	810	Crissium	280
Frigoris	750	Humorum	250
Imbrium	680	Vaporum	200
Serentatis	420	Nectoaris	190

Never mind the strange names. Just look at the numbers. They correspond closely with those on earth from the smallest at 190 mi. The upper limit at 1300 mi lies in the mid-range of all terrestrial examples listed in Table A.1. Megacrater Sizes. Thirty on earth, however, are larger that South-Pole Aiken. They are up to three times larger at 3640 mi. Perhaps the weaker gravity of the Moon has accelerated the incoming missiles to a speed less than on earth. Differences in geological structures might also account for the smaller megacraters on the moon.

Objects falling from space to the surface of earth and moon will have different, terminal velocities. Stronger gravity of earth will draw in an initially static object at infinity to the surface at 25,060 mi/hr. Weaker gravity of the moon would pull in such an object to strike at 3,331 mi/hr. Thus the ratio of impact speeds at earth's surface relative to the moon would be 25060 / 3310 = 7.6. Since kinetic energy delivered by objects involve the squares of velocities, objects of any size striking the earth would have about 7.6^2 or 57.8 times the corresponding figure for lunar impacts. That factor could easily explain the disparity of megacraters sizes on the earth and moon. However, the question may be moot as shown later.

Mercury

A fruitful line of inquiry is to extend the comparison of crater sizes to other neighboring planets. The situation on Mercury is quite simple. Taking 311 mi as an arbitrary but convenient cutoff, one finds that Mercury has only a single, larger crater at 388 mi. As a small planet its gravity may be handicapped in attracting comets. Or the powerful gravity of the sun may deflect most prospects of collision.

Venus

Dense, opaque clouds have completely blocked all views of this planet's surface for eons. Modern technology of radar and spacecraft, however, has overcome the problem and mapped the surface. Size distribution of its 900 asteroid craters is typical. This small fraction of those on earth is probably due to the same factors affecting Mercury. In addition, both are far removed from the Asteroid Belt that lies outside of earth's orbit. Twenty two, prominent scars range in diameter from 350 mi to 3400 mi in close agreement with the bandwidth on earth. They are also rare compared to earth, amounting to only 10%.

Vast mountain ranges and equally expansive plains have been interpreted by the author as evidence of aerial explosions. Details are much to complex to treat here so they are shoved back into Appendix C. Megacraters On Venus.

Earth

The number of craters that have been found on earth is certainly a small fraction of the total number of asteroids that have hit it. Nearly 71% of the asteroids and comets would have landed in the oceans and all but the largest would have left no permanent craters. Of those striking the continents, an unknown number of craters have been washed away, filled with sediment, or totally buried under thick layers of sedimentary rock. And that loss would hold for all sizes. Megacraters would be more easily preserved than their smaller siblings. And, like the lunar mares, some of them could be relatively young thus more likely to be found. Otherwise, they would have been planed down to grade level, buried, and lost forever.

Geomorphologists have contended that most of the modern landscape had its origin within the past 65 million years. On the other hand, it appears that many of the megacraters are much older. As the earth's megacraters were not previously recognized, they have not been studied and the question of their ages has never been raised. It seems that discovery of the large number of huge craters on Mars and Moon would have triggered suspicions of similar scars on earth and ignited a frantic search for them. It didn't happen that way. In fact, a learned and clairvoyant scientist has strangely proclaimed that nothing like mares has ever been created on earth and never will be.

Mars

The opposite extreme is found on Mars with 90 craters larger than the cutoff, half of them greater than 620 mi! The proliferation of craters on Mars is obvious. A statistically significant group of craters is clustered around 620 mi. Small clusters at 1240 mi and 3110 mi may be purely accidental or indicate minor tendencies in nature. It should be clear at this point that something fundamental distinguishes the large, multi-ring structures from the simple, bowl-shaped depressions.

The size distribution of Martian craters is compatible with those on earth and the upper limit for both is comparable. A progression in the density of craters is roughly proportional to the distance of the planet from the sun. Using the number of craters per billion sq mi of planetary surface as an index, the numbers are 5, 16, 71, and 240. That trend toward a great maximum on Mars should be expected because its orbit is just inside the asteroid belt making a favorite target.

Jupiter

Thirteen moons of Jupiter display the typical pattern of craters in the solar system although none exist on the planet itself. Being composed of gases, no solid surface is available to retain a record. Actual impacts by comets, however, were closely observed and photographed in July 1994. Astronomers found the comet, Shoemaker-Levy 9, a year earlier and watched it break into 22 pieces during a flight through Jupiter's gravitational field. A modified orbit shifted the string of fragments into a collision course the next time around. Calculations proved to be very accurate. The whole world watched as the fragments, one after another, struck just over Jupiter's rim. The impact sites were hidden but resulting plumes reaching enormous heights could be seen over the horizon. Spacecraft in fortuitous positions were able to photograph the actual impacts. Debris clouds quickly rotated into view of terrestrial telescopes. These dark spots on the flaming surface expanded up to the size of earth. These matters are so complex that they must be deferred to Appendix D. Lessons From Jupiter.

Chapter 4. Multi-ring Structures.

"Many discoveries have been missed because scientists ignored data that didn't fit into their established mode of thinking."

- Richard Muller

The weight of earth's crust at any particular location must be balanced by buoyancy of material that is submerged in the underlying, molten but viscous magma. Continents simply float on molten rock with a portion below the surface like the draft of a ship. Having elevations in the thousands of feet, they ride on top of solid rock of comparable thickness. Ocean floors, on the other hand, have a thin crust supporting only the weight of water overhead. This requirement dictates the thickness of rock as it is created when large blocks of the crust drift apart at mid-oceanic ridges. Thick continents, therefore, can withstand assaults from cosmic attacks more effectively than thin, ocean floors.

Chapter 1 illustrated the various results of impacts that were confined to a single continent, those confined to an ocean, and others where the craters overlapped both. Crustal damage varied according to the location of ground zero and, of course, the power of the event.

Basic features of those surface scars are entirely different from the bowl-like depressions of ordinary impacts by asteroids. The latter have been intensively studied for decades and their formation is well understood. An asteroid smashes into the surface with an explosion of enormous energy depending upon its mass and velocity. Detailed mechanics of crater formation have been worked out on the principles of physics so that dimensions of a crater can establish the mass and speed of the asteroid. The known density of different types of asteroids leads to determination of their size. Accuracy of the theoretical model has been verified by data from bomb craters including those from nuclear devices. Material excavated from the bowl is blasted into the environment while a rim of moderate height remains, bending upward away from the explosion.

More than 150 asteroid craters have been identified around the world whose diameters average about 12 mi with a distribution skewed toward the smaller sizes. Even so, a few exceed 100 mi. Scanning of the data on asteroid craters emphasizes their diminutive sizes and their locations around the world. Asteroid craters are found as far north as 75.70 ° in Russia but the southern range is restricted to 32.87 °. Most impacts farther south would have been lost in the great expanse of oceans after producing tsunamis of Herculean heights. It is not known if predominance of craters in the northern hemisphere is related to asteroid orbits. But the global distribution of asteroid craters closely echoes that of megacraters.

Creation

The title of *megacraters* emphasizes their primary distinction from familiar craters. Usually, they have two concentric rims closely spaced relative to the crater size. Circular faults are found a) inside the inner rim, b) between the rims, and c) beyond the outer rim. In oceans, higher elevations of the rims are revealed as perfect arcs of islands. Those on the inner rim are typically volcanic whereas those on the outer rim are not. The latter are also smaller and more widely separated. Magma oozes up from the depths through fracture zones sometimes filling the shallow basin out to the base of the inner rim. Even flow through saddle points can fill the annular space between rings. And some of it may escape the structure entirely to flow into adjoining craters.

Immediately outboard from the structures one finds deep trenches down which ocean floors dive under the bedrock of the impact basins. This process, known as subduction, advances very slowly at the rate of tectonic displacement of participating plates. Creation of island arcs and their associated trenches have often been described in the literature. Drift of tectonic plates has been proposed as the driving force to create island arcs. But explanations have been missing, tenuous, and unconvincing. That issue has survived until the present expose'.

Linear rift systems from megacraters can run for thousands of miles without interruption or they may become offset by tectonic drift or disrupted by other craters. These complex faults may also exude magma to form straight chains of volcanoes, volcanic islands, and coral reefs. These distinguishing features of multi-ring structures are illustrated in Fig. 4.1. Event Diagram.

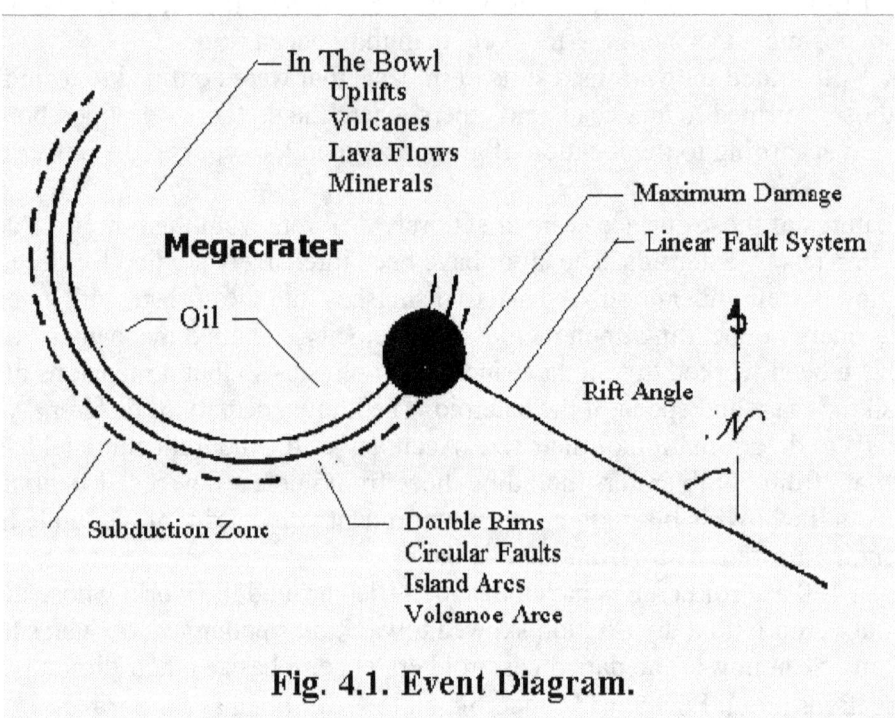

Fig. 4.1. Event Diagram.

Crustal damage would be compounded at intersections of the radial and linear faults. Instability at such locations could produce earthquakes for millions of years as broken fragments of the crust jostle one another. Also, earthquakes are prominent along the arcs as ocean floors descend under adjoining plates in a halting motion. Great forces drive the subducting plate with little effect until resistance between the plates is periodically overcome in devastating jerks. Of such movements are earthquakes born and they are exclusively located on the faults from megacraters.

Initial Basins

Whereas asteroid craters involve contact with earth's surface, megacraters give every indication that the explosions occur at high altitude. Tests of nuclear weapons in the atmosphere have shown the growth of a giant, fireball immediately rising as cooler air near the surface rushed in toward ground zero. Dust, dirt, and debris are lifted in a swirling column toward the rising fireball. Heat of extraordinary intensity radiates in all directions at the speed of light, incomparably faster than a powerful, blast wave expands away from the site. Damage on the surface to all kinds of targets is horrendous as observed in tests and demonstrated in 1945 in the Japanese cities of Hiroshima and Nagasaki. No craters were formed by the aerial explosions and a demolished building at ground zero in Hiroshima has been preserved as a memorial to the dead and wounded.

With powers many millions of times greater than atomic or even hydrogen weapons, a cosmic missile would produce two types of damage; a) asteroids striking the surface and penetrating some distance compared to b) objects exploding in the atmosphere at high altitudes. The former blast out large cavities from which all sizes of rock are hurled into ballistic trajectories. Aerial explosions exert enormous pressures that depress the surface into basins that rebound from their original depths. Sedimentary rock would yield as energy from an aerial explosion of a very large object as a distributed source and last many times longer than the instantaneous ignition of a bomb. A source many miles across would depress the surface over a very large area. This process can be visualized as squeezing an orange with emphasis on the thumb. If strong enough, the orange skin will break in an arc under the edge of the nail. And, in a delayed motion, spring back toward its original position.

Intense heat would vaporize or melt rock over a wide area while a shock wave blew away unimaginable quantities of rock and fractured whatever remained. Depression in the earth from an extremely powerful explosion in the atmosphere would be strongly resisted by the thick layers of rock at the site. That extremely stiff rock does not bend easily to produce a typical crater of limited diameter. Instead, the vertical forces are spread out far beyond ground zero to produce an enormous depression.

Shock waves striking the surface some distance away from ground zero quickly develop a horizontal component while the vertical component becomes weaker. Fifty miles away from an explosion at an altitude of 50 mi, for example, the vertical and horizontal components have become equal. At a distance of 200 mi, the vertical component amounts to only 25% of the horizontal. At 500 mi, the vertical component is reduced to 10%. Much beyond that, the vertical component vanishes altogether. So the extreme stiffness of bedrock below ground zero governs the response of the earth by distributing the vertical component over very wide areas.

Velocities

Instead of an asteroid dredging out a bowl in seconds, growth of megacraters is ponderous.

An asteroid strikes the surface at 40,000 mi/hr like a BB shot into an orange peel. Megacraters are created by the shock wave and trailing pressures that expand at the speed of sound, only 770 mi/hr. Megacraters grow slowly. The first indication of an explosion 500 mi away would be a blinding, sky-filling light brighter that the sun. Two minutes and eleven seconds later, an observer would experience a frightful earthquake as the primary seismic wave reached his position. A blast wave in the atmosphere would not arrive until 40 minute after the explosion followed by whip lashing winds 1000s of mi/hr that would sweep the surface clean. One sees that formation of a megacrater would take an hour or so.

With these points in mind, the formation of megacraters is examined in Fig. 4.2. Growth Of Megacraters where the vertical scale is greatly exaggerated.

69

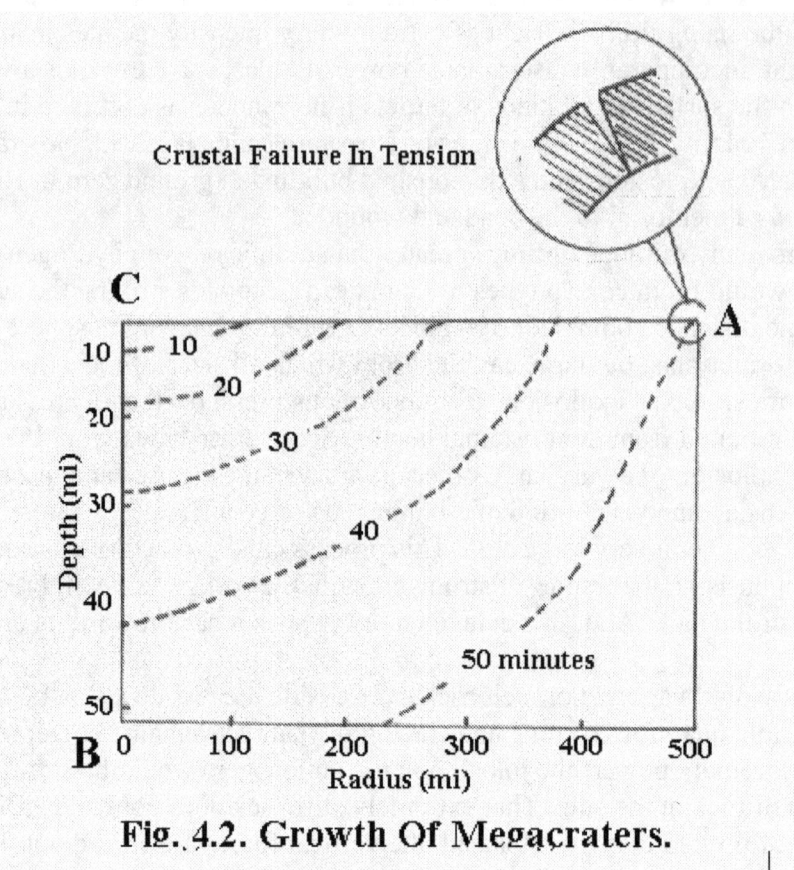

Fig. 4.2. Growth Of Megacraters.

While the suggested profiles at different times are highly speculative, some qualitative aspects of the crater growth are probably correct. Stiffness of the crustal rock layers quickly spreads the downward force of the blast tending to maintain a horizontal orientation for great distances. The flat bottom expands as the crater gets deeper. At the point where the rim meets the undisturbed grade in the mature crater, bending of the rock increases to the breaking point as shown in the inset.

After reaching a maximum diameter and depth, pressure below the basin will tend to raise the bottom. The pressure would be equal to the value that prevailed at the particular depth before the disturbance. At first, the rebound rate would be rapid because of the extreme pressures from below compared to the resistance to one atmosphere above. As the bottom rises, it will slow down as the driving forces are gradually reduced. The final approach toward the original grade will become very slow as the pressure difference approaches zero. Complete recovery may take millions of years.

Overthrust And Basin Depth

The expanding basin stretches the rock along the bottom as clearly illustrated in Fig. 4.2. Growth Of Megacraters. The distance from A at the rim along the dotted contour to B at the center is greater than the distance along the initial grade A to C. What happens during rebound of the depression? Irresistible pressures attempt to force the bottom disc upward into a hole that is too small. The stretched rock experiences no forces of compression back

70

to the original size. Curvature of bedrock at point A exceeds its tensile strength so it fractures from the top like breaking a pencil in two.

The net result is obvious. At point A the layer along the basin bottom must be driven *upward* on the left side of the fracture. At the same time, the rebounding bottom expands *outward* over the right side of the fracture to allow the space required in its return to grade. The extent of the overlapping layers can not be calculated precisely because the exact profile of the maximum basin is not known. But values can be approximated. The length of the final profile in Fig. 4.2 is seen to be longer than the diagonal, straight line from A to B. It is also shorter than the sum of the radius and depth. Half the sum of these two lengths then provides a reasonable estimate of the extent the rock layer was stretched. The stretched length indicates the spacing between the outer and inner rims. In the present example, the amount is $(502.49 + 550.00) / 2 = 526.25$ mi. So the overlap in this case would be about $526.25 - 500.00 = 26.25$ mi or 5.25%. This problem can be solved backward so that knowledge of the overthrust for a specific megacrater permits calculation of the maximum depth.

Considering the great expanse of some of the megacraters, one suspects that the original basins must have been quite deep. Some calculations indicate that the depth would approach half the radius, an estimate that must be treated with caution. Even so, some very deep holes are implied as follows

Diameter (mi)	Depth (mi)
1000	230
2000	460
3000	690

Very deep, indeed. Something like these values may be verified by thorough analysis of crater dynamics or by geological evidence at the sites.

The temporal sequence of overthrusting is shown by sketches in Fig. 4.3. Rim Ratio And Island Arcs in which the vertical scale is exaggerated. The top sketch in the figure represents the circumstances when the basin has reached a maximum radius and the inner fault system is created by fracture of bedrock. Rebound pressures would be enormous at the basin center. Even below a 6-mi slab of bedrock, it would amount to 70,000 atmospheres. The second sketch illustrates the point in time when the edge of the basin floor has been offset just enough to allow slippage over the bedrock at grade. Overthrust has begun in Sketch 3 where weight of the top slab drives the underlying slab downward. A block designated "**B**" becomes too heavy to be supported as a cantilever of the basin floor. It breaks off under gravity and settles down with the subducting slab. Notice the broad arrows in each sketch representing the applied forces that drive the entire process. Elevated points identified by the letter "**I**" become the inner arc of islands. Points at "**O**" become the outer arc of islands that settle to lower levels. The bottom sketch shows the geometry of double arcs of islands after the whole system has achieved stability. The only exception is the extremely slow motion of the subducting plate.

Rim Ratios

A convenient method of recording the geometry of twin arcs has been adopted. Values of Rim Ratios, whenever available, are found in Appendix A in the 7th column of the tables. An appropriate tool for probing that data is a scatter plot. Complications, however, require that detailed investigation to be handled in Appendix E. Analysis Of Rim Ratios. But a brief summary is provided here. About one quarter (26%) of all craters displayed closely-spaced, double rims. The average value for the whole group amounts to 1.114 that can also be expressed as 11.4%. In other words, radii of megacraters averaged an expansion of 11.4%. Values do not vary greatly as all of them are confined between 3.8% and 21.1%. Half of all Rim Ratios fall between 7.0% and 15.8%.

Pacific Plate

The present separation of North America from Africa and creation of the Atlantic Ocean could have happened in two ways. Blockage by the Pacific Plate prohibited motion until excess floor of the Pacific Ocean found some outlet. The first option was for the continent to slide over the Pacific Plate along the fault system on the west coast. The second option was for Pacific floor to dive under Asia. Both processes must have been active during some periods.

A 5000-mi front of subduction trenches runs from the southern tip of the Philippine Island to about Anchorage, Alaska as shown in Fig. 1.18. Disappearing Ocean Floor. Loss of bedrock down the trenches contributed to the mobility of North America and still does. Material was submerged under the Asian Plate where it is eventually melted back into the magma from whence it came. Islands arcs of volcanic origin are found west of all the trenches. In addition to the eight major trenches previously depicted, two smaller ones lie southwest of *Mariana* with familiar names of Palau and Yap. They will be neglected.

Island arcs are formed by outward expansion of crust within the impact basins coupled with elevation by pressure from below. A deep trench is thereby created with the approaching plate plunging to great depth. Trenches define the outer boundaries of their megacraters. Diameters of the eight trenches protecting Asia range from 710 mi at Japan to 1980 mi at Kuril. These values are in perfect accord with diameters of all the megacraters. It appears that weaker events are incapable of breaking through the crust but a reliable threshold has not been found. Although clues from *Marianna* are helpful.

To bring the trenches into sharper focus, islands in their respective chains can be mentioned. Most are widely known but some obscure specks of land will be noted whenever their presence bears strongly upon the subject; 1) Philippines - all islands from southern Mindanao to Manila on Luzon, 2) Ryukyu - all islands from Taiwan to Japan's Kyushu including Okinawa, 3) Japan - major islands of that country excepting Hokkaido, 4) Kuril - all islands from the northeastern corner of Hokkaido to the East Coast of Kamchatka where it ties into the Aleutian Islands, 5) Aleutian - all islands from Kamchatka to Yukutat Bay in Alaska including the Alaska Peninsula, Kenai Peninsula, and Kodiak Island. [The last two items clearly illustrate a double-rimmed crater with a ratio of 1.108, quite close the average value.] 6) Marianna - all islands including Guam, Saipan, and Tinian, 7) Bonin - Iwo Jima, and 8) Izu - a straight line of islands inclluding Miyake leading to the Izu Hanto peninsula between Yokohama and Nagoya then inland to the volcanic Fuji San. Benin-to-Izu, being the only straight trench, strongly suggest that it was created by *Mariana.* When subduction was initiated under the islands, it would naturally proceed northward along Bonin-Izu. While it was instantaneous under the arcs, progress to the north may have been very slow. If, however, compression of the Pacific Plate was substantial at that time, it could have sprung westward briskly under Bonin-Izu. In any event, compression in the Pacific Plate would have contributed to subduction everywhere along the chain.

Loss of Pacific Plate by overthrusting of North America is treated separately. Naturally, prominent chains of earthquakes and volcanoes back up all the trenches. Overlapping plates creep past each other in jerks that are most often minor but sometimes devastating. The fracture zones between plates provide a pathway for lava to feed volcanoes along the inner rim and build seamounts and islands. Lava simply oozes out of fissures and caldera over eons but occasionally releases energy in cataclysmic eruptions. Examples in this part of the world are the ongoing activity of Mt. Mayon on Mindanoa, island building by Myosin-syo in Izu, and colossal explosions of Mt. Bezymianny in Kamchatka and Mt. Katmai near Anchorage.

Island Arcs And Subduction

Treating the configuration of earth's crust at discreet moments after the rebound starts will be helpful. The first sketch in Fig.4.3. Geological Processes depicts the fracture caused by excessive bending at the edge of the crater.

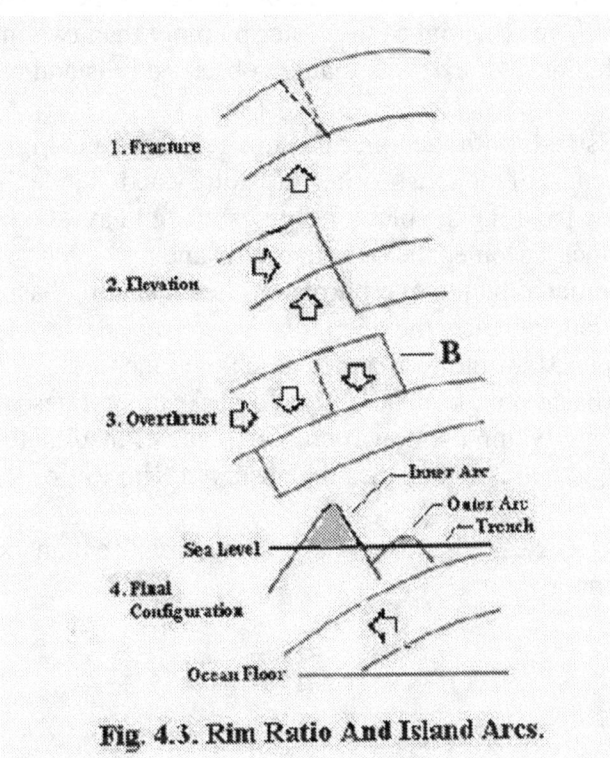

Fig. 4.3. Rim Ratio And Island Arcs.

The second sketch represents an intermediate time when the vertical motion is great enough to allow the edge of the bowl to begin expanding outward. Driving forces for this motion are the deep, rebound pressures that make the basin more shallow and drive the crust toward the initial grade. The third sketch indicates the arrangement when substantial overlapping has been achieved.

Weight of the top layer drives the bottom layer farther downward and initiates subduction at the time of Sketch 2. The lower layer sinks deeper as the overthrust grows. Stiffness of the top layer may cause a separation between the layers as indicated by the gap in Sketch 3. At some point in the increasing overlap, the weight of a portion labeled "**B**" can no longer be supported by the crustal rock. It is possible to estimate the maximum length of that portion before a rupture. In a highly idealized model, the layer will fail when the weight of the 6-mi thick slab exceeds the shear strength of rock. Actually, the breaking

73

length should be shorter because a bending force would help to start the failure at the top surface.

The final configuration of this system is shown in the last sketch where many details are noted. At lower elevations, peaks on the outer rim protrude less frequently than those on the inner. So the outer arcs of islands are smaller and more widely spaced.

The fall of block "**B**" is directly linked to the depth of the resulting trench and provides a clue for estimating the elevation of its starting point. An ocean trench should be about as deep as the change in elevation of the block. That movement should correspond with *the depth of the trench minus the depth of the ocean.* Careful examination of Sketches 3 and 4 should make this point clear. Depth of subduction trenches vary over a rather narrow range from 25,300 ft for the Aleutians to 35,600 ft for Mariana. The average depth of the Pacific Ocean is 13,400 ft. So the change in elevation at the Aleutian Trench must have been about 25,300 - 13,400 = 11,900 ft. For an island on the outer arc with an elevation 3,000 ft, it would have reached up to 11,900 + 3,000 = 12,200 ft. It and others in the arc must have stood far above sea level, sticking out of the water by more than two miles!

The last sketch contains all the details that are observed in island arcs as follows;

a) twin chains of islands with the inner most being larger and higher than the outer,
b) the original fracture of the crust forms the fault system inside the inner chain,
c) the fracture zone isolating the block becomes the fault system between chains,
d) the tip of the block becomes the outer arc of islands,
e) subduction is initiated by the overburden of the expanding basin floor,
f) collapse of the block creates the deep trench,
g) a pathway is opened for molten rock to reach the inner arc,
h) rebound of the basin produces the observed peaks and ranges,
i) extreme fracturing within the bowl permits escape of magma to make lava lakes, and
j) spaces between two arcs establishes the observed Rim Ratios.

Formation of island arcs can, therefore, have no other cause than explosion of a cosmic body creating a megacrater.

Chapter 5. Linear Rift Systems.

Science has become so specialized that each branch must be left to its oracles. Generalists like Aristotle, Leonarado, Hooke, and Darwin belonged to past ages.
- S. Warren Carey

The second major aspect of megacraters is the appearance of linear rifts that have rended ancient continents but are more frequently found in the ocean environment. Their piercing through the heart of the multi-ring bull's eye was initially very puzzling. What was the relationship between megacraters and the rifts? Why did they seem to aim toward ground zero? The answers were unexpectedly found in the gravitational link between the earth and the sun.

Preliminary Analysis

While swinging around in its orbit, the earth experiences balanced forces. Attraction toward the sun by gravity equals the centrifugal force tending to sling it away into space. Hence, a stable orbit. This model of treating earth as if all its mass were concentrated at the center works well enough on the cosmic scale but ignores the forces of gravity experienced by the earth itself. A refinement of the model requires that the earth be conceived as two hemispheres with the flat faces together. The center of gravity of the half away from the sun, the dark side, is a little more distant than the closer half in full sunlight. The outer half experiences an excess of centrifugal force against solar gravity while the inner half experiences an excess of gravity over its own centrifugal force. The halves are in a tug-of-war. So earth attempts to fly apart with the proximate half falling toward the sun while the outer half tries to fly away. The plane of separation would be perpendicular to the line toward the sun separating day from night. This dividing line on the surface is known as the terminator.

Earth's own gravity, however, is much stronger causing compression of all its constituents. The net result is a very slight relief of that compression along the terminator. It is very small but very real. That weakness dictates the location of crustal rupture when cosmic missiles apply massive stresses. Apparently an impact might occur at any instant anywhere on earth far away from the terminator. Here intuition simply fails as the data indicate otherwise.

This puzzle can be cleared up in Fig. 5.1. Impact At Dawn. Earth is shown in its orbit moving counter clockwise to the point of impact with an approaching object. The dark side is shaded. Rotation of the earth on its own axis, also counter clockwise, means that the transition from night to day is always on the leading edge as earth flies through space! Earth runs into the object head on and always on the morning terminator. Any other location would violate the observed facts that linear rift systems are aligned with ground zero.

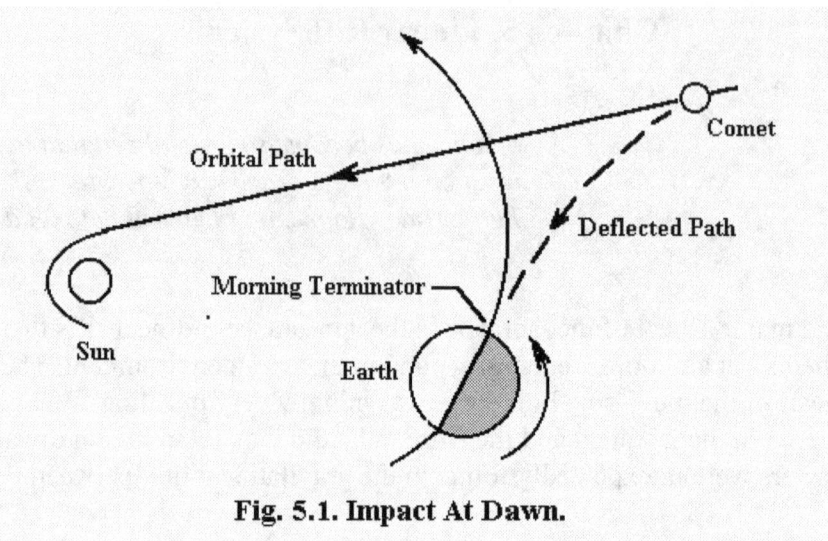

Fig. 5.1. Impact At Dawn.

It is not any easy target. The terminator remains perpendicular to sun's rays while earth rotates 360 ° every 24 hr. In relation to earth's surface, therefore, the terminator speeds along at 1000 mi/hr at the equator or 810 mi/hr at the latitude of Washington, D.C. But global maps and satellite images insist that the linear rifts pass through the epicenters. Missiles must have struck at 6:00 o'clock in the morning because the sun always rises on the leading edge where that is the local time. Any deviation by a matter of only minutes would have caused an offset from the epicenter. Suppose the event took place a 6:10 or 5:50 a.m. Even these minor difference in time would displace the terminator by 2.5 ° corresponding to a surface displacement of 173 mi at the equator. At the latitude of Washington, it would amount to 135 mi. Any offset by these distances would be immediately noticed. Measurement of the offset and consideration of the correct latitude would yield the exact moment of impact within a fraction of a minute before or after 6:00 a.m. In others words, the object approaching under the pull of earth's gravity always hits very close to the morning terminator even though it moves along the surface at hundreds of miles per hour.

Terminator Motion

Linear rifts exhibit a wide range of azimuths from the impact point, that is, the angle of the rift relative to true north. Why this should be the case is rather obscure so it must be examined carefully. Fig. 5.2. Rocking Terminator shows how the dark side of earth depends upon its position relative to the sun or, the equivalent, the time of year and exact location on its orbital path. Because earth's axis is tilted about 23.5 ° from a perpendicular to the orbital plane, seasons are created.

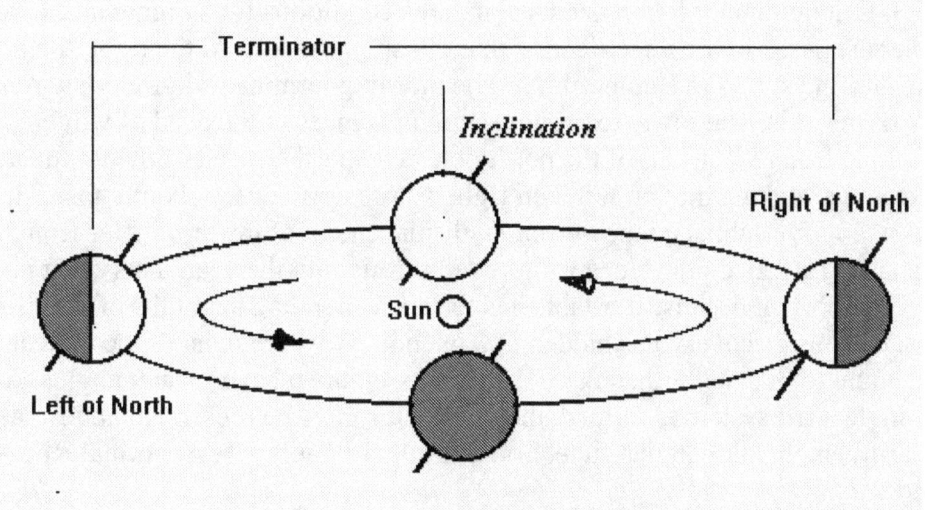

Fig. 5.2. Rocking Terminator.

Consider the northern hemisphere. The north pole at the extreme left in the figure tilts toward the sun producing summer. Conversely, at the extreme right it tilts away from the sun producing winter. Intermediate locations in the center represent the autumnal equinox on September 21st with night in full view. The sun illuminates the globe in the background at the vernal equinox on March 21st. Every location on these dates experiences 12 hr of day and 12 hr of night.

Orientation of the terminator is thus seen to be a lively phenomenon. At the summer and winter positions, it is observed to be left of true north as viewed from space. When observed from earth, however, a different pattern arises. In the left image, the earth spins left to right so that the terminator depicted there is the position at dawn. In the position at the right, the viewer faces the terminator at dusk as earth moves into the paper. A proper perspective and complete understanding of this subject can be achieved by imagining earth at this point from *behind* the paper. That is not easy so another drawing is required.

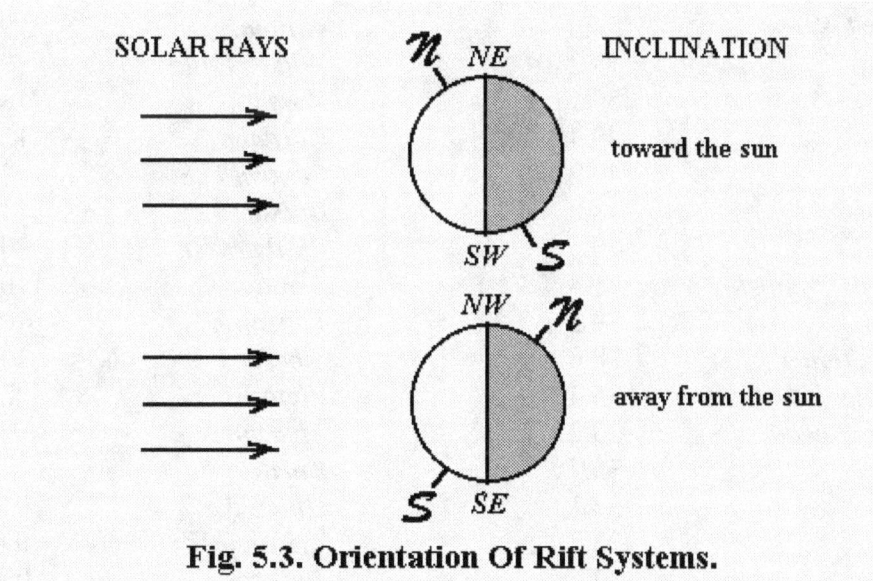

Fig. 5.3. Orientation Of Rift Systems.

Fig. 5.3. Orientation Of Rift Systems compares conditions for summer and winter in the northern hemisphere. In the upper icon with earth spinning toward the right, the north pole is tilted toward the sun as in summertime. The morning terminator in full view runs from NE to SW of the axis. The lower icon shows circumstances in the dead of winter when the terminator runs from NW to SE of the north pole. The northern end of the terminator is thus seen to execute a rocking motion left then right of true north every six months. The tilt angle of the axis at 23.5 ° established the *maximum* displacement of the terminator from true north at the solstices. Intermediate values will prevail throughout the year. This point is exceedingly important because it establishes the corresponding azimuths of the rift systems.

A most curious discovery lies hidden in Appendix A where most of the indicated rift angles are equal to or smaller than 23.5 °. But a dozen or so have greater angles. At whatever angle a rift system is formed, the terminator must be rocking to and fro at the same angle as earth's inclination at that time. Some results in Table 5.1. Abnormal Rift Angles are simply shocking.

Table 5.1 Abnormal Rift Angles.
(Sorted by absolute values.)

Name	Angle (°)	Name	Angle (°)
Campeche	84.9	Brisbane	32.7
West Pacific	82.6	Iran	32.3
Yukon	80.8	Indian Ocean	32.1
India	78.6	Great Western	29.9
Mexico Gulf	76.0	Chad	29.7
Arabian Sea	72.9	Kalahari	29.1
Guatemala	72.8	Arabia	27.9
Indonesia	71.5	Chicago	24.0
South China	69.9	Champaign	22.8
Chaco	60.5	Moscow	22.0
Caroline W	57.8	Sardinia	21.9
Madagascar N	56.3	Timbuktu	20.6
Bismark Sea	54.6	Hungary	19.5
Artesian	53.2	Wilkes Land	19.0
Klondike	50.1	Ross	18.2
Venice	49.4	Cordoba	15.1
Bengal	49.2	Central U.S.	13.5
Angola	48.0	Baikal	11.9
Kara	46.5	Khabarovsk	10.7
Noril'sk-Yakutsk	41.4	Lena	10.7
Aegean	39.2	Tanzania	10.3
St. Lawrence	38.6	Siberia 1	10.0
Ethiopia	35.4	Victoria	8.6
Sascatchewan	35.2	Amazon	5.8
Alberta	34.7	Kuril	4.7
Texas	34.1	Himalaya	0.75
Angkor	33.7	Aleutian W	0.66
Bothnia	33.4	Jodhppur	0.14
		Great Basin	0.0

Twenty seven percent of the craters (57) were found to have rift angles greater than 23.5°. In rocking across the north polar axis, any one of them could have created a crater with a rift within the expected range of 23.5° as the terminator sped through that zone. Similarly, a listed rift does not necessarily indicate a maximum tilt. The resulting value could have been produced by any greater angle at some time during the year as the terminator executed its now familiar rocking motion.

Tabulated values do indicate minimum conditions at the time of impact. The axial tilt at the time of *Aegean* had to be at least as large as 39.7°. And so on for even greater tilts. The value of 84.9° for *Campeche* approaches an astonishing 90° when earth's axis would lie flat in the orbital plane with the spin vector pointing in the direction of travel part of the year and in the opposite direction six months later. Weather during such times would be chaotic. Imagine the temperatures of polar regions swinging every six months between full radiation of the tropics to continual, frigid darkness. The range of temperatures and duration of extreme conditions would threaten both flora and fauna. Periods of day and night would vary wildly everywhere.

Another aspect of Table 5.3 must be addressed. Recent research and mathematical modeling of the highest order by others has shown that the moon stabilizes the earth. If it were not present, earth would reel in contrast to its placid behavior today. Maximum tilt angles up to 85° were calculated in the absence of the moon. That finding is in remarkable accord with the physical evidence of the rift systems on earth as presented here.

The moon is traditionally believed to have appeared several billion years ago but the craters with extraordinary rift angles are no where near that old. If the ages of earth's craters can be determined with confidence, a remarkable possibility arises. One should find the youngest crater whose rift angle was more than 23.5° indicating when a transition occurred to the present stability and *when the moon appeared in the sky*. The result would show that the origin of the moon is relatively recent.

Much effort by scientists has been focused upon reconstructing the birth of moon out of debris from a Mars-sized impact with earth. A resulting ring of debris around the world would require eons to coalesce into the moon and leave no trace in the lunar orbit. These worthy efforts have not been completely satisfactory so the research continues. A mathematical model is being developed to track the gravitational interactions of 20,000 fragments of matter ejected by the collision.

Evidence presented here strongly implies a youthful moon probably less than 500 My old. If that proves to be correct, then the assumption that it was created by cosmic impact must be re-examined. And more attention should directed toward the prospect that it had a long life before becoming attached to earth and that it was simply captured intact by earth's gravity. Such is known to be true for certain moons of Jupiter.

Practically nothing is known about crater ages in Table 5.3 with the exceptions of *Kalahari* and *Noril'sk-Yakutsk* at about 250 My and *Aegean* at about 26 My. Only *Noril'sk-Yakkutsk* is known with precision because of dating the huge Siberian Traps by laboratory methods utilizing radioactive decay of Argon. An enormous amount of work will be required in new field investigations and assembly of related data already in the geological archives. It is expected that such data could answer key questions like a) by what means did earth acquire its moon, b) when did earth's axis stabilize to the present inclination, and c) were all of the largest craters formed in the early history of earth?

Rift Formation

Available space on single pages in Chapter 1 did not allow full descriptions of the subjects. So treatment of the Emperor-Hawaiian chain of seamounts, shoals, and reefs in Section 1.20 omitted some important details that must be examined now.

Very straight rift systems from craters would take some time to develop. If the process were slow, they would display a gradual curve toward the west following the lively movement of the terminator. Suppose that a rift were created on a terminator at 50° north latitude and ran southward to the equator at 0 °, a distance 3470 mi. To extend that far at the speed of primary seismic wave at about 15,000 mi/hr would require an elapsed period of 0.230 hr or about 13 min 47 sec. During growth of the crack, the terminator would have moved westward by 3.45° or 239 mi. Thus the gradual progression of the crack following the local position of the terminator would slowly bend to the west producing a curvature along the way. This scenario would apply to an impact at one of the equinoxes when the terminator passes through both poles, however, it conflicts with the geographical facts.

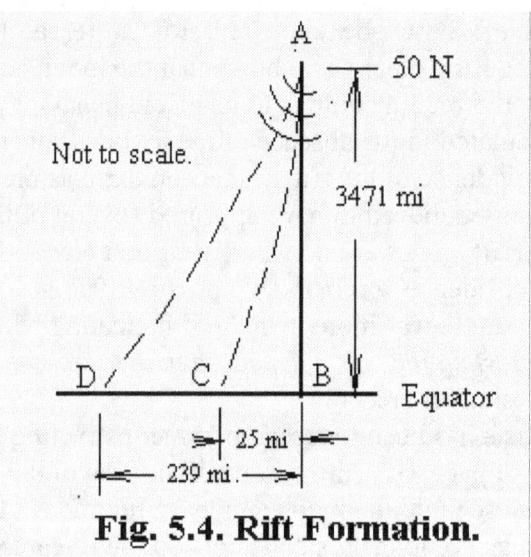

Fig. 5.4. Rift Formation.

Fig. 5.4. Rift Formation can put these details in perspective. Instead of arriving at the equator at point B, movement of the terminator would have forced arrival at point D some 239 mi west. Such large-scale bending of rift systems around the globe would certainly be noticed. They just don't happen.

Longitudinal displacement would be increased by slower speeds and decreased by faster speeds. But what phenomenon can transmit an influence faster that 15,000 mi/hr? With computer software based upon satellite photos it is possible to detect offsets as small as 25 mi. So that amount would require a rate of 139,000 mi/hr. Outrageous but true. Recall that only a weakened zone is created along the path. The front of the crack could progress at an enormous speed compatible with the size of the enormous craters in the earth when they have grown to maximum diameters and depths. Pressures pushing the walls apart act upon the entire crust like a wedge in splitting wood. The crack progresses much faster than movement of the wedge.

Marching Volcanoes

While defining the eventual path of an island chain, transverve forces from the expanding crater must open up the crack to considerable width. Estimated widths of such trenches are 50 to 100 mi. Magma is so lethargic that it can not fill the rupture immediately. Side walls would be more rapidly forced toward their original positions leaving a permanent gap of, say, 2 to 10 mi. Extremely sluggish lava can then ooze through the crack as it builds the volcanic island closest to the explosion. A long chain of islands may develop over the tens of millions of years. This sequence for the Emperor Chain was noted in Chapter 1 but deserves a closer look.

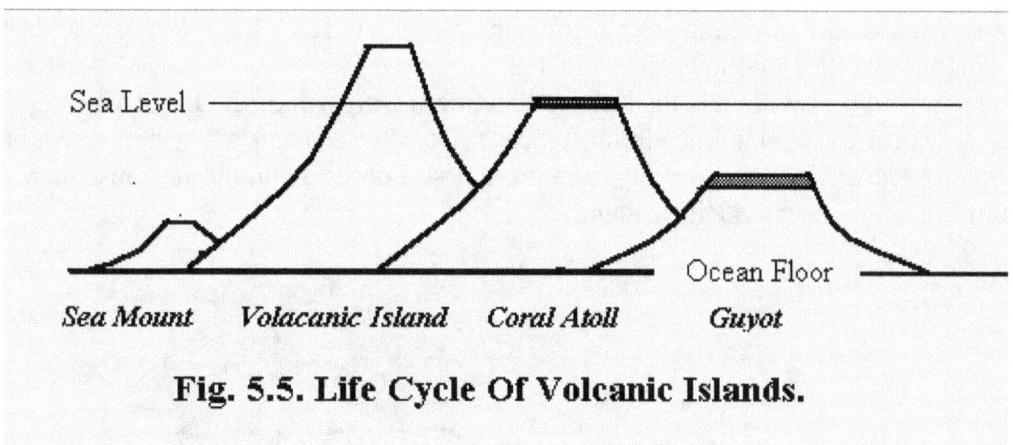

Fig. 5.5. Life Cycle Of Volcanic Islands.

Birth of a volcanic island begins when lava leaks through the ocean floor as illustrated in Fig. 5.5. Life Cycle Of Volcanic Islands. Topographical features at the site then undergo profound changes. The sketch on the left represents the initial appearance of an underwater volcano. It continues to grow until breaking through the ocean surface as a young island. Additional growth will then widen the base on the ocean floor and drive the island to such heights that snow may cap its peak even in the tropics. The island tends to settle by its own weight pressing down on the supporting layer of rock. Rain and waves scour the island and, over several million years, destroy it completely leaving only an atoll at sea level. A balancing act then ensues between the growth rate of coral versus the sinking rate of the submerged mountain. If coral can keep up with its loss of footing, the atoll is preserved. If, however, the sinking rate is too fast, the whole ensemble disappears in the depths forming a flat-topped mountain known as a *guyot* (pronounced GEE-oh).

From birth at the first flow of lava to demise upon sinking into the sea takes millions of years and the number can be calculated for any particular example. The first island forms outside the crater rim not far away that, of course, becomes the oldest guyot in the eventual chain. The process is replicated down the rift. It continues down the rift to the youngest one at the opposite end. The number of seamounts along the chain can easily be counted along with the total length of the chains. Upon determining the ages of the first and last structure, much information becomes available. This method was applied to the Emperor Seamounts in Section 1.20 to find the progression down the chain to be 7.8 cm/yr. This submerged ridge is nearly continuous so the number of islands is hard to estimate. Say there were 15 islands created during the period of 38 My. Then the average lifetime of volcanic islands would amount to 2.5 mil yr.

Several adjacent seamounts may become visible islands as in the present Hawaiian group. Kawaii, the oldest island in the group, dates back 4 mil yr. And a new island along the chain southeast of the Big Island, Liohi, is being born as a young seamount reaching for the surface. In due time, it will become an adult island when Kawaii has been reduced to an atoll. It appears that 4 to 5, adjacent islands may be the standard squadron in the parade of volcanoes.

Some useful deductions follow this line of inquiry. Most, if not all, straight chains of seamounts were created by massive, cosmic impacts defining the avenue down which volcanoes grew. In addition, the center of a megacrater can be found by projecting the alignment of the chain in the direction of its oldest member. Any offset between that projection and the center is a true measure of the tectonic drift of the ocean floor since the explosion. Finally, the date of the explosion can be estimated as 5 to 10 My earlier than the fisrt seamount that took that long to be created.

As yet not fully explained is the knee redirecting the growth of the Emperor chain near Midway toward Hawaii. Unique circumstances arise when a linear rift system encounters another at a large angle that, naturally, was there first. Forces dictating the consequence are shown in Fig.5.6. Truncated Rift System.

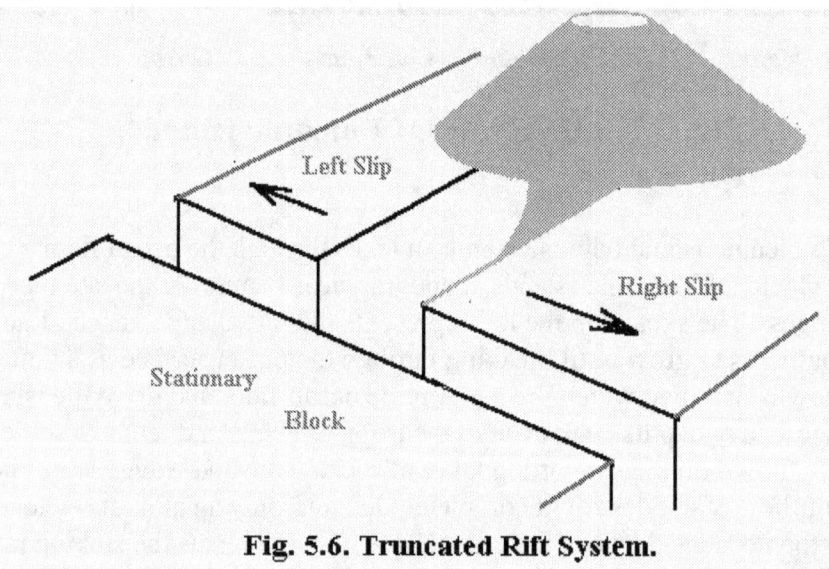

Fig. 5.6. Truncated Rift System.

The last volcanic island in a chain growing at the upper right is shaded along with a simplified channel accommodating the lava flow. Weight of the island wedges upon the rift ahead as it approaches another fault across its path. Lateral forces that served well in the past now become ineffective. As blocks separate, slippage begins as shown by arrows in the drawing. Someone standing on the stationary block would see segments across the young fault moving in opposite directions. The only forces transmitted to the stationary block come from resistance along the slip faults and they are balanced. There is no way that the lengthening chain of volcanoes can leap the gap so it must turn off onto a side street, one way or the other. The direction taken will be along the side closer to the original path.

Ninetyeast Ridge

The large Bay of Bengal between India and southeast Asia, being notably circular, could be a megacrater. If so, what supporting evidence can be found? Is the possible crater associated with a major, rift system? Yes and yes. Radiating southward is a straight fault 2310 mi long. Because it closely parallels the 90° east longitude, oceanographers have named it Ninetyeast Ridge. This ridge is covered with massive lava flows analogous to the Emperor Seamounts. It is then terminated at Broken Ridge where it turns east toward Australia along a line of ancient, submerged islands. The similarity with the knee at Midway must not be overlooked. A megacrater in the southern Indian Ocean is identified as *Bengal.* Furthermore, Broker Ridge may well have been formed by *Peak Hill*, the dominant megacrater in western Australia as depicted in Fig. 1.17 of Chapter 1.

Projecting Ninetyeast Ridge northward toward the presumed crater shows an easterly offset of only 30 mi from the center of the Bay. Not only is the probability of the Bay of Bengal enhanced by that fact but it might indicate the exact time of the explosion. At the latitude of the center, one degree of longitude equals 49.84 mi or 30/49.84 = 0.60°. Earth's spin rate of 0.25 °/min converts the distance into a time interval of 2.4 min or 2 min 24 sec. The terminator had just passed the site so the object exploded at 6 hr 2 min 24 sec.

These calculations illustrate a point and may be off base if the geography has been disturbed by tectonic motion. However, no such distortion can take place when the crater and rift ride on the same plate.

Chapter 6. Comets And Surprises.

"There are powers next to which nuclear
violences are but faint puffs in the air.
- Richard Bach

Much can be learned from an aerial explosion in Siberia in June 1908. Early on the morning of the 30th, a great flash rent the sky. Forests were flattened for miles around. It was such a remote location that scientists did not begin a local investigation for years. The site was near the Podkamennaya Tunguska River from which the event's name derives. It should be distinguished from a megacrater in the same region dubbed *Tunguska* by the author for the paucity of suitable landmarks.

Siberian Explosion

Cosmic impacts millions of years ago are too remote to seem real. But nature recently hurled an awesome missile at the earth that some people can still remember. A mere pipsqueak compared to mega-events, it's damage was nevertheless massive. It could have easily wiped out the entire city of Los Angeles. Fortunately, the object struck in the vastness of Siberia where 1400 sq mi of forest were leveled. Whether anyone was killed is not known. The region is exploited for the fur trade so some hunters may have perished. The closest surviving witnesses were 125 mi away. One unlucky fellow had his horse knocked over by the blast and his clothes charred by the radiant heat. The site was about 62 mi to the northwest of Lake Baikal at 60.92 N 101.95 E. Shock waves in the atmosphere and seismic waves in the ground were detected and recorded by sensitive instruments at remote laboratories. Thus the time of the explosion was reportedly 7:17 a.m. local time. But local time refers to a particular time zone in which deviation from true time can be as large as 30 min either way. The region in question is time zone - 7, that is, 7 hr before Greenwich Mean Time. However, the exact location of the event at 101.95° E lies close to the western boundary of the zone at 100° E. The true, local time must have been nearly one half hour earlier. The numbers work out with the answer of 6 hr 44 min 48 sec, a considerable distance from the terminator at that time. Russian scientists recognized it as a "morning meteorite."

Prominent scientists and authoritative sources today refer to the missile as a comet, meaning that it was a dirty snowball perhaps with a thin skin of cosmic dust. The constituents would be essentially the ices of water, carbon dioxide, and methane.

In a magnificent book on giant meteorites, a Russian scientist devoted 142 pages to this event. Author E. L. Krinov provides a comprehensive history of planning expeditions, seeking financial backing, leading the field team, mapping the damage, and laboratory analysis. The slow growth of understanding just what had happened becomes clear and the final conclusions of a scientific committee are stated.

Because of the Russian Revolution at the time, no expedition to Tunguska was launched until 1921, thirteen years later. The primary purpose was to locate witnesses and record their testimony that was extensively quoted by Krinov. Precise location of ground zero and collection of physical evidence were completed in a series of five more expeditions until 1962. Careful observations and mapping disclosed the extent of the damage that has been refined by recent research. The pristine forest had been lain down in a radial pattern with an uneven periphery. Some trees were protected from the blast on the leeward side of hills. Several fires had been started. But there was no crater! A swampy area about 3 mi across was found at ground zero where trees remained standing. Their limbs had been sheared off but they were not killed. The evidence was unmistakable: The explosion had been at high altitude directly overhead.

Soil samples from the devastated area and a zone of comparable size in the direction of the missile's flight produced new findings. Microscopic spherules up to millimeter size were composed of silicate and iron oxide. Those compounds could not have come from the earth because there was no excavation. They must have originated in the comet itself. Evidently, extreme heat had vaporized the object. Silicon and iron atoms then reacted with oxygen in the atmosphere to produce the observed compounds. Lowering temperature in the explosion cloud solidified or froze them. Studies under a microscope showed silicate particles as transparent, or nearly so, while iron-oxide particles were black. In some instances, they were attached. In others, a smaller iron-oxide spherule could be seen inside one of silicate.

A few puzzling questions about Tunguska go unanswered. What caused the object to explode? And why didn't it explode several seconds earlier or later?

Cosmic Sparks

In electrical circuits for television antennae, one end reaches for the sky while the opposite end is attached to earth that is treated as a universal, reference datum. Normal, electrical strength is measured in voltage from this datum known as *ground.* Artificial grounds are established in the metal frames of vehicles to which all circuits return after fulfilling their functions of starting, lighting, air conditioning, etc. Both types of grounds are convenient but are simple fictions. Far from being electrically neutral, the earth has a great surplus of electric charge that creates a powerful field in its vicinity. It is everywhere perpendicular to the surface and quite strong. It drives a weak current of positive ions downward through the atmosphere and a counter current of free electrons upward. The atmosphere does not easily break down in a violent discharge because air is an excellent insulator.

Cosmic rays from the sun are simply charged nuclei of hydrogen atoms called protons that shower earth in a great flood. Naturally, a strong charge builds up on earth over its entire lifetime of 4 to 5 billion years. A comet spending its lifetime far removed from the sun would not be exposed to so strong a flow of protons and those reaching it would be minimal because of its negligible size. Any charge buildup can be ignored and the comet used as a cosmic ground. As the object eventually approaches earth, an electric field between them grows stronger.

Bombardment of earth by cosmic rays also generates a conducting region high in the atmosphere known as the ionosphere. So it prevents a discharge until the approaching body breaks through the bottom at an altitude that is usually about 3.4 mi. The best estimate of the explosion at Tunguska places the altitude at 3.7 mi! Energy release has been compared with a large H-bomb of about 50 million tons of trinitrotoluene, 50 MT of TNT. Some Russian scientists suggested that Tunguska was a nuclear explosion. That concept was abandoned, perhaps prematurely, because tree rings failed to register an expected change in concentration of Carbon-14.

Data from Tunguska strongly suggest that megacraters are produced by explosions of comets at the bottom of the ionosphere triggered, in all likelihood, by a cosmic spark. It would be effective only if the missile had already achieved high pressure and temperature.

Composition Of Comets.

A concept of dirty snowball was introduced to the scientific community in the early '50s by Fred Whipple as the "…icy-conglomerate model." Components were thought to include water, methane, ammonia, carbon dioxide, and hydrogen cyanide (H_2O, CH_4, NH_4, CO_2, and HCN). While seemingly impossible, molecules present in comets can be identified. Their light carries unmistakable signatures that can be read by astronomers and modern research has revealed dozens of constituents. With a density about the same as water, they display a gaseous corona, a nucleus with a dusty skin, and a great, gaseous tail. It is not critical to the present discussion to account for all the details so a simplified model is adopted as follows - water ice 25%, methane ice 25%, and solids 50%. Following the clues from Tunguska, the solids will be assumed to contain substantial quantities of silicate and iron.

A hypothetical comet approaching earth with these constituents and relative abundance would present an interesting mixture of atoms. For any arbitratrary volume of the gases, the atoms would be in the ratios hydrogen 6, carbon 1, and oxygen 1. Corresponding percentages would be 75%, 12.5%, and 12.5%. So the gases look very much like a cloud of hydrogen. While rarer, carbon from methane will be found to have a prominent role in the explosion and its aftermath.

Nuclear Fusion

The enormous power required to generate megacraters seems to be much greater than could be delivered by the largest known asteroid. Comets are also too small to deliver the requisite amount via their kinetic energy. One suspects another source of energy and the Russian scientists may have been right all along.

A well known, nuclear reaction converts hydrogen into helium with the release of energy about twice as large as that from breaking apart an atom of uranium. This fusion of hydrogen allows the sun to radiate its energy as it "burns up" hydrogen. All the stars in the universe do the same. So nature not only embraces nuclear fusion as all of creation is powered by it. Igniting the process, however, requires enormous temperatures and pressures coupled with some means of containing the reaction.

Everyone is familiar with the general concept of H-bombs based upon nuclear fusion that have played the dominant role in international politics for decades. Not so many are acquainted with the details that must be understood in the present context.

Pardon My Equations

The U.S. Atomic Energy Commission published the essential information in June 1957 that involves a rather complex chain of events and different forms of several elements called isotopes. Any element may have an excess of neutrons in its nucleus giving it a different weight on the atomic scale while retaining the basic properties of the element. To be encountered here are isotopes of hydrogen and helium that are traditionally denoted by superscripts, such as H^2 (deuterium), H^3 (tritium), and He^3 in contrast to the normal He^4. The nuclear reactions converting hydrogen into helium can be written as

$$H^2 + H^2 = He^3 + n + 3.2 \text{ Mev}$$
$$H^2 + H^2 = He^3 + H^1 + 4 \text{ Mev}$$
$$He^3 + He^3 = He^4 + n + 17 \text{ Mev}$$
$$He^3 + He^3 = He^4 + 2n + 11 \text{ Mev}$$

where n is a neutron. Mev indicates the amount of energy released in millions of electron volts (1.6×10^{-6} erg). Fusion of He^3 contributes most of the explosive energy at 86%. The first two reactions lead to two atoms of He^3 that then fuse in the last two reactions.

When the nuclear reactions cease in the plasma, an extremely hot cloud contains unburned nucleii of normal hydrogen and helium nucleii with atomic number 3. Also present are nucleii of normal helium as the end product. As the plasma cools free, negatively-charged electrons attach themselves to positively-charged nucleii resulting in a cloud of very hot gases. Thus, the ratio of He^3 to He^4 would remain much higher than the value found on earth. During the reaction taking microseconds, there is an abundance of He^3. These facts are of utmost significance because He^3 hardly exists on earth.

The amount of helium of both kinds in the atmosphere is zero. But substantial quantities are found in the natural gas fields of Texas. While being delivered to consumers by pipeline, a stream of gas is side tracked into an industrial plant operating at cryogenic temperatures. Most of the helium is removed and the processed gas is sent along.

In other words, it is extremely rare. Yet He^3 is found elsewhere under the most unusual conditions. Carbon atoms from methane and carbon dioxide aid the nuclear reactions but are not consumed.

Proof In Buckyballs

Recent research has discovered a new form of carbon molecules in which 20, 40, or 60 atoms arrange themselves into tiny spheres. They are linked together on the surface much like the appearance of a soccer ball or structures designed by the famous architect, Buckminster Fuller. In his honor, these molecules have been named fullerenes and are affectionately referred to as buckyballs. They are stable up to a temperature of 600° C where their bonds fail. At that time, they release an atom of noble gas that has been trapped inside. The reverse process means that carbon atoms in the presence of any noble gas at high temperature and pressure would capture an atom of the noble gas as it cooled past 600° C and formed a buckyball.

What, then, might be expected in the cloud of a hydrogen explosion containing the specified quantities of ordinary ice, frozen methane, and undetermined amounts of silicate and iron? The answer is too elementary. Upon cooling down to 600° C, the carbon atoms would arrange themselves into a stable fullerene while capturing atoms of a noble gas, in this case, helium of which a significant fraction would be the rare isotope of He^3. Fullerenes, incidentally, are produced in large quantities by electric arc welding between carbon electrodes in a helium atmosphere and at the site of lightning bolts.

Fullerenes containing He^3 atoms have been found widely dispersed in the layers of rock separating the various geological periods. Since the isotope barely exists on earth, experts in planetary science and astrophysics have concluded that fullerenes must have been brought to earth from outer space. An improved perspective suggests that they were manufactured during fusion explosions of comets at the bottom of the ionosphere.

For a nuclear fusion reaction to be ignited and sustained, hydrogen must be compressed to millions of atmospheres and held at a temperature of tens of millions degrees Fahrenheit. Magnetic fields or inertial conditions just before the explosion can provide the containment. With considerable widths measured in miles, air can not escape around the sides of a comet during its plunge through the atmosphere. Most of the atmosphere, 99.9%, is below the altitude of 54.8 mi. So a comet reaching the bottom of the ionosphere at 3.4 mi would travel 52.8 - 3.4 = 48.4 mi. If traveling at 100,000 mi/hr, the elapsed time to explosion would be the distance divided by the speed or 13.4 milliseconds. In that brief flash like a meteor, escape of air around the edges would be impossible. The front of a comet, like a piston, compresses the air and its own gases to the requisite conditions utilizing inertial containment. After the onset of fusion, additional hydrogen is driven into the nuclear furnace sustaining the reaction.

The nuclear fusion is thereby spread out in space across the face of the comet and also sustained in time as fuel is added to the fire. The resulting effect upon the surface of earth becomes a sustained force over a wide area compared to shock wave from a point source. Multi-ring structure produced by exploding comets is a different breed than the craters excavated by impacting asteroids.

Most of the time comets simply hover at great distances from the sun while traveling along with it in galactic orbit. At some time, the sun urges them out of their lethargy to begin a long descent. After picking up speed all the way they approach the sun at about 1,380,000 mi/hr. Their path of descent will sometimes be bent toward earth by the strong gravity of Jupiter. As their maximum speed is nearly reached at earth's orbit, the final approach could be even greater than discussed above.

Rather than relying upon kinetic energy from the mass and velocity of the missile, it generates energy from nuclear reactions. Variation in the sizes of megacraters apparently depends upon the amount of hydrogen available as fuel instead of the projectile size and velocity. Also, the perfectly circular scars fail to indicate any direction of approach. The explosions are isotropic so that forces, applied equally in all directions, produce scars on the surface that are close to perfect circles.

Becker (2000) and colleagues have recently found fullerenes with abnormal ratios of He^3 and He^4 in meteorites and at the Sudbury site of an asteroid impact. They have also discovered the link between fullerenes containing noble gases and extinctions at the end of geological periods. Their conclusion was that the fullernes were of extraterrestrial origin in contrast to the suspected birth in earth's atmosphere.

Further details concerning nuclear explosions are presented in Appendix F. Kinetic Versus Nuclear.

Chapter 7. Chains Of Craters.

*"Every man takes the limits of
his own field of vision for
the limits of the world."*
- Arthur Schopenhauser

To a first approximation, a stream of comets is considered to travel along the same trajectory. So impacting a rotating globe would be similar to shooting it with a machine gun as the bullets spaced themselves out along a common latitude. The globe is also assumed to be stationary in space.

Astronomers have reassured the public the last few decades that earth is perfectly safe from impacting asteroids. The theory said that such events were common in the distant past but our cosmic neighborhood has been swept clean of wandering debris. Recent years, however, have seen major changes because asteroids of frightful size have been zooming too close to earth. NASA established the Near-Earth Asteroid Tracking program to find all asteroids larger than 1 km in diameter and identify any that might threaten earth. Plans are being laid for some sort of defense against disaster, such as, sending up spacecraft to rendezvous with the intruder then blow it to pieces with a nuclear bomb or nudge it off its path with a powerful rocket. These methods may be effective in saving earth someday.

The potential damage could be devastating on a global scale but unlikely to terminate civilization or wipe out most of the fauna and flora. Asteroids certainly could vaporize a metropolitan area and the tsunami raised by a strike in an ocean could wipe out dozens of coastal cities. But physical damage in remote areas on continents might be limited to a few hundred square miles of downed trees or scoured sand except for dust clouds that might circle the globe and partially block sunlight for many months. An impressive crater would mark the site. The real problem here is that asteroid threats are trivial compared to the final approach of a comet against which there would be absolutely no defense. Unfortunately, comets can come into view in less than 2 yr before the big black out. It might be argued that the asteroid responsible for *Chicxulub* destroyed the dinosaurs but that crater is negligible compared to the megacraters under study.

Greatest Show Off Earth

The summer of 1994 was an exhilarating time for astronomers around the globe and common folk too. A new comet had been jointly discovered and confirmed by David H. and Carolyn Levy of Jet Propulsion Laboratory and Eugene M. Shoemaker at Palomar Observatory. Dim at that time but destined for greatness. Astronomers in every country exchanged information in a frenzy. It came under scrutiny by 28 observatories, 8 spacecraft, and an uncountable host of amateurs using their own telescopes.

Fireworks on Jupiter were astonishing. The point of impact just beyond the left edge (limb) of the planet was not visible from earth. But fiery plumbs shooting up 2000 mi were detected. As the planet turned, the scars came into view and marched in a column to the east. One of them was as large as the whole of earth, some 8000 mi in diameter. Gases at 200,000 °C rushed outward at 8,000 mi/hr.

Such violence can hardly be imagined. As the fragment diameters were judged to be 3 to 5 km, about the same size as many asteroids, a question comes to the fore. Was the extraordinary power of the fragments based upon nuclear reactions in contrast with their mechanical energy based upon weight and speed?

Answers are not easy to find and the hunt involves some tedious effort. Some principals of physics and some knowledge of nuclear reactions must be called upon. It would be burdensome to explore this area in this point so it is addressed in Appendix F. Kinetic Versus Nuclear. Conclusions of that study indicate that nuclear reactions must be the source of energy from comets. The amount so far exceeds that from asteroids of the same size that the latter can be neglected. Kinetic energy is responsible for compressing air ahead of the comet to create the temperature and pressure required to ignite a fusion reaction in hydrogen, in particular, deuterium, whose presence on S/L9 was not only detected but measured. Asteroid impact craters are smaller than *Chicxulub* with the possible exception of *Aegean* that is somewhat bigger.

Lessons From Jupiter

The comet was first spotted in March 1993. It swung in a loop close to Jupiter on July 7, 1993. The planet's strong, gravitational field stretched out the loosely agglomerated material that broke into 8 major fragments. Later subdivision brought the total to 20 plus a few minor pieces that were eventually lost. Further observations established the precise orbit that allowed astronomers to predict a collision with Jupiter. The leading fragment was expected to plunge into Jupiter's atmosphere in mid-July 1994. It struck on July 16th.

Scars on Jupiter left by S/L9 reveal the weakness of the idealized model. The following analysis is based upon data forwarded to the astronomical community by West (1994). Table 7.1 shows the individual fragments, their impact dates, times, and latitudes expressed as decimal fractions to the second place. The event started with fragment A on July 16th at 20.18 hr Universal Time (UT same as Greenwich Mean Time) and lasted until fragment W struck on July 22[d] at 08.10 hr UT. The simple model would predict all fragments landing along the same latitude as A but substantial differences are noted. Latitudes between A and W were a whole degree. Taking the radius of the planet at that latitude as 85,750 mi, the surface displacement would be 747.65 mi, say rounded to 750 mi. There was a pronounced trend for each fragment to hit farther south than the previous one, as easily seen in the table and illustrated in Fig. 7.1. Latitude Drift.

Table 7.1. Data For S/L9 Impacts.

Fragment Number	Day July 1994	Latitude S Minus 43.00°	Cumulative Time (hr)
A	16	0.15	0.00
B	17	0.17	6.70
C	17	0.38	10.84
D	17	0.46	15.72
E	17	0.48	19.00
F	18	0.55	28.37
G	18	0.60	35.37
H	18	0.74	47.35
K	19	0.80	62.06
L	19	0.92	74.10

N	20	1.39	86.30
P2	20	1.64	91.20
Q2	20	1.26	95.55
Q1	20	1.05	96.04
R	21	1.07	105.39
S	21	1.16	115.09
T	21	1.99	117.99
U	21	1.43	121.74
V	22	1.44	128.20
W	22	1.15	131.92

The gradual increase in south latitude becomes more obvious by plotting the decimal fractions starting at 43° S instead of the values directly. Numbers above 44° S appear as one plus a decimal fraction. The bargraph in Fig. 7.1. Latitude Drift shows the strong and nearly uniform trend toward more southerly impacts from **A** to **U**, amounting to 1.43° over a period of 121.74 hr. If all fragments had landed at 43.15° as established by **A**, the vertical bars would have all been the same height.

Fig. 7.1. Latitude Drift.

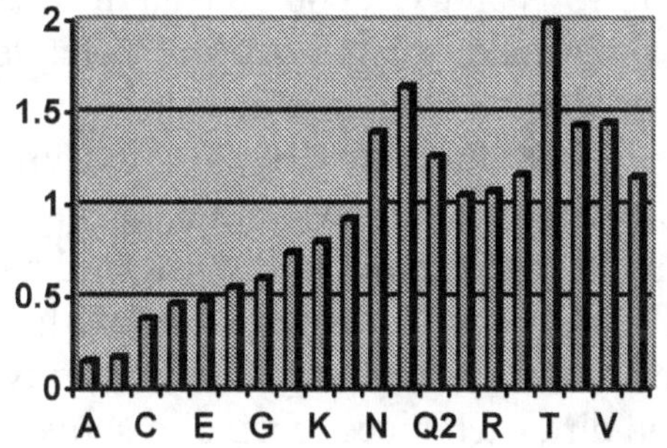

Several rogue fragments in the string of flying comets were substantially offset from the main axis. Among them were **N**, **P2**, and **P1**. Fragment **T** was traveling with the greatest offset of 1.99 - 1.35 = 0.64° of latitude that translates into 480 mi off the track. On the surface of Jupiter its displacement from the simple prediction would be 1.99° x 750 mi/deg = 1490 mi.

As to longitudes, the scars could have been created anywhere. The huge planet completes a rotation on its own axis at the amazingly short period of 9hr 50 min 30 sec. So during the period of the entire event, several Jupiter days passed. Specifically, the number of rotations was 131.92 hr / 9.842 hr/rot = 13.404 rotations, that is nearly thirteen and a half Jupiter days. If the scars had been permanent craters, it would have been impossible by studying them to establish the order of their positions in the original chain.

Why, one may ask, was the pattern created? Answers appear to lie in two areas.

Jupiter was not a stationary target but moved along in solar orbit at the good clip of 29,520 mi/hr. During the period of bombardment, it moved a total of 29,520 mi/hr x 131.92 hr = 3,894,000 mi! So the landing point on the surface of any delayed fragment would not be likely to hit exactly on the same latitude as the previous one.

Secondly, longer exposure to the gravitational field of more distant fragments probably contributed to the observed drift. But complications in this area are too great to address.

A short string of asteroid fragments landing on earth within an hour would leave a chain of craters less than 730 mi long. Within reasonable limits, they would be on approximately the same latitude depending upon their initial positions relative to the axis of the aerial chain. Similar results should not be expected of a chain of comets stretched out far enough to rain down for a few days. Two fragments flying only 0.5 hr apart, however, would land close to the same latitude as *Aleutian E* and *Aleutian W.*

Chains On Earth

There is every reason to suspect that one or more streams of comets laid down a chain of megacraters on earth. Finding them would be a worthwhile effort.

A section in Chapter 1 described the proximity of two impacts in the Bering Sea related to the Aleutian Islands and their subduction trench. Separation in longitude amounted to only 109 mi that could be accomplished by earth's rotation in 38 min 46 sec. Another clue that the two were related was found in the retarded response of the surface in the region of the Kenai Peninsula. Full development of a double rim appears to have been disrupted by a closely following fragment. These events were labeled *Aleutian E* and *Aleutian W* whose essential data can be easily found in the alphabetical listing of Table A.3 in Appendix A.

How close were they flying together? The answer, of course, lies in the tabulated values of latitude, specifically, **E** at 62.15 N and **W** at 60.58 N, a difference of 1.57° At that location on earth, one degree of latitude equals 69.25 mi. Relative to the first impact, the second was 1.57 deg x 69.25 mi/deg = 109 mi too far north. Such dispersion should be expected in light of the known pattern laid down by S/L9 on Jupiter. In other words, an offset of that magnitude does not preclude the two craters being sister fragments. If they had been traveling at 80,000 mi/hr, their separation in flight would have been 80,000 mi/hr x 0.767 hr = 96,760 mi apart. This proximity in flight is rich with implications.

When comets are stretched out to the breaking point in a gravitational field and crumble into discrete bundles of ice and dirt, each piece receives a shove. Without any resistance in the vacuum of space, the whole chain continues to lengthen as the gaps between each piece widens. Consider fragments **A** and **B** of S/L9 and the data in the previously presented in Appendix D. They were 6.7 hr apart upon impact. But the mother comet broke up in its near miss of Jupiter on July 7, 1993, looped around a complete orbit, then crashed on July 16, 1994. In just over 1 yr the objects drifted apart to a distance than can be deduced from the record of scars.

Both flew at 134,000 mi/hr with **B** landing 6.7 hr later. They were both joined with other fragments on July 7, 1993. Then by the time of impact, they became separated by 134,000 mi/hr x 6.7 hr = 898,000 mi. Duration of the entire sequence of impacts, lasting 131.92 hr means that all of the beads had been strung out on a necklace an amazing 134,000 mi/hr x 131.92 = 17,700,000 mi!

These details are brought out because most people would have no idea of the facts and they are important in understanding the Bering Twins. If the Aleutian craters were caused by fragments, they had achieved a very limited separation. The time since their breakup was necessarily short and the most likely, if not only possible cause was a brief tour through the strong gravitation field of Jupiter. Because it is by far the largest planet in the solar system with the strongest gravity short of the sun, there appears to be no alternative. If the Bering Twins were broken apart by Jupiter and separated at the same rate as **A** and **B**, the time since breakup can be calculated. They drifted apart 898,000 mi in 374 days so the spread rate must have been 898,000 mi / 374 days = 2400 mi/day. That result then permits a calculation of how long they traveled . It is the distance of separtion divided by the available time or 104,000 / 2300 = 43 days! This astounding figure suggests strongly that the breakup was actually caused by Jupiter. Apparently, the change of course near Jupiter also threw the comet onto a collision course with earth. Throughout the history of the solar system, Jupiter has performed as a guard against devastation on earth by taking the punches as happened in 1994. This recent date completely shatters the feeling of safety on earth that has dominated the astronomical community. Comet crashes are thankfully rare and Jupiter may perform its traditional role in the next threat by taking the punch.

Chains of craters will be generated when a string of celestial objects strike the earth. Because of its daily rotation to the east, they fall approximately on the same latitude. Some slack must be allowed in positions of craters to accommodate vagaries in relative positions in the approaching stream. These observations are highly relevant to the effects of such impacts upon earth. Megacraters that have been described could have easily been produced by an isolated comet or a string similar to S/L9.

Nuclear Fusion

The task at hand is to examine the composition of comets at the atomic and nuclear levels to see how much energy would be available. A starting point is learned from Glastone (1957). Fusion of one pound of hydrogen is equivalent to an atomic bomb at 26,000 tons of TNT abbreviated to 26 KT with TNT implied. The maximum energy release by a single fragment on Jupiter was estimated at 200,000,000 million MT! The numbers get so large that scientific notation must be used. The convention is to write a number close to unity times 10 to some power. For example, 200,000 000 would appear as 2.0×10^8. Also the straightforward calculations can become a bit messy. Skipping over them can still reveal the logic and conclusions.

The energy in a pound of hydrogen would be 26×10^3 tons or

$$26 \text{ KT} \times 1 \text{ MT} / 10^3 \text{ KT} = 26 \times 10^{-3} \text{ MT} = 2.6 \times 10^{-4} \text{ MT}$$

and the amount of hydrogen required is the total energy released divided by the energy from the fusion of 1 lb of hydrogen or

$$2.0 \times 10^{14} \text{ MT} / 2.6 \times 10^{-4} \text{ MT} / \text{lb} = 0.769 \times 10^{18} \text{ lb}$$

Is there that much hydrogen in a comet? Can it be ignited?

Equations presented in Chapter 6 showed fusion reactions starting with the heavy isotope of hydrogen, deuterium (H^2). That isotope was observed on SL/9. Meier (1998) and a team of other scientists using highly sophisticated techniques with the James Clerk Maxwell Telescope on Mauna Kea, Hawaii for the measurements. They found the ratio of deuterium to normal hydrogen in water to be

$$D / H = (3.3 \pm 0.8) \times 10^{-4}$$

where the uncertainty is the standard deviation. This value closely matches previous observations of Halley's Comet and Hyakutake. So H^2 was, in deed, present in S/L9 but only as a weak contaminant in normal hydrogen. It would appear in the same proportion in the gaseous state of water and methane and doubtless in their ices. Their fusion would be spontaneous when exposed to proper levels of temperature and pressure.

Crater Chains

A search among all the megacraters can be much simplified by arranging the database according to latitudes as in Table A.2. Crater Latitudes. Of special interest here is the likelihood of *Aleutian E* and *Aleutian W* being the first two impacts of a longer string. Running down the latitude column, one finds several examples between 62.15 N and 59.00 N that include the twins and deviate from the mean value of 61.36 N by no more than 1.36° or about 94 mi. That data is brought forward in Table 7.2. Candidate Fragments.

Table 7.2. Candidate Fragments.

Name	Latitude N	Diameter (mi)	Deviation (deg) *
Aleutian E	62.15	1140	0.79
Klondike	61.60	360	0.44
Noril's Yakutsk	61.78	1650	0.42
Aleutian W	60.43	1340	0.78
Anchorage	60.43	670	0.93
Siberia	59.85	710	1.41
Hudson	59.50	200	1.86
Ungava	59.50	150	2.00

* From the mean of Bering Twins.

Recall the first seamount in the Emperor-Hawaiian Chain, named Detroit, whose age has been determined as 85 My. It evolved out of a crack made by *Aleutian W* and took 5-10 million years to grow to maturity, weather away, and sink. *Aleutian W* must then be that much older, say 95 My setting the age of any associated chain of craters. The much older *Noril'sk-Yakutsk* at 250 My could not possibly be a member so must be discarded as indicated by the shaded row. *Klondike* lying between the twins is a strong candidate. Also notice the gradual reduction in sizes of *Anchorage, Hudson,* and *Ungava*.

Longitudes have little meaning in the discussion thus far because several days could have passed during completion of the chain. A string of fragments as long as those on S/L9 would take 5.5 days from the first to last on earth. Craters could be at any longitude and completely out of sequence in the original string. But the proximity of the Bering Twins suggests that a chain including them was comparatively quite short. While longitudes can be set aside here they become a factor in another context.

In only 43 days of travel, separation of the twins was minimal and a similar limitation must apply to all the fragments. They probably landed with delays not longer than a few hours between each impact. Hence, fragments following the twins should not be very far to the west of the Bering Sea.

Of immediate interest in the table is *Siberia* that was off track by 1.41° or 98 mi, well within the acceptable range for offsets. This 710-mi crater is centered near the town of

Tomot. The eastern rim lies on the shore of the Sea of Okhotsk south of the city of that name. Lying west of the International Date Line requires a bit of juggling to add the angular distance from Aleutian W to the date line then to the shortfall of east longitude for the site. The longitude of 130.37 E corresponds to a delay of 3.42 hr, not unexpectedly. No candidates were found in Russia nor northwestern Europe.

In Canada, the next candidate is *Ungava Bay* in the extreme north of Quebec followed by the prominent *Hudson Bay*. Their longitudes indicate that Ungava was followed by Hudson in only 49 min 12 sec, strongly reminiscent of the short period between the twins. The pattern in Canada involves two additional craters that may be related, Belcher Bay on *Hudson's* southeastern shore and a great swamp surrounding James Bay to the south. All of these, if related, would represent a tight cluster of fragments with limited offsets that landed within 1 hr.

The final two craters in Table 7.2 are *Klondike* in Yukon and *Anchorage* in Alaska. Their small sizes, 360 mi and 670 mi in diameter are appropriate for smaller fragments tagging along behind the big boys. Here is where consideration of longitudes must be examined thoroughly based upon data in Table 7.3. Possible Chain.

Table 7.3. Possible Chain.

Structure Name	Diameter (mi)	Longitude (deg)	Cumulative Time (hr)
Aleutian E	1140	168.68 W	0.00
Aleutian W	1340	178.37 W	0.646
Siberia	710	148.92 E	2.83
Ungava	150	67.20 W	17.2
Hudson	599	80.50 W	18.1
Klondike	360	131.97 W	21.5
Anchorage	670	141.30 W	30.9

The total time of 30.9 hr at the bottom of the last column shows that the earth would have made more than a complete rotation and scrambled the location of some craters in this possible chain. It is, therefore, not possible to determine an exact sequence of these impacts within the assumption that they are related. Values in the last column would require adjustments if an undiscovered structure were found east of *Aleutian E*. The long time interval between *Siberia* and *Ungava* of 14.8 hr is highly suspect so Ungava along with the following craters are probably not part of the package.

It should be noted that *Ungava, Hudson, Belcher,* and *James Bay*, while unlikely members of the present chain, could represent the impacts of another short string of comets with limited offsets only if their ages prove to be the same. The probable finding here is that the *Bering Twins* and *Siberia* belonged to the same comet chain but others in Table 7.3 did not. Smaller craters in the string may yet be found west of *Siberia* but are below the threshold of the present research, have been obliterated by later events, or buried under deep sedimentary rock.

A number of tiny pieces in a comet string may not have left ring structures but may have been devastating nonetheless. Such a fragment flattened a forest in Siberia but did not land until 1908. All traces of that event will disappear in a few thousand years. Appendix G. Size Of Tunguska shows how the size was estimated.

This process of exploiting the database would certainly aid in the search for other chains of megacraters. For that purpose, a convenient tool has been made available. Table A.2. Megacrater Latitudes of Appendix A arranges the crater locations from the north pole to the equator thence to the south pole. It should be used with caution and awareness of the shifting continents over the geological eras. That complication can be ignored if the craters are on the same tectonic plate.

Scanning down the column of latitudes in Table A.2 raises some possibilities regarding *Yukon, Tunguska,* and *Siberia* all between 66.73 N and 63.45 N. *Aldan Plateau, Moscow,* and *Albert* lie between 57.45 N and 57.00 N. Finally, scattered craters share the narrow band of latitudes from 44.87 N to 44.00 N. And so on. These possibilities stray too far to explore now so there is plenty of homework to go around. Everyone is invited to participate.

Realizing the locations of all the craters mentioned thus far is hardly possible without reference to an atlas or world map. But essential details can be found in the alphabetical listing of the database in Table. A.3. Short descriptions of each megacrater are presented in Catalog Of Megacrater that is also in Appendix A.

Asteroids

Three chains of craters from known or suggested asteroid impacts have been reported in the literature. While completely unrelated to comets, the same method of research as used above can be helpful in finding others.

Aerial Explosions

Widely reported sonic booms with no apparent sources have challenged the curious and become a popular mystery. Explanations were known to a select few but were hidden from the public. Air Force satellites designed to detect heat from launched rockets and nuclear explosions had registered many explosions at high altitude. The information was classified until 1994. An average of 8 events per year was detected from 1975 to 1992. The grand total up to that year was 136. It was believed in official circles that the phenomenon was under reported and that the actual number might have been 10 times greater. Energy releases ranged from 500 tons of TNT to 15 KT comparable to an atomic bomb. They must have been isolated, small comets whose size can be estimated by the method in Appendix G. Size Of Tunguska. It remains a mystery why the Air Force report was silent about people reporting bright and sustained flashes of light during night or day.

Chapter 8. Volcanoes-Earthquakes-Lava.

"The problem is not just to say something might be wrong, but to replace it by something—and that is not so easy."
- Richard Feynman

Among the most famous volcanoes about which anything is known at all blew its top in the Aegean Sea sometime in 1400 BC. The island of Thira (Santorini) in the Aegean Sea threw ¼ cu mi of rock, lava, and dirt into the sky with much of it settling far into the Mediterranean Sea in the direction of Cairo, Egypt. The energy of the explosion has been estimated as equivalent to 4,000,000 H-bombs, a concept that has now become familiar. Tourists come there today from all quarters to inspect the town high above a precipice, walk the quaint streets, scan the caldera filled with blue water, bask in the sun, and savor the modern Greek culture that began so long ago. What geological factors selected such an unlikely spot for the cataclysm?

Aegean Crater

Among the megacraters, *Aegean* is the most recent at 26 My. The region has not yet stabilized from a bowl-shaped crater with rebound forces approaching nil as elevations approach the existing grade. Many of the islands are still rising. A dense and nearly continuous ring of earthquake epicenters surrounds the entire sea. The floor is subducting to the northeast down the Hellenic Trench that has lifted western Turkey hundreds of feet. The Roman port of Ephesus now resembles a mountain village. Powerful earthquakes rack that region. In addition, the linear rift from the explosion weakened the crust to the southeast that opened into the Red Sea. Hot springs along the fault collect in pools on the bottom in which minerals, precipitating in contact with cooler water above, fall to the bottom forming new deposits. Could any further evidence be required to demonstrate that the Aegean Sea is a water-filled crater of cosmic origin? Being among the smallest under consideration and its shape suggest that the offending missile might have been an asteroid instead of a comet.

Delos Island at the center of the Aegean Sea was the holiest spot in deep antiquity. Apollo and his twin sister were born there thousands of years later. People from the known world assembled there to participate in festivities and to worship Apollo. The entire cultural aspect of Delos and related shrines is examined in the author's earlier book (McCampbell, unpublished).

Volcanoes

Hundreds of volcanoes dot the global landscape. Most of them are completely dormant, some just rumble, others leak lava, and many explode at any time with or without warning. All are of interest because the primary question is their location relative of other geological features, in particular, the rims, linear faults, and central uplifts of megacraters. Any reference book on this subject will show nearly all of them associated with subduction zones that was explained and illustrated in Chapter 4. Potential pathways for lava exist at the interface of the undisturbed bedrock and the subducting slab. Some 400 border the Pacific

Ocean known as the rim of fire. Others have formed great chains of seamounts, islands, atolls, and guyots along arcs of large radius and in straight lines many hundreds of miles long. A few are scattered randomly throughout all the continents. In short, their locations correspond exactly with the principal features of megacraters from which they were spawned.

It is quite impossible to cover this gigantic subject or even to select a number of examples on the basis of their explosive power, volume of lava ejected, or year of eruption. Those categories would not provide the variety of points that need attention. A few will be used to develop the argument primarily ones that are most likely to be familiar. More details on all of the megacraters can be found in the alphabetical listing by assigned names in Appendix A. Reference Data.

The Kamchatka Peninsula hangs down from eastern Siberia into the western Pacific 850 mi in an arc of long radius that continues south another 800 mi through the Kurile Islands to, and through, Japan's northern-most island of Hokkaido. Both the peninsula and islands are saturated with volcanoes. All of them lie along the rim of *Kuril* whose diameter is 1980 mi. One of the most active volcanoes in the world, Bezymianny, is of special interest. For nearly two years, its rumblings generated up to 100 earthquakes per day until March 30, 1956 when it let go. Half of the mountain was blown away. Trees were knocked flat for many miles and dry wood was set ablaze out to 18 mi. Curious effects came from violent lightning in the cloud. Telephones kept ringing and loudspeakers burned out.

This mountain is situated very close to the intersection of faults on the rims of *Kuril* and the westward extension of *Aleutian W* as was illustrated in Fig. 1.19. Bering Twins. Bezymianny lies less than 20 mi south of the primary peak, Klyuchevskaya.

The entire chain of the Aleutian Islands is highly volcanic with a notable example all the way to the other end on the mainland of Alaska opposite the large Kodiak Island. There, in the Valley of 10,000 Smokes (fumerols) rises Mt. Katmai. Its eruption on June 6, 1912 is ranked among the most violent on record. In contrast to Bezymianny, Katmai exploded without warning. Not much farther along the arc is Mt. McKinley, the highest peak in North America.

The most recent volcanic tragedy to hit the news was a massive, lava flow from Mount Nyiragongo when the caldera full of molten rock drained out of a fissure. Some 40 people were killed, refugees fled the destruction of 10,000 homes by the creeping monster and an international relief effort was mounted. The mountain lies on the rim of *Tanzania* about 20 mi north of Lake Kivu west of Lake Victoria. Kivu is one of a string of major lakes in east Africa created between the double rims of the megacrater. These lakes continue into the East African Rift that cuts through the lava plateau of Ethiopia. The central uplift of the crater includes Mt. Kilimanjaro made famous by African safaris and background in wildlife films on the Serengeti Plains. That absolutely flat expanse of grassland clearly developed on top of lava flows. It should be pointed out that a smaller impact site, *Victoria*, was created later inside of *Tanzania*.

The floor of the Indian Ocean slowly plunges down a deep trench 2,000 mi long in front of the Indonesian island chain. These features record a powerful explosion over the China Sea whose diameter is 3420 mi. The inner rim is plainly visible as the Malay Peninsula plus an island arc that together run 2,000 mi. Between Sumatra and Borneo there used to be a large island named Krakatoa. Volcanic, of course, it again became restless after a considerable history of eruptions. Smoke plumes attracted curious people in boats who should have known better. This island with an estimated elevation of 6,000 ft exploded in 1883 essentially removing the entire mountain down to the water line. It ranks among the most powerful eruptions. Three small islands remained around the edge and new ones have been formed inside the gap.

A different kind of volcano periodically throws lava into the air or ejects lava calmly over very long periods. Mountains of extraordinary size are built in the process. The "big island" of Hawaii is a prime example. The peak of dormant Mauna Loa at 13,796 ft offers prime viewing for the cluster of major observatories on the crest. On its flank some 4,000 ft lower, a river of lava and smaller streams flow down the southern slope into the sea. The vent of Kilauea is not accessible to the public but the moving lava is quite a show from shipboard. Great flashes of light are seen as the cooling crust breaks open and exposes the molten rock. From a distance, the river appears like a bracelet of tiny, sparkling diamonds. Upon entering the ocean, clouds of steam are raised and blown away in a wedge of ever-increasing width.

This slow release of lava at various places around the island has been more or less continuous for millions of years. Starting at the bottom of the Pacific Ocean, a new cone developed and grew in height to break through the surface. It continued to grow to the present height. But this "big island" of Hawaii is only the peak of one of the largest mountains in the world. When measured from the bottom of the ocean its elevation becomes 31,000 ft. Several theories have been advanced concerning creation of Hawaii and each may hold some truth. A new theory says that they are exposed peaks on a rim of a megacrater called *Mid-Pacific* and are a snapshot of a long track of similar islands.

A compelling example of a megacrater and its linear rift is displayed by the west coast of South America. The northern half is the rim of *Amazon* that carved the northwestern bulge of the continent. The linear rift sliced the straight southern half down to Tierra del Fuego. Westward drift of the continent since separation from Africa overthrust the floor of the Pacific Ocean and, as expected, volcanoes formed along the entire length of 4,500 mi from Colombia to Tierra del Fuego. Such peaks form the backbone of the mighty Andes with Aconcaqua at 23,085 ft near Valparaiso, Chile.

The continent of Africa is nearly a carbon copy of South America as their west coasts followed the arcs of megacraters and the straight lines of their rift systems. *Timbuktu* shaped Africa whereas *Amazon* shaped South America. One difference is noticed. There are no volcanoes on the African coast. The reason is clear. Africa is attached to a plate that is escaping from a *spreading*, mid-ocean ridge. South America, moving in the opposite direction, slides over the Pacific Ocean floor along a *compression* fault. The only volcanoes near the African coast are on the straight line of the small islands of San Tome and Principe. A few more volcanoes on the same line penetrate deep into the Sahara Desert. Along this alignment ashore, their cause is found to be the crater as *Chad* with a more modest diameter of only 1100 mi.

Antarctica is so inconvenient to show on world maps that it is frequently omitted. And relatively little is known about that continent except the international missions that have set up base camps where scientists winter over to study the conditions. Fossil leaves of deciduous trees and ferns are indisputable evidence that it was not always at the bottom of the world. That is, of course, known by its former connection to south Africa. Cut free by *Kalahari,* it drifted southward taking along some of the Drakensburg Mountains. That portion now is the range of the Antarctic Peninsula plus its continuation inland. Twenty-seven major volcanoes are included. So these volcanoes in Antarctica were sired by the crater, **Kalahari**.

This line of discussion would be endless if volcanic activity of all the 208 megacraters were accounted for. Instead, these few examples demonstrate the point that may be pursued over the surface of the globe by studying the Catalog Of Megacraters in Appendix A. They are arranged in alphabetical order according to their assigned names.

Earthquakes

Volcanoes and earthquakes are siblings that travel together. It is virtually impossible to discuss one while ignoring the other. But examples will be presented while several aspects of the same megacrater are deferred to following sections of this chapter. So much has been written on the San Francisco earthquake of 1906 that little else is needed. A sudden slippage along the San Andreas Fault offset a causeway on the San Francisco peninsula by 16 ft. The average movement of the western side relative to the eastern side has been about 3 cm/yr. Over 40 million years, the total slippage has been about 750 mi. A recent event on the fault near Loma Prieta nearly destroyed downtown Santa Cruz, ignited frightful fires in San Francisco, and knocked down a section of the San Francisco Bay Bridge. Huge sums will be required for reinforcement. It appears that a fault system running north from *East Pacific* opened the Sea of Cortez and continued under the name of San Andreas.

The most powerful earthquake in North America was centered at an unexpected location. Near Cape Girardeau, Missouri, the New Madrid fault slipped in 1811 to generate a quake of 8.0 on the Richter scale. Very few people were killed because no one was around. The fault runs from northeastern Arkansas to southwestern Indiana and is still active. A 4.6 quake in September of 1980 rattled a wide area with little damage. It did cause the Mississippi River to run upstream and create Reelfoot Lake in Tennessee. A repeat of the big one would destroy portions of five developed and densely populated states. The location is believed to be linked to a massive explosion.

India suffered severely from a strike on January 20, 1991. In the city of Bhuj, with a population of 150,000, high rise buildings toppled and 90% of the other real estate shattered. Death toll, 2,300; injuries, 14,000. This region is highly prone to earthquakes because it is situated at the intersection of the rim fault and linear fault of the huge *Iran*, whose center is 740 mi to the northwest near Mashhad. Such regions are exposed to the maximum hazard as was emphasized by a shaded circle in Fig. 4.1. Event Diagram. Several other aspects of this megacrater will be addressed shortly.

A 7.2 R hit the Tonga Islands on January 8, 2000. Hardly surprising because they lie along the Tonga Trench where the ocean floor dives into the depths toward the west. That linear fault links up with the Kermadec Trench running an additional 1500 mi to North Island, New Zealand. Near Auckland, the fissure produces a fine display of heat from underground sources in the form of fumerols, boiling pools, and bubbling mud pots. The fault continues down through the whole of New Zealand close to the highest mountain in the country, Mt. Cook. And it does not end there. The origin of this lengthy fault was the Antarctic megacrater named *Ross* as shown in Fig. 1.9. Birth Of Australia.

It would be well worth the effort for someone to look into the entire history of earthquakes and relate them to megacraters with which they were associated. In addition to earthquakes associated with megacraters rifts, very great activity is concentrated along the mid-oceanic ridges. The origin of most of them has also been traced to the aerial explosions and, in all likelihood, comets are responsible for the whole lot.

Further research on earthquakes might begin with those in Guatemala on April 19, 1902, China on December 16, 1920, Japan on September 1, 1923, Pakistan on May 30, 1935, and China on July 27, 1976 when a Richter 8.0 killed between 255,000 and 655,000 people. The list is endless but hundreds of primary events like those above deserve investigation.

Lava Plateaus

Massive lava flows are known by a variety of names, such as, *lava plateaus, flood basalts, traps*, and *provinces*. Shield volcanoes should also be included because they are built by a steady production of lava over geological periods. Some of these formations spread out over vast areas on land, on the ocean floors, and on the bedrock of oceans underneath the sedimentary layers. Many can be attributed to damage from specific, comet explosions and no other mechanism has been suggested. All major lava flows, therefore, were probably caused by those explosions whatever their size, shape, or location. This aspect of geology is as big as the sky in Montana because lava flows are so numerous and scattered around the world. A review of this subject may be found in an authoritative treatment by Mahoney (1997).

Those accounted for in Chapter 1 were Siberian Traps, Ontong Java in the South Pacific, Rio Grande in the South Atlantic, Parana-Entendeka with part in South America and the remainder in south Africa, Ethiopian highlands with the bulk near the source in Ethiopia and a portion in southeastern Arabia, Ninetyeast Ridge in the Indian Ocean, and Broker Ridge west of Australia. Traps by all names became prominent in Chapter 2 regarding life extinctions. Since they can be dated with impressive accuracy, they become the stopwatches for major geological events.

Instances where a lava bed has been spilt and separated are most instructive. They were, of course, completed and hardened before something sliced through them like scissors through paper. The forces and power required were monumental. Since these cases are unequivocally tied to specific megacraters, there is little doubt of the cause. Entendecka was formed from crustal damage of **Kalahari** then sliced by the linear rift from **Timbuktu**. The western fragment drifted away on the South American plate where it is today in Argentina. The sequence of events requires **Kalahari** predating **Timbuktu** and the breakup of Pangea.

Ethiopia is especially interesting because it was cut into pieces on two occasions. First came the opening of the Red Sea by **Aegean** that spit off the fragment at the heel of Arabia. After the sea had attained considerable width, the Ethiopian Highlands were cut again by the East African Rift from **Tanzania**. This sequence is assured because the Afar Triangle in northern Ethiopia is clearly a river delta deposited from drainage down the fault. **Ethiopia** must have followed the origin of the Red Sea by **Aegean** that is dated about 26 My ago. So the Ethiopian explosion must have been geologically recent but old enough for erosion to destroy its volcanoes except for cores rising above the plateau. These denuded cores are the primary features of a national park.

Ninetyeast Ridge south of the Bay of Bengal is an underwater, mountain range that is a flood basalt. The same nomenclature applies to Broken Ridge west of Australia. Another example covers the ocean floor east of New Zealand with about the same area as that country. Still another one covers a large area of the Indian Ocean that will be considered in some detail later.

Among the largest flood basalts, one covers the entire bottom of the Caribbean Sea. The eastern end is marked by the Lesser Antilles islands including Grenada, Barbados, Dominica, and Antiqua. The western end reaches to Nicaraqua and Costa Rico in Central America. With dimensions of approximately 1500 mi long and 450 mi wide, the surface area amounts to 675,000 sq mi. Thicknesses determined in the Deep Sea Drilling Program were found to vary from 3.7 mi to 12.4 mi. Taking an average thickness of 8 mi yields an estimated volume of 5.4 million cubic miles. That's a lot of lava unknown to most people.

Volumes of lava at other large sites vary considerably but 500,000 cu mi may be taken as typical.

A point touched upon earlier can best be understood with the aid of Fig. 8.1. Caribbean Subduction. The 560-mi crater called ***Antilles*** produced the island arc and thrusting bedrock out and over the Atlantic floor. Subduction then followed as the Americas continued an inexorable drift to the west. Local portions of the sea floor also continued the westward drift as new floor was created along the Mid-Atlantic ridge. A relatively narrow strip diving down the trench continued on the same motion but was invisible. It flowed under the Caribbean Plate.

The drifting floor of the Atlantic Ocean is indicated by the broad arrow that applies to all latitudes. Loss down the subduction trench renders the westward part of the ribbon invisible as suggested by the shaded area. Vast damage to the crust and new passages for lava must have initiated the flow near the islands. As time progressed, lava flows opened to the west following the hidden floor. If this scenario were true, ages determined for drill cores along the axis of the Caribbean would display increasing ages from east to west.

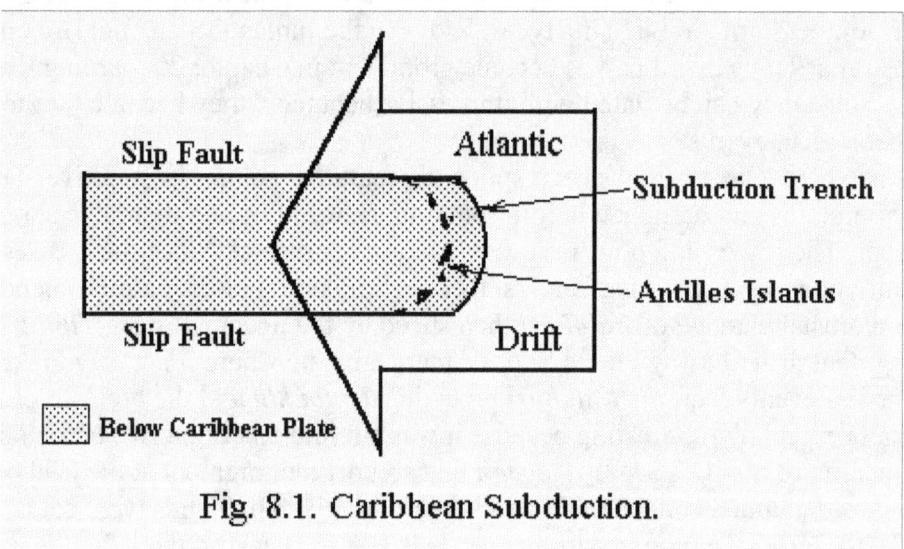

Fig. 8.1. Caribbean Subduction.

The generally accepted view is that the islands originated in Central America and drifted eastward to their present position along with the Caribbean Plate. Such swimming upstream against the flow of Atlantic drift seems improbable. More likely, formation of the megacrater anchored the island arc in place at the present position. Creation of the Caribbean Plate progressed and elongated westward as growth was accommodated by slip faults along the north and south boundaries. An exact duplicate of this geology is found at the South Sandwich Islands off the southern tip of South America as shown in Fig. 1.15. More Latin Examples.

Chapter 9. Global Landscapes.

"...he well understood that most of us can digest only so much unorthodoxy at one swallow."

- Eric Larrabee

Extremely powerful explosions in the atmosphere over hundreds of millions of years have plowed earth's surface into a huge, patch-work quilt. Major rifts cut continents loose from their motherlands defining tectonic plates. The stage was set for continental drift and, in conservation of land area, some were lost from sight in returning home. Shapes of the continents were determined. Secondary events added nooks and crannies along the shorelines. Fault systems provided the foundation upon which mountain ranges grew from tectonic forces. Islands of every sort were born whether continental fragments, straight chains, arcs, or isolated. Volcanoes grew from lava passages provided at subduction zones on crater rims and rifts and within their basins. Major rivers flowed in the rift zones. Fossil deltas formed from rivers now extinct. Strings of lakes remained in their shallows during dry periods. Depending upon the amount of local rain, various basins turned into inland seas, lakes, swamps, deserts, and salt flats.

Ridges And Plates

It has been shown that some of the more powerful explosions fractured the original land mass and subsequently some of the pieces. The great fault systems established the boundaries of continents and smaller pieces. A great continental rumba began as the dancers clasped, jostled, slid, dove, and separated. Water quickly poured into the voids while lava leaked out of the fracture zones freezing into new ocean floor. When the separation was slow, parallel mountain ranges grew up on the bottom. So all oceans have primary cracks down their middles where lava has flowed continually since they were created. Dates of creation can be established by radio-active analyses of drill samples. Oldest ocean floors are, therefore, close to the continental margins while ages at the ridges are nil.

Major fragments of the primordial landmass are known as tectonic plates and carry the continents. Prominent among the small pieces are New Zealand, Madagascar, British Isles, Caribbean, and Cocos. Other dance steps produced crushing, slippage, and subduction at plate boundaries depending upon relative motions between adjoining plates. All of these geological processes are possible only along the explosion-created rifts. Some examples are assembled in Table 9.1. Geological Features.

Table 9.1. Geological Features.

	Name	**Location**	**Crater**
RIDGES			
	South Atlantic	South America - Africa	*Timbuktu*
	North Atlantic	North America - Europe	*Arctic*
	Carlsburg	India - Africa	*Iran*
	Austral - Antarctic	Between continents	*Ross*

TRENCHES

Atacama	West coast South America	*Amazon*
Kermedec	South Pacific	*Ross*
Java	Indian Ocean	*Indonesia*
Mariana	West Pacific	*Mariana*
Island Arcs	World Wide	*Various*

MARGINS

San Andreas	Western North America	*South Pacific E*

Birth Of Continents

Earth can take tremendous beatings with little permanent damage beyond welts that, by ordinary standards, are impressive. Only the most powerful attacks can bite off continental-sized chunks and spit them out. Table 9.2. Creation Of Continents accounts for four out of seven.

Table 9.2. Creation Of Continents

Continent	Event	Diameter (mi)
Africa	*Iran*	1470
Antarctica	*Kalahari*	1120
Autstralia	*Wilkes Land*	2200
South America	*Amazon*	1800

A crater as small as Kalahari at 1120 mi is capable of creating a continent under favorable conditions. But a crater around 2000 mi seems to be the general rule for fracturing a major land mass.

Mountain Ranges

The study of how mountain ranges are formed is the subject of *orogeny,* a special discipline in geology. It addresses the distortion of sedimentary rock into folds like a curtain or even loops. Tectonic forces drive these processes that are countered by erosion. Other types of mountain ranges based upon escaping lava are treated in a separate category although they should be combined. Details have been thoroughly studied for all the ranges and their growth is well understood. However, an important issue remains open. Little attention has been addressed to explain why any given range selected its site. The mechanism has become obvious. All mountain ranges, whether arcing or linear, arose where rift systems were created by the megacraters. Let that point be emphasized. *Cosmic explosions gave birth to all mountain ranges.* Many grew from overshooting slabs during crater rebound. Others grew from complex fault systems. Once initiated, the locale was subject to the normal geological forces that are so well understood. Some of the major mountain ranges are shown in Table 9.3. Mountain Ranges to be associated with particular craters.

Table 9.3. Mountain Ranges

Range	Location	Megacrater
Alaska	inland extension of island arc	*Aleutian E*
Atlas	northern Algeria and Morocco	*Timbuktu*
Big Horn	northern Wyoming	*Big Horn*
Carpatian	north central Europe	*Hungary*
Elburz	Caspian shore of Iran	*Caspian S*
Great Dividing	southeastern Australia	*Brisbane*
Kjolen	Norway-Sweden border	*Bothnia*
Sangre de Cristo	Colorado and New Mexico	*Navajo*
Serra do Mar	coast range at Sao Paula	*Timbuktu*
Sierra Madre Oriental	eastern Mexico	*Gulf*
Stanovov-Dzhugdzur	eastern Asia	*Yakutsk*
Verkhoyanskiy	central Siberia	*Lena*

Many more are described in Appendix A. Catalog Of Megacraters where mountain ranges are seen to augment other clues to the existence of crater rims or linear rifts.

Continental Drainage

Craters heal themselves because pressures far below grade drive the floor upward. Recovery may be incomplete because these forces become weaker and weaker as the previous grade is approached. The final result is a shallow basin of enormous extent whose mountainous rims and faults dictate drainage of the entire area. Very large craters break a channel to the sea along their linear rifts while smaller ones, confined within the continental boundaries, produce a variety of landscapes.

A large portion of central Africa drains into the Atlantic Ocean via the Congo River. The basin of more than a million square miles, caused by *Congo*, is laced with waterways that converge in the vicinity of Brazzaville for the final run of 250 mi to the sea. It is a mighty river. The mouth is strewn with an array of jungle-covered islands that are most impressive when viewed from a small airplane just off shore.

Smaller *Chad* at the edge of the Sahara Desert does feed the Niger River but retains much of the limited moisture. A similar pattern prevails at *Timbuktu*. Both have extensive swamps or shallow lakes at their centers. Deba Swamp just south of Timbuktu covers about 25,000 sq mi. Lake Chad and associated swamps have a comparable area. Swamps, therefore, can be used as clues to locating new craters.

A lacework of rivers flows in the Amazon basin like veins of a leaf converging to the stem near Manaus with others joining farther down stream. The mouth is so wide that the opposite shore can not be seen. *Amazon* produced this river system that is replicated at other locations.

St. Louis did the same thing in the center of the North America. The upper Mississippi, Ohio, and Missouri rivers join near St. Louis then run to the Gulf of Mexico accepting contributions from the Arkansas and Red Rivers. All the landscape south of the intruding point from the state of Mississippi is alluvium. Sand, dirt, tree limbs, and trash have been deposited there with a great deal more in the alluvial fan of the Gulf of Mexico. Above that point from the neighboring state, the river runs down a linear rift created by *St. Louis.* The river meanders through the silt between the barriers where it stays most of the time.

An interesting variation on this pattern was mentioned regarding the Himalayas. The rivers involved are the Ganges, Brahmaputra, and Indus. The first two arise in proximate lakes high behind the frontal range then set off in opposite directions. They are guided by a valley between a double rim of *Himalaya*. Ganges runs to the northwest half the length of the range and makes a 180° turn around the western end. After reversal to the southeast, it flows in front of the Himalayas then turns toward the Bay of Bengal. Discharge is accomplished via multiple outlets 250 mi wide.

The Brahmaputra took a short cut around the eastern end of Arunachal Pradesh. After cutting back in front of the mountain range, it joins the Ganges near the coast. The Indus drains the Tibetan highlands behind the Himalayas. After making a turn through a broad gap near Islamabad it runs along a rather straight course down the axis of Pakistan. The mouth at the Arabia Sea is near Karachi.

Most of these points have been mentioned previously but further attention must now be focused upon specific megacraters and their consequences. The world is too large a camel to inspect every flea. So, selected specimens will be studied to illustrate the nature of the whole bunch. Because many examples might strain the geographical knowledge of the reader, this is a good time to lay hands upon a world atlas. Additional information on individual craters can be found in the alphabetical listing in an appendix.

- Mackenzie, northwestern Canada through Alaska to Arctic Ocean, *Alberta*
- Yangtze, eastern China into East China Sea, *South China*
- Mekong, in Cambodia through Vietnam to South China Sea, *Ho Chi Minh*
- Mississippi, central United States into Gulf of Mexico, *Mexico*
- Lena, arcing through Siberia to Laptev Sea, *Lena*
- Upper Nile, great arc and offset, see Fig.1.3 and description, *Saudi*

Smaller craters are less successful at opening paths to the sea with a variety of consequences. Bays are formed when the sea floods a crater too close to shore. Other landmarks depend upon the amount of rainfall and whether it is seasonal or varied over long periods. Depressions in dry locations produce deserts. Swamps and shallow, seasonal lakes are sustained by limited rain. Lakes and seas are kept full by adequate downpours and overflow as the headwaters of streams and rivers. Long intervals of dry and wet climates produce salt flats and salt lakes. The world is pockmarked by all of them.

It is necessary only to point out some obvious examples along with their crater names as in Table 9.4. Terminal Depressions.

Table 9.4. Terminal Depressions.

Location	Country	Crater
Hudson Bay	Canada	*Hudson*
Kalahari Desert	Botswana	*Kalahari*
Lake Victoria	Ugnada	*Tanzania*
Lake Chad	Chad	*Chad*
Caspian Sea	Kazakistan	*Caspian*
Lake Baikal	Russia	*Baikal*
Bonneville	Utah	*Bonneville*
Sahara Desert	Mali	*Timbuktu*

108

Notice the variety of landscapes in the first column and crater names in the last column. Every effort was made to assign names to craters that indicated their global location. Plenty of other examples will be found in Appendix A.

Waterfalls

Hundreds of waterfalls dot the landscape attracting tourists like moths to a flame. Among the most dramatic is in Yosemite National Park in California. Water cascades down three levels for a total drop of 2370 ft into a canyon of breath-taking beauty. A glacier scoured the canyon during the last ice age that was not so long ago. Rising temperatures melted the ice and created the topography. In other words, Yosemite Falls is an infant on the geological time scale.

More significant here are those where great rivers cascade down rifts of continental proportions. Victoria Falls on the Zambizi River near Livingstone, Zimbabwe is a tourist attraction rivaling the pyramids of Giza. The gorge lies on the northern rim of ***Halakari*** and the site is on its linear rift. The latter is a wide, geological feature of southern Africa running northeast from the epicenter along the western border of Zimbabwe. Once again major structural changes are seen at the intersection of a megacrater rim with its own linear rift where maximum damage is predictable.

Before creation of a fault, the surface defines the grade level. After a fault rips through it, elevations on both sides are frequently different. This vertical offset creates sharply defined escarpments that may be thousands of feet high. If a river previously ran through the region, its final course would include a waterfall if the downstream side were lower. Whenever the downstream side is higher, the course must change to flow along the fault. These conditions appear to prevail at Victoria Falls whose existence can probably be attributed to the megacrater.

Waterfalls are frequently judged on the basis of their height alone because they are most spectacular. An alternative rating would be their power, taking full account of their height and the volume of water flowing over the precipice. On this basis, the most powerful waterfall on earth is, or was, the Guari Falls in Argentina. Alas, it was inundated by construction of the Itaipu Dam on the Paean River so is now dormant and invisible. The confluence of this tributary of the Parana River is near the latter's mouth into the Atlantic Ocean. At the time of completion in 1982, Itaipu Dam was the largest hydroelectric facility in the world, providing 25% of the energy needs of Brazil and 78% of Paraguay's. It dutifully earned recognition as one of the Seven Wonders of the Modern World. Tourists came in droves. Nine million people from 162 countries visited in 1995. Their interest included the famous Iquasu Falls only 12 mi downstream of the Parana River.

The Parana flows in an arcing trough to an encounter with a wide, transverse fault, falls over the edge, then runs down that rift after a turn approaching a right angle. Part of the year, several streams make the plunge but, in the wet seasons, the entire river drops 215 ft into the rift on a 2-mi front.

Courses of the Parana and Paean Rivers and sites of their waterfalls reflect some fundamental geology. Even though obscured, the ultimate cause is recognized as the far away ***Kalahari***. Recall the original position of South America vis-a-vis South Africa. The Parana flows in an arc of that crater's rim while Paean drains the lava plateau of that name formerly linked to the Etendeka plateau in south Africa.

Below the Iquasu Falls, the Parana makes a sharp turn to the east through a broad swamp, clearly a portion of the rift that has been filled with silt. It then broadens abruptly into Rio de la Plata (Plate River), reaching the Atlantic Ocean between Buenos Aires, Argentina and Montevideo, Uruguay. The term, Rio de la Plata, applies only to a wedge-shaped bay 130 mi long that gradually widens to 60 mi at the mouth. This pattern is not typical of a delta so something else is afoot.

Answers can be found in Fig. 1.15. More Latin Craters. The transition from the Parana to Plata lies at the point where two megacraters are tangent with their respective rims touching. The north shore of the bay was formed by the 410-mi *Uruguay* while the south shore was formed by the 1020-mi ***Buenos Aires***. The lower Parana back to Iquasu can be attributed to the combined effect of these two megacraters.

A charming, little game is suggested by these findings. Players select a country in which a national park contains a major waterfall. Then study of the local geology seeks to clarify whether it is or is not associated with a known megacrater. Scoring would be 10 points to prove or refute the proposition and 5 points for uncertain results.

Escarpments

Primary rifts from megacraters split bedrock and everything lying above. Frequently, sides of the fault settled at two elevations leaving a nearly vertical and bare face on the higher block. This is, of course, the same manner in which waterfalls were created if a river was present at that time or developed later. This idealized concept may exhibit some variations such as a series of cascades or secondary faults sloughing into stair steps. Sediment washed down from the heights over eons form gently slopping ground on the lower block and beyond.

In the eastern United States, this configuration is expressed as a coastal plain. Narrow at Long Island, New York, it progressively widens to the south at Philadelphia, Washington, and Richmond where it reaches 100 mi. A dividing line between the continent proper and the coastal plain is a unique feature known as the fall line. That name comes from modest rapids and waterfalls created by the different elevations at the transition line between the continent and its coastal plains. Obviously, the latter is geologically younger. Also apparent is the link between the fall line and the rift system that broke North America off of Pangea. Despite severe erosion, evidence of the original rift is still present.

In particular, part of this rift is revealed by the precipitous face of an escarpment up to 500 ft high, namely, the Palisades in New Jersey. This formation along the Hudson River is of lava that existed when Pangea was fractured. The correct age was established by discovery of a fossil in 1910. Bones of a reptile resembling a 35-ft crocodile, classified as Clepsysaurus manhattanensis, were excavated and are on display at the American Museum of Natural History in New York. The animal lived about 210 My ago. Also found at the site were fossils of a large fish known as coelacanth that had been proclaimed extinct until the mid-1960. A living specimen was caught off the coast of South Africa. Several others have come to light since then. Also found at the Palisades were remains of Icarosaurus, a flying reptile of the correct age. A conclusion is hard to suppress: The Palisades were born with the creation of the mid-Atlantic ridge.

More Lava

Recent research has identified a lava plateau that, not considered thus far, is probably the largest known. The volume has not been determined but the area amounts to 2,700,000 sq mi. Being several times the area of the typical example takes it into a class by itself. Before the breakup of Pangea, lava spread out from what is now France to southern Brazil. Something cracked it apart releasing North America on its long, slow journey. This massive formation is known as the Central Atlantic Magmatic Province (CAMP). A mass extinction associated with CAMP about 200 My at the boundary of the Triassic-Jurassic periods has been generally recognized. Suggested here is the source of energy for the breakup and extinction. A comet must have approached the earth with its inclination almost 90° when the North Pole pointed in the direction of its orbital motion. Consequently the megacrater, located near the North Pole and now filled with water, is the Arctic Ocean. This crater has been named Iceland because that island lies squarely on the mid-Atlantic Ridge and is noted for its intense vulcanism.

On the other side of the Atlantic Ocean on the coast of southern Africa one finds a geological formation comparable to the Palisades. The Congo River reaches the sea at the north end of Angola except for a small enclave on the Right Bank. That country has a rather narrow and uniform coastal plain that backs up to a major fault scarf. Rising abruptly several thousand feet, the rock face is best observed from the air. Flying close to the cliff, a famous geologist pointed out to the author, many rock formations whose counterpart in South America were well known. *Timbuktu*, in this instance, was the culprit. It cut South America loose forming the southern portion of the Mid-Atlantic Ridge.

Major escarpments around the world may have been formed along the linear rift systems of other megacraters. They should all be identified whether wet or dry. Their alignment may associate them with specific megacraters that have been pointed out or lead to discovery of new ones.

Grassland

Many plains and farmlands depend upon loose material washed down from nearby hills and mountains. Gravity assures that they are nearly level and suitable for irrigation. A motorist can enjoy cruising through the Central Valley of California observing its rice patties, vineyards, orchards, vegetable farms, and cotton fields. Seldom will the driver think about the bottom of the real valley about 2.5 mi under the pavement. That much silt has filled the original fault zone.

Other plains are known to rest atop lakes of molten rock that froze eons ago. An additional layer of sediment provides the footing for vegetation. This type of plain is demonstrated in the Ngorongoro Crater in Africa about 1 mi west of Mt. Killimanjaro. Internal collapse of a volcano created a haven for an incredible variety of wildlife. The level floor 10-12 mi across can be observed from the rim 2,200 ft higher. The adjoining Ngorongoro Conservation Area of 3,200 sq mi consists of varied terrain including lakes, streams, outcroppings, and numerous craters. The latter, resembling the surface of the moon, have been assumed to be volcanic in origin. Au contraire, they may be asteroid craters on the smooth surface of lava duplicating the conditions on the lunar mares. In addition, the Serengeti National Park 100 mi southeast of Lake Victoria is very flat. This grassland of 5,700 sq mi also supports innumerable grazers, predators, and birds. The controlling factor in this geography must be an underlying layer of lava.

Even Serengeti is miniscule compared to the lunar mares. Their equivalent structures on earth are much larger. So much so that their existence has not been recognized. Their areas can be very great. A modest megacrater of 1,000-mi diameter would cover 785,000 sq mi. Apparently limitless plains on earth are, therefore, suspected to be lava-filled, multi-ring structures covered with soil. Probable examples are

- Plains of west Texas by **Great Plains** (1060 mi diameter),
- Steppes of central Asia by **Moscow** (3,330 mi diameter), and
- Wheat farms of northwest Canada by **Alberta** (1330 mi diameter).

Erratics

Strange boulders are lightly sprinkled about the earth. They are huge and their composition is unlike any geological formations within many miles. Various theories have been proposed to explain how these *erratics* were extracted from their homes and transported great distances. Perhaps the continents had been overwhelmed by oceans that sloshed out of their basins. Or they were carried into new positions on the backs of ice sheets and glaciers. None of these concepts, however, proved to be satisfactory. So the erratics are still a mystery.

A 10,000-ton boulder of granite stands near Conway, New Hampshire with dimensions of 90 x 40 x 38 ft. No natural formation of granite has been found for miles around. Another boulder of similar size on the beach at Humbolt Bay, California may also be geologically distinct from its immediate surroundings.

The great explosions from the cosmos could easily have overturned huge slabs of rock and projected giant boulders over long distances. If this explains the puzzling phenomenon, research should begin at the nearest megacrater and compare the local geology with the composition of the great stones.

Overview

Features of the landscape are so closely intertwined that none can be discussed in isolation. As many have already been treated in previous contexts, a summary in Table 9.5. Crater Formed Geography can provide a proper perspective.

Table 9.5. Crater-Formed Geography.

Mid-oceanic Ridges	Subduction Trenches
Oceans	Island Arcs
Continents	Island Chains
Coasts	Seamounts
Rift Systems	Coral Reefs
Mountain Ranges	Continental Fragments
Volcanoes	Plains
Earthquakes	Seas
Lava Plateaus	Bays
Rivers	Lakes
Waterfalls	Swamps
Escarpments	Deserts

All of these items have been found in association with a limited number of megacraters. It seems to be apparent that a general pattern has been established. For megacraters not specifically investigated, similar results can be expected. It follows that all the megacraters under study and those not yet discovered have sculpted the landscape by creating all of the major feature of earth's geography.

114

Chapter 10. Minerals.

"Speculations are an important component of science...but they must eventually answer to the facts."

- Charles B. Officer

Damage to earth's crust from explosions was far greater than surficial. Not only shaping the global topography, they dug holes of unimagined proportions. Previously horizontal layers of sedimentary rock were tilted, folded, and distorted into grotesque shapes. Results from a particular event were often scrambled by another one nearby. Much of the rock down to the igneous basement was crushed, melted, vaporized and/or blown away. Thousands of cubic miles of material fell back into the crater. Refilling the cavities from below brought material from the depths to the surface or close to it. Maximum crater depths probably exceeded 300 mi in imitation of dimples on golf balls. Gigantic slabs of rock and smaller fragments brought all kinds of elements close enough to the surface to be reached by whatever mining technology happened to be available. Faults and broken rock provided pathways for intrusion of molten materials and water carrying dissolved minerals.

Oil And Gas

Petroleum geologists have long recognized a relationship between basins on the surface and sources of oil. Also domes in geological structures that are sometimes uplifted on the surface have been a boon to wildcatters. Certain conditions must prevail for oil to accumulate in marketable quantities. It must be held in a trap that is closed by imperious, hard rock. Also, layers of porous sandstone must be inclined to the horizontal allowing migration of oil into the trap. In domes, a sealing cap can gather oil into a commercial pool. The most valuable sources, however, occur when oil percolates through sandstone until encountering an impenetrable wall. Such arrangements are found at faults where a layer of vertically displaced, impervious rock blocks the flow. Lighter fluids in the trap separate according to density producing horizontal zones of natural gas, oil, and water, respectively.

Numerous basins cover earth's surface like a carpet with a pattern resembling the moon. Craters of all sizes lie close together, abut, or overlap a neighbor. A few of them have been suspected to be asteroid impacts and most probably are.

Oil has been found near the center of the basins and along their peripheries. Notable examples are

- Ames Crater, reserves, northern Oklahoma 16 mi west of Fairview (36.25N 98.17W), 10-mi diameter
- Avak Crater, reserves, immediately south of Barrow, Alaska (71.25N 156.63W), 7.5-mi diameter
- Red Wing Creek Crater, production for decades, west central North Dakota 18 mi from Montana border (47.60N 103.55W), 5.6-mi diameter.

At diameters of only 10.0, 7.5, and 5.6 mi in diameter, these craters are mere chicken pox but a point is sustained. Oil and impact craters go together because the required conditions were established by the explosions.

U.S. Bullies

Three primary oil-producing areas in the United States are the Gulf Coast, West Texas, and the Midwest. Maps of oil fields in these areas show a remarkable pattern. As shown in Fig. 10.l. U.S. Fossil Fuels, they are clustered along great arcs that are directly related to three megacraters, namely, **Gulf, Texas, and Michigan!** The general patterns are unmistakable and even the details of field sites are significant.

Concentric arcs through southern Texas and Louisiana show how fields developed in conjunction with a double-rimmed crater. Separation between the arcs ranges from 18.8 mi due north of the epicenter to 37.7 mi to the west. An average value leads to an approximate, rim ratio for this crater, the outer radius divided by the inner radius, $(440 + 28.3) / 440 = 1.064$. The fields are clearly related to three, concentric faults through the region. Starting at the Gulf coast, they are Lower Miocene Fault Zone, Vicksburg Fault Zone, and Frio Fault Zone. They are equally spaced and reflect common kinks. A more accurate rim ratio of the crater can be determined from a map. Spacing in the north is 15.9 mi whereas in the west it is 27.2 mi in agreement with corresponding data for oil fields, namely, 18.8 and 37.7. A rim ratio based on an average value for the faults is found to be exactly the same as the value calculated for the fields. There can be no doubt that the oil fields were formed in conjunction with the faults created by **Mexico Gulf**. It must be presumed that twin mountain ranges between the three fault zones were lost to erosion and blanketed by sediment. Their remnants may be found in the folded layers of rock below the sediment.

While oil fields on the western end of the arc are not shown extending into Mexico, they must surely exist whether developed or not. On the eastern end, large quantities of oil are obtained from offshore platforms. Their locations indicate a direct extension of the onshore fields. Drilling beyond the present locations should be high-priority for prospecting. In May 1991, Shell Oil Company and British Petroleum found a new field, called Mars prospect, 130 mi southeast of New Orleans. Analysts consider this strike to be the most important one in the U.S. in the last 20 yr. Estimated reserves of 2 billion barrels in a string of fields know as the Mississippi Canyon could rival the North Sea. This discovery could have been predicted because it lay on the rim of **Mexico Gulf** arc extended out to sea.

Another prominent crater in the present catalog is **Texas** whose center is near Amarillo. Its rim was found to have two arcs of mountains in New Mexico that are separated by the Rio Grande River. It continues down through Big Bend National Park then proceeds through an arc near Austin and Ft. Worth to the Oklahoma border at Lake Texoma. Evidence on the surface beyond the lake fades away.

Distribution of oil fields in Fig.10.1 closely follows the topography. This massive collection of fields begins in the southwestern corner of New Mexico. It swings toward the northeast in a wide band all the way to Lake Texoma. That is not the end. The fields continue in full density north through Oklahoma City and Tulsa then, turning to the west, into central Kansas. A central uplift of **Texas** was certainly responsible for a large cluster of fields ranging north from Amarillo through the Oklahoma panhandle well into southwestern Kansas. A roughly rectangular zone is about 175 mi N-S and 120 mi E-W. This large region covers about 21,000 sq mi but is not out of scale for a crater 990 mi across with an area of 770,000 sq mi.

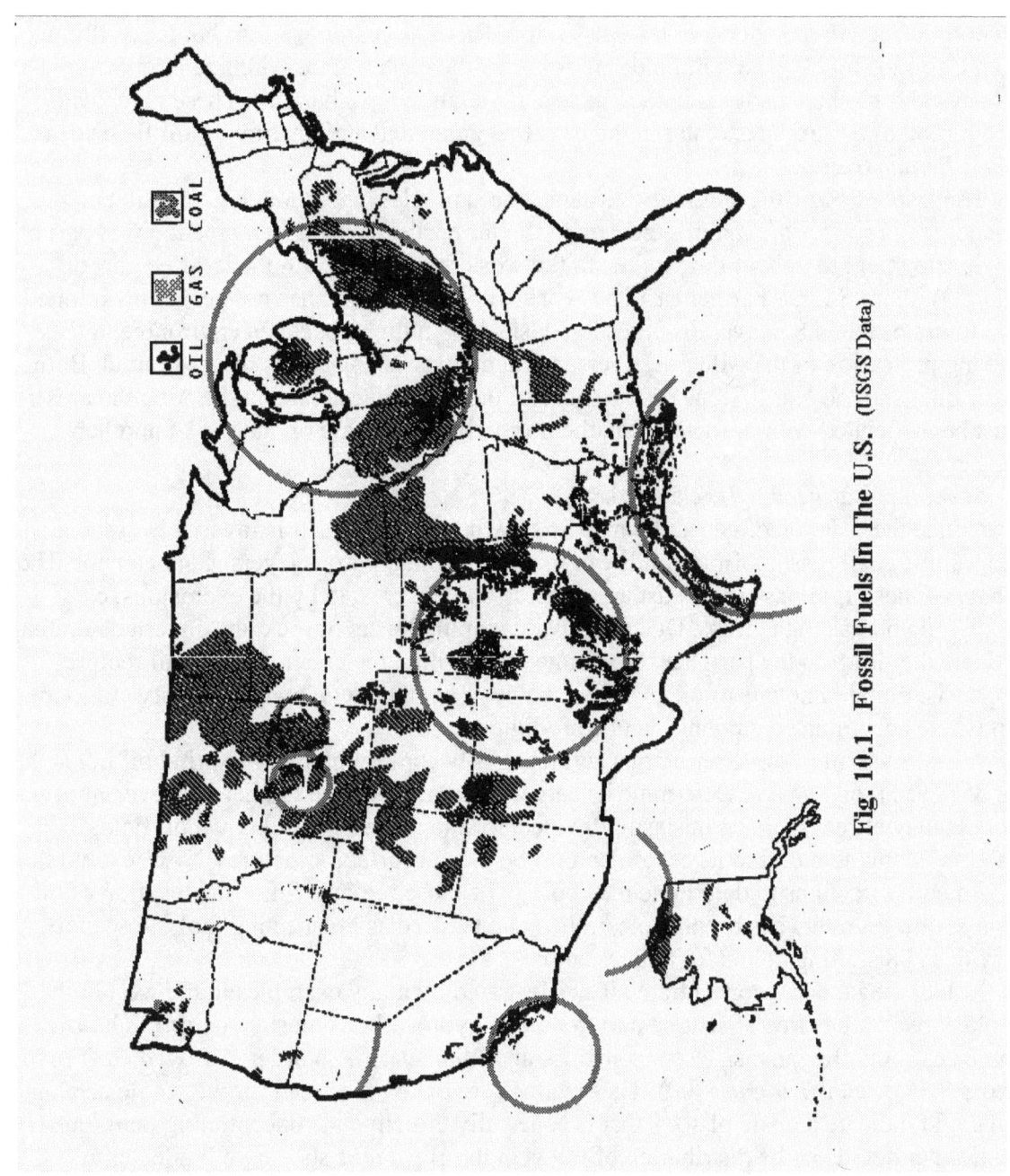

Fig. 10.1. Fossil Fuels In The U.S. (USGS Data)

Oil seeping out of the ground and collecting in small pools has been used as medicine and flaming arrows for 5000 yr. Only in August 1859 was oil obtained by Edwin L. Drake from the first well drilled for that purpose. It is unlikely that he could have envisioned the explosive growth of a new industry with all of its economic, chemical, transportation, and military implications. Western Pennsylvania became a major producing area as wells spread in a narrow band north from Pittsburgh to the New York border and south through West Virginia. No one suspected at that time that the oil deposits defined a great arc 600 mi in diameter centered on the Michigan Peninsula between Lake Michigan and Lake Huron. After reaching the border of Tennessee, the arc turned to the northwest through Kentucky and into Southern Indiana.

There are few, obvious clues on the surface that identify this arc as a crater. But the lake district in southern Kentucky and Mammoth Cave come under suspicion. Also drawing

117

attention is the curved course of the Mississippi River along the western border of Illinois from Cairo to Rock Island. Oil fields are found at the center of the Michigan Peninsula exactly echoing the structures in the Texas and Oklahoma panhandles. These facts justify recognition of a third megacrater in the U.S. that generated a giant array of oil fields. It is called *Pennsylvania*.

The inset of Fig. 10.1 shows the circumstances in Alaska. Although Cook Inlet at Anchorage produces oil, major fields are in the far north. The Alaskan pipeline was part of the development to deliver the product to the world market through the ice-free port of Prince William Sound. Farther east lays a primary resource that has not yet been exploited. The future of the U.S. Naval Reserves in Alaska lies in limbo between environmentalists seeking protection of the wilderness versus pragmatists seeing riches below ground. Both Cook Inlet and the North Slope on the shore of the Arctic Ocean lie on the Aleutian arcs so must be associated with it. In addition, the northern reserves are on the rim of another megacrater.

Several points need to be clarified.

Notice the location of gas in the map as shown by stippling. It is invariably associated with oil for good reason. Both fluids can migrate through porous layers. Coal can not. The beauty of these findings is that vast areas inside craters, created by the explosions, can supply oil and gas to the traps. Due to outward sloping layers of rock, an "inverted" drainage delivers the fluids to the periphery from anywhere within the crater. This detail requires emphasis. Small amounts of oil anywhere below a giant crater slowly drift outward to the rim where accumulated amounts can be prodigious.

A new method of discovering megacraters has become apparent. First find oil fields that are disposed along an arc. Determine its center. And measure the diameter. Reversal of this process may attract more enthusiasm. Prospectors should search for known oil fields disposed along an arc, find its center, and probe the subsurface structures. Next, extend the arcs in both directions to determine the geology of those regions. Since the location of oil vis-à-vis megacraters has become clear, they can be used as prospecting tools. Appendix A provides 208 starting points for this research.

Not all megacraters concentrate oil as illustrated by many examples in the western U.S. Explosions in areas free of sedimentary rock and hydrocarbons must be barren. The Big Three are about the same size so a comparison of diameters is in order - *Mexico Gulf* (810), *Texas* (640), and *Michigan* (700). These values are below the mean for the whole catalog and could indicate a range of sizes that is especially effective in concentrating petroleum.

Further details in the distribution of fields in the Big Three show two hidden craters that are relatively small. At the western end of Texas a dense circle of fields is quite evident. Its diameter is only 130 mi. A similar, hidden crater about 100 mi in diameter appears in eastern Ohio. Neither of them is included in the master catalog.

Two clusters of oil fields in Wyoming are clearly associated with craters that had been identified independently, *Bighorn* (120 mi) and *Blackhills* (520 mi). Still another weak explosion must have struck just east of the Four Corners at the junction of New Mexico, Arizona, Utah, and Colorado creating a ring of oil fields about 60 mi across. While of little consequence on the petroleum scene, the region is an important source of uranium. A landmark in the region, Shiprock, is a former volcano nearly lost to erosion with radial dikes of rock formed by molten intrusion in linear fissures.

California makes a substantial contribution to U.S. oil requirements along the southern shore. Fields extend from Santa Barbara down to Long Beach with many offshore rigs. This arc along with the Channel Islands is part of a 300-mi crater in the Pacific Ocean called

Catalina. Another source in southern California is related to the 170-mi *Bakersfield* that carved the arc of mountains at the south end of the San Juaquin Valley. A linear fault to the north of the epicenter set the stage for recovery of limited oil and substantial gas in the San Joaquin and Sacramento Valleys.

Foreign Sources

Space does not allow as much detail on the global scene as presented for the U.S. The purpose of this section must, therefore, be limited to showing some of the principal sources and their relationship to megacraters. Regions to be included are the Near East, South America, Indonesia, Caspian Sea, and Siberia. Major oil sources in the Middle East are addressed in Appendix A under the heading of *Iran*.

Near East

Oil from the Near East dominates the industry by producing about 40% of the global demand with Saudi Arabia handling the lion's share of nearly one-half. Regarding reserves, the Near East enjoys an even stronger position accounting for over 60% of the world's future supply. Nations of the region in descending order of their reserves are Saudi Arabia, Kuwait, Iran, Iraq, and Abu Dhabi. The oil fields of all these countries span a range of 1200 mi in a narrow band from Turkey to the Gulf of Oman where the Persian Gulf enters the Arabian Sea. This vast collection of deposits can be recognized on a library globe as a distinct arc. The fields lay on the rim of a megacrater that has already been introduced, namely, *Iran* centered near the city of Masahd with a diameter of 1470 mi. Iraqi president Saddam Hussein attempted to commandeer the resources of Kuwait and threaten the take over of Saudi Arabia. These actions were intolerable to the specific countries and to all that depend upon them as sources of oil. Both foreign nations and companies with local investments could envision huge losses. So a blitzkrieg was unleashed to drive the Iraqis back home in only six days.

An obscure feature of Iran's geology along the southern coast is noteworthy. On a flight from Bandar Abas, the principal city in that region, to Jask on the Pakistan border, the author observed a strange formation. Not far inshore, formerly horizontal layers of rock rose vertically out of the desert. Layers of hard rock alternated with softer sandstone that has been essentially washed away. Remaining walls of hard rock are separated from their neighbors by a distance somewhat greater than their thickness. Tops of the walls have been scalloped irregularly like vertical plates on the back of a dinosaur. A ready explanation for this spectacle depends upon a vanished mountain range. An ancient range was folded by tectonic pressure then worn away by the elements. That concept can not be correct because there has never been sufficient movement of the Arabian Peninsula relative to the Asian continent. A more palatable scenario is related to the megacrater that carved valleys for the Tigris and Euphrates rivers, opened the Persian Gulf, and implanted so much oil. The standing walls must be the broken rim of the crater where bending stresses fractured the sedimentary rocks to great depth. Their vertical orientation at that time has been retained. If this formation has a geological name, the author does not know it.

Discovery of huge deposits of super-light oil in remote areas of Saudi Arabia was announced in September 1990. The oil has the color of gasoline and the fluidity of water. It could even be burned in automobile engines. But it will be refined for that application at much lower costs than normal. The new field is located 150 mi south of Riyadh at Hawtah.

119

Four other wells produced the oil and another one produced gas. A drilling area of 1400 sq mi has been laid out having the potential of the biggest field in the world comparable to all of Texas. Can such a find out in the desert be explained? Certainly. On the rim of **Saudi** centered close to Jiddah on the Red Sea. The crater of 610-mi diameter is about the right size. The oil field is linked to the mountain range of Jabal Tuwayq on the eastern rim. This crater, it will be recalled, was sliced in two by the fault system creating the Red Sea as illustrated in Fig. 1.3. Afar And Environs.

South America

Almost all the oil in South America comes from the northwest shores of the continent. The primary role is played by Lake Maracaibo that fills the deepest part of a depression surrounded by a ring of mountain ridges 190 mi in diameter. Such a small crater is certainly capable of providing oil traps but its location is unique.

The Andes Mountains were raised when the 2010-mi crater of **Manaus** depressed the entire Amazon basin. The comet creating the Maracaibo basin struck squarely on the rim where damage was already severe. The new crater raised the circular pattern of Sierra de Perija to the west and Cordillera de Merida to the east. The central depression, now flooded by the sea, is called Lake Maracaibo. It is the primary source of oil in the continent with most being recovered from offshore platforms.

Neighboring Colombia is also a significant producer while new exploration proceeds in the adjacent Ecuador. Both of these locations are linked to the Andes.

Indonesia

The Empire of the Rising Sun sought to share world domination along with Nazi Germany. To do so required conquest of Southeast Asia to obtain oil, in short supply on the home islands, for military applications, domestic energy, and refinement. Expansion began with invasion of China in 1933. Fully recognizing the implications, the Council of Foreign Relations made an attractive offer to the U.S. State Department. With no charge, CFR would undertake studies of regions of the world known as Grand Areas. The purpose was to find out what portions of the world the U.S. must control to sustain WWII if Hitler should conquer all of Europe. Indonesia was found to be essential as a source of oil, tin, and rubber in direct conflict with the Japanese plans. Their expansion into the Pacific Islands, Singapore, and the Philippines had to be stopped. A threatened embargo of Japanese shipping triggered an "infamous day" when the U.S. fleet was attacked at Pearl Harbor. Invasion was never an objective. Destruction of the fleet at the docks would block interference with the Rising Sun. The ensuing struggle in the Pacific theatre with horrendous loss of life and property was conducted for national security and access to precious commodities. The primary one was, of course, oil.

It is found in fields on a concave arc on the concave side of the Indonesia-Sumatra-Java island arc. The 3430-mi megacrater of **Indonesia** was clearly responsible. As for oil fields near the epicenter, consider Brunei. A country the size of Long Island has the highest per capita income in the world. Free education and medical care are provided to the citizens by the King who was the richest man in the world until surpassed by a computer expert with business acumen. Revenues from oil go to the King who disburses it as he wishes. Most of the oil is obtained from platforms close offshore in the East China Sea near the epicenter of **Indonesia**.

Caspian Sea

A large body of water dominates western Asia between Iran and Kazakhistan being 700 mi long by 200 mi wide. While especially noted for its beluga caviar, it is the source of other products including oil. It appears to be a compound depression from two explosions that overlap, *Caspian N* (960 mi) and *Caspian S* (340 mi). Where would oil be found? At the places of most violence from the combined explosions. These points are near Baku to the west and Krosnvodsk to the east where the huge Tiblesi field is located. Much wrangling accompanied development of the field and a means of delivery to the world market. Changing political climates have opened an acceptable route for a new pipeline to the Black Sea through the former Soviet satellite of Georgia.

Siberia

Vast reserves of gas have been identified in frigid wilderness of Siberia. A prominent field parallels the Lena River that has already been treated. Exploration for oil proceeds apace and great discoveries will likely be found on the rims of craters with diameters of 600 to 800 mi. Larger ones tear up the region so badly that the basin is flooded by lava. Smaller ones, as seen in the U.S., could be fruitful but of limited capacity. All the megacraters in Siberia should be used to direct the exploration. Naturally, regions that are already known to harbor gas reserves will be strong candidates for oil.

Without attempting to account for all the oil in the world, the observed pattern may be generalized in that direction. Explosions of comets make hiding places for oil and gas. If the catalog of craters is nearly complete then most of the sites should be linked to a crater. Large deposits are very unlikely to be found dissociated from all craters but may exist in conjunction with any that are still hidden. Data presented in this book should play a major part in expanding the petroleum reserves of the world.

Coal

Distribution of coal resources in the United States is also show in Fig. 10.1. Fossil Fuels. Deposits are generally found in conjunction with oil and gas although it is notably absent in the Gulf arc. Large regions where it is found in isolation include the mid-western states of Illinois, Missouri, and Iowa. Also, vast deposits independent of oil are common in the western states east of the Rocky Mountains.

Coal is normally found in veins only a few feet thick that reflect the amount of carbonaceous material available at any given locale at the time of formation. Fossilized tree trunks still stand in some veins. Their feet are at the bottom of the vein whereas the trunk is cut off at the top of the seam. These circumstances call to mind the conditions at the 1908 event in Siberia. Trees left standing to full height at the epicenter had their limbs sheared off. A shock wave from the aerial explosion clearly did the damage. Away from the epicenter, a shock wave acquired a gradually increasing, horizontal component that felled the trees in a radial pattern. Portions of standing trees above the accumulated debris will rot away as loose rock or volcanic ash caps the area. Heat and pressure convert the carbonaceous material on the ground into coal. Similar circumstances are also observed in the absence of coal. Trunks of former forests are fossilized within layers of broken rock and volcanic ash. Twenty three such layers of successive forests have been found in Yellowstone

National Park with a total thickness of 1230 ft. Tree ring counting has shown that some of the trees lived for 500 yr. Volcanic activity there has been prominent for a long time.

Major coal deposits around the world are relatively scarce. Little or none is reported in South America nor Africa. On the other hand, any existing deposits may have been poorly reported or overlooked in the emphasis upon other resources. Some fields lie in eastern Australia that could be related to *Artesia* and *Brisbane* as shown in Fig. 1.17. Same Old Thing. Similarly, vast deposits in Siberia are probably related to the several craters found in that region. A large fraction of northern China is underlain by coal. It is randomly scattered within a rough circle about 1200 mi in diameter with the center 400 mi east of Beijing. This pattern displays the earmarks of another megacrater for which substantiating evidence has not been found. It can not, therefore, be cataloged among craters whose reality is not in doubt. It is simply recognized as a possibility and listed under the title of Shanghai in Appendix H. Candidate Megacraters.

Other Energy Sources

Uranium is often mingled among many other ores that are traceable to megacraters. Hope for uranium as a major source of energy was high at the beginning of the nuclear age. Electric power plants were built by the dozens in the U.S. and Europe. Japan, facing shortages of fossil fuels, adopted policies strongly supporting nuclear energy. But the U.S. experienced a setback in the mid-60s for several reasons. Environmental issues delayed construction along with endless hesitation by regulatory agencies. Construction of new plants ceased and some, already in process, began converting to fossil fuels. A lethal blow was delivered by uranium producers when pricing was shifted from the cost of production to the energy content in competition with alternative fuels. Enthusiasm for nuclear power has waned and may never be revived.

Geothermal sources have been tapped to advantage at a limited number of sites. They are typically linked to megacraters such as the use of natural steam for heating in Iceland sitting on the mid-Atlantic ridge. Commercial generation of electricity in California began on a small scale but, over time, the available heat dwindled away. Some promise is found for sustained usage on a large scale by deep wells removing energy from very hot rocks. Prospectors could well use knowledge of megacraters to find places where magma has moved up toward the surface and shared its heat locally.

Extraction of energy from winds remains promising, especially with new windmills of advanced design and large diameters. Enron corporation has recently advertised a windmill whose hub is 140 ft above the ground and whose blades are 250 ft in diameter. Work is underway in northern Europe to construct fields of monsters on the continental shelf. Security is increased and environmental issues are diminished. But how does wind relate to the megacraters? It will be recalled that faults from the explosions opened up all the oceans as continents drifted away from the mid-ocean ridges. Global weather is driven by the temperature differences between the oceans and the continents. Land areas are heated up by the sun while the ocean waters remain cool. Differences between temperatures and corresponding pressures drive the global winds. Even wind energy can be attributed directly to cosmic impacts in the distant past.

Industrial Minerals

Most mountain ranges of the world have been linked in the present research to specific megacraters and extension to all of them seems to be justified. Because mineral deposits are strongly correlated with mountain ranges, the implication is clear. Ores are emplaced by cosmic events. These relationships hold for a remarkable range of crater diameters.

Mini-craters

Much of the scientific research today is based upon cooperation of several scholars from one or more universities and laboratories. Activities focus upon an elaborate and expensive apparatus that must be constructed or rented at considerable expense such as the Stanford Linear Accelerator and Hubble Telescope. Funding is provided by grants from government offices and foundations.

By way of contrast, rare instances of significant discoveries are made by an individual conducting simple, inexpensive experiments. An example of such poor-boy work by Saul (1988) who used a three-dimensional map of Nevada with exaggerated elevations. This plastic map was lain on a table with intense lighting from a nearly horizontal beam. The arrangement simulated the terrain exposed to solar rays at sunset. Only the highest regions of the map were illuminated while lower elevations remained in shadow. A startling pattern was observed. All the elevated ridges closely resembled the pattern of impact craters on the moon. Arcs and circles lay nearby, abutted, overlapped or were contained within larger ones. A likely explanation of this topography is that the basins are craters of small asteroids that have been preserved in a dry climate. Photographs of the illuminated map were zoomed to match the scale of a USGS map showing the location of mines in the region. Mines were found to be preferentially distributed along the rims of these basins and, especially, at their intersections. Maximum damage at such locations has been pointed out.

Middle Craters

Some asteroid craters have produced significant amounts of valuable metals that came directly from the impacting object. The Barringer Crater of Arizona, less than 1 mi across, has been universally recognized as an impact explosion of an iron asteroid or meteorite. Fragments were blown all over the desert. Untold numbers of iron pieces were collected and shipped to market in a commercial operation. Extensive efforts, however, failed to find a large remnant of the iron meteorite under the bottom of the crater. It would have been very valuable. Evidently the explosion totally shattered the impacting object.

A nuclear explosion may have released most of the energy. In parallel with fusion of hydrogen as previously discussed, the likelihood of an atomic explosion must not be overlooked. A plasma created upon transit through the atmosphere and contact with the ground would have been composed of unattached ions of iron and electrons traveling at enormous speeds. Two iron atoms would coalesce if their relative speeds overcame the repulsive force from positrons in their nuclei. The same process could capture a third ion of iron. In that event, a trans-uranium element would be created that would undergo spontaneous fission as in an atomic bomb.

Since this event dates back less than 50,000 yr, some evidence of the nuclear reactions might remain in the desert or the bowl. Radioactive fission products would have decayed away except for those with very long half-lives. Whether or not Barringer involved a nuclear explosion could be a straight-forward although difficult task. A suitable fission product must be selected that

- is rare on earth
- produced in large quantity by fission of uranium not far from atomic numbers of 95 and 140
- decays by loss of an electron with emission of an identifiable gamma ray
- has a suitable half-life.

A cursory review of isotopes shows possible choices of rubidium-87 and samarium-148. Upon cooling in the atmosphere, droplets of molten rock would freeze into small, glass nodules in the shape of spheres, teardrops, and hamburgers. The age of these tektites could be established by the radioactive decay of tracer isotopes embedded in the glass. Any samples collected within ballistic range from Barringer would yield the age of the explosion.

Further indication of a nuclear source would be clear if induced radioactivity were found. Vast floods of neutrons escaping the explosion would activate atoms that captured them. There is quite a bit of work begging for attention at Barringer.

Nature is fond of fusion reactions as illustrated by all the stars and probably the explosion of comets in earth's atmosphere. Even *fission* reactions have been found in nature although uranium in ore deposits would appear impossible to trigger. But remains of a natural reaction in Gabon, Africa have been analyzed. Isotopes and fission tracks at the site showed that the geological structure went critical about 1.7 billion years ago and baked for 200,000 years. The site still contains detectable amount of rubidium and samarium plus telltale cesium and barium.

A unique, geological structure in Ontario, Canada lies 40 mi north of Georgia Bay on Lake Huron. A circular formation 155 mi in diameter is especially rich in minerals. Recovery of nickel from the site accounts for 20% of the production of the Western world. All of Ontario's copper comes from the basin. Also produced in great quantities are gold, silver, platinum, cobalt, sulfur, and iron.

The origin of Sudbury has been hotly debated among experts that has been so common in the study of many other circular structures. Some argue that volcanic activity was responsible. Another faction proposes that an asteroid impact created the structure. One group has suggested that all of the minerals came from the asteroid body itself. The amount of ore and the great variety of minerals at Sudbury argue strongly against this hypothesis. The impactists are winning the debate that is also the case for basins elsewhere.

What seems apparent is that ores are dredged up from great depths by the asteroid impacts. Nearly instantaneous removal of overburden exposes the bottom of the original crater to a pressure of 1 atmosphere. The rocky bottom, however, remains at the pressure prior to the disruption. These values are established by the weight of the overburden that may be as high as 6 million atmospheres. Very high temperatures at depth soften the rock and material at the bottom will begins to move rapidly. By the time the basin is filled, vast quantities of ore that are present in a particular event reach the surface or within the reach of miners. Remember that Sudbury only 155 mi in diameter is a mere pimple on earth's face and does not qualify as a megacrater.

Waters of the Mediterranean flooded a somewhat larger crater 370 mi in diameter. It is known as the Aegean Sea, the birthplace of Greek civilization in antiquity and a vacation Mecca today. Formation of the sea by an impact is clearly established by several factors. A dense pattern of earthquake epicenters encircles the basin and the bottom is subducting to the northeast via the Hellenic Trench. Rebound in the original crater brought mineral ores from great depth carrying valuable resources that sustained ancient Greece. Examples are gold and silver from Sithnos, marble from Tinos, lead from Khios and Antiparos, iron from Milos and Serifos, barium from Khios, and magnetite from Lesbos. Such mineral wealth would be unlikely on small islands unless they rose from great depth to the present elevations above sea level. On a recent visit to the eastern islands, the author encountered local traditions that they rose up out of the sea. And they may still be rising ever so slowly.

Greek mythology supports these concepts. In brief, Zeus became enamoured of Apollo's aunt, Asteria. He chased her until she changed into a rock and was cast into the sea that eventually became an island bearing her name. Poseidon, the god of the sea, anchored the island of Delos that became the most holy shrine referred to as the Center of the Earth. The heart of this story is easy to understand. *Aster* means star and the terminal *ia* merely indicates a female name. By crashing into the sea it was an evil star or *disaster* in modern English. Hence, *Asteria*. It seems odd that someone in antiquity would have known about it.

Arcs of submerged mountain ranges in the eastern Mediterranean suggest that ***Aegean*** may have been as large at 640 mi in diameter with internal walls sloughing off to form concentric ranges. If so, opening of the Red Sea by this explosion would be more palatable.

Most mountain ranges, whether linear or arcuate, have been linked to specific megacraters. So the probability of all mountains having the same origin is very high. They were raised along fault systems created by cometary explosions. And mines of every sort are almost exclusively found in mountainous terrain. So varied are the minerals in any region that discussion of each one throughout the world is impractical. It is better to illustrate the point and leave the rest of the investigations to others.

In Siberia, for example, a half a trillion tons of ore have been found in the vicinity of Noril'sk containing 2.7% nickel, 3.9% copper, 3 parts per million (ppm) platinum and 12 ppm of paladium. These treasures must have been associated with rising material from one of the several impacts in that region although available data does not permit a specific assignment.

Diamonds

Diamonds were mentioned earlier regarding megacraters in Africa and South America. They are formed by carbon under very high temperatures and pressure, conditions that prevail at great depth. Lava flows bring them to the surface through channels that build volcanic mountains. When the volcanoes die and rain washes away the mountain, diamonds are flushed away into riverbeds, gravel banks, and ocean beaches. Large numbers are retained within the lava pipe below ground that are recovered by hardrock mining. Of the eight major deposits of diamonds in the world seven have been associated with megacrater as follows

South America - ***Venezuela, Brazil***
Africa - ***Timbuktu, Congo, Tanzania, Kalahari***
Europe - ***Bosnia***
India - ***India***
Siberia - ***Lena, Aldan, Baikal***
Far East - ***Khabarovsk***
Australia - ***Kimberley***.

The only major diamond field in the world that has not been associated with a specific megacrater is located close to the Arctic Circle about 200 northeast of Yellowknife in the Northwest Territories of Canada. Gemstones discovered in 1991 at Lac de Gras prompted the Dia Met Mineral Company to file a prospecting claim with the government covering 400,000 acres. They were soon joined by Australia's Broken Hill Properties and hordes of prospectors whose claims covered a region the size of Texas. In the original claim area, 107 kimberlite pipes were found. Production began at the Etaki mine in 1998. This mine quickly accounted for 5% of the world production based upon value. Increase of production to 10% has placed Canada fourth in diamond mining behind Botswanna, Russia, and South Africa. The cutting and polishing affiliate uses a laser, appropriately enough, to inscribe a tiny, polar bear on each stone. Reserves at Ekati are estimated at 72 million carats.

Meanwhile only a mile to the east, several new kimberlite pipes were discovered in a shallow lake. Great expansion is underway at that site known as the Diavik Diamonds Project. Submerged pipes will be exposed by construction of surrounding dikes and pumping out the water. Production is scheduled to start in 2003 and continue for about 20 yr. This massive operation will require a dozen, huge trucks working around the clock to transport ore to the processing plant. Each truckload will be 177 tons. When fully developed at an investment of US$1.3 billion, the value of diamonds extracted will amount to 40% of the world production. The value of all the diamonds to be produced will approach US$10 trillion.

The similarity of kimberlite pipes in Canada with those of the De Beers complex in Africa can not be ignored. Magma simply does not flow to the surface anywhere without a cause so a megacrater in the Northern Territories is suspected. One is, in fact, found whose center is only 25 mi west of Lac de Gras with a diameter of 760 mi. Other evidence of a crater consists of a nearly complete circle of waterways including Coronation Gulf, Hottah Lake, North Arm, and Artillery Lake.

This investigation has unveiled a hidden megacrater that is posted in the master catalog as *Territories*. This Canadian example completes the roster of all major sources of diamonds!

An interesting question arises in this context. Where are all the rest of the diamonds in the world near enough to the surface to be recovered economically? They must be associated with some of the numerous megacraters that have been identified. Naturally, special conditions must prevail for creation of diamonds so all the craters are not candidates. Also, world-wide prospecting has been thorough enough that epicenters and rims in remote areas are especially promising. However, exposed pipes of extinct volcanoes should be examined. And anywhere on earth, undiscovered pipes may be underwater or covered with sand and/or sedimentary rock. Any alluvial diamonds would be a valuable clue to a 'mother load' upstream.

Mineral Summary

Space does not allow a comprehensive review of worldwide production of fossil fuels, other energy sources, industrial minerals, precious metals, and gemstones. Evidence described to this point, however, may justify a generalization. Deposition of all these products essential to society became available through the agency of comets in the solar system that ended up colliding with earth. All of them are concentrated at or near the epicenters, along the linear rifts, or in the mountains raised by tectonic forces around the crater rims.

Chapter 11. Total Dependency.

"That's the most exciting thing I ever heard."
- Harold E. Puthoff

All topography, whether exposed to view above sea level or hidden under the ocean waves, was created by cosmic impacts. Extremely ancient craters have been lost and some others probably remain hidden. But a few hundreds of millions of years have not completely obliterated many that have been discovered from visual clues or geological inference. Since the whole world has been molded by these impacts, they have profoundly influenced mankind. Every aspect of human existence is ultimately traceable to the great explosions. Components of these relationships are so numerous and intertwined that adequate explanations seem impossible. The best one can hope is that some of the relationships, hitherto unsuspected, can be revealed. Others may then become apparent without further explanations.

Oceans

Fractures of earth's crust caused by impacts widened into the existing oceans. That fact alone is overwhelming because all existing flora and fauna are dependent upon them or their global consequences. Changes in weather originate above the oceans as the sun heats the continents while the oceans remain frigid, cool, or tepid depending upon the latitude. The sun evaporates water molecules from their surfaces. Thus winds and rains in complex patterns saturate some areas leaving others arid. High mountains catch and preserve snow that provides reliable sources of water for irrigation and all other uses. The same mountains created river channels that deliver water year round. Rivers flow to sediment-filled valleys where they can be tapped for irrigating crops. Dams, tunnels, and canals control flooding, generate electricity, and shift this precious resource from areas of plenty to areas in need. Every snowflake and drop of rain contains molecules of water that rose out of the oceans and other large bodies of water. Oceans and seas were, in turn, derived from the explosions. In a unique sense, therefore, weather can be considered as the grandchild of megacraters.

Consider these points in relation to food supplies. A great variety of fish, crustaceans, and octopods has been a primary source of protein for mankind from the beginning and persists in some countries today. Oceans, seas, and rivers are the breeding ground where the supply is replenished almost without limit.

Key events in the increase in human populations are the first uses of water for irrigation. Hunting animals and collecting wild vegetation could only sustain a very limited number of people. They had to spread out over large areas to find their sustenance. But artificially watered crops allowed many more people to live close together. The basis was thus set for communal living in greater numbers. Cities were built of necessity to house the people. These developments led to political structure, division of labor, and trade. The earliest civilizations developed along the Tigris-Euphrates, Ganges, Indus, and Nile Rivers. Irrigation today is common on a grand scale throughout world. A worthy example is the water management developed in California. A complex system delivers water from the wet, northern regions to the thirsty south. Irrigation became the basis for flooding large areas to grow enough rice for export. All sorts of fruits and vegetables could be nurtured through hot, dry summers to satisfy a significant fraction of the national demand.

Expanses of plains became grasslands with nearly unlimited food for all grazing animals. Huge herds could be sustained, such as, antelopes, zebras, gnus, and giraffes in Africa and buffaloes and caribou in North America. They became the source of protein for all the predators - lions, leopards, and wolves. Scavenging birds consumed the leftovers. Birds without number congregated at the lakes and rivers to feed on fish and vegetation. Thus nature has provided food for all creatures in the plains, mountains, valleys, woodlands, jungles, and tundra. In short, food sources for all creatures are the long-delayed consequences of comets exploding in the atmosphere.

A little thought about the following items may clarify the linkage between megacraters and the current diet in several countries—

- Grains – wheat, corn, rice, barley, and rye
- Meats - cattle, sheep, goats, poultry, fish, and crustaceans
- Vegetables - potatoes, beans, peas, lettuce, tomatoes, and carrots
- Berries - strawberries, blackberries, blueberries, and raspberries
- Dairy Products - milk, cream, cheese, ice cream, and yogurt.

Food sources of every kind for every creature are traceable through intermediate steps to the megacraters. For example, the linkage between comets and rice pilaf is;

- a comet explodes in the atmosphere
- a gigantic crater is formed with many faults
- mountain ranges are raised at the faults
- moisture from the oceans precipitates in the mountains
- water drains down channels between the mountains
- the river is tapped downstream for irrigation
- rice fields are flooded during growth
- people harvest, clean, cook, season, and eat the pilaf.

Modern Civilization

It is not adequate for the world's population to have enough to eat. A few countries face famine because they have been swindled by nature. On the whole, however, people survive at the borderline of hunger or with an ample diet. Lucky ones have surpluses that can be shared with the less fortunate by trade and generosity. But modern society needs a great deal more. A cornucopia of mineral resources must be extracted from the earth either directly or through intermediate roles of plants and animals. Certain mines were discussed in the previous chapter but society's requirements are much broader. They must provide the basis for every component of civilized living, such as, housing, energy, transportation, vehicles, communications, and health. A partial listing is

- Energy (repeat) - oil, gas, coal, nuclear, hydroelectric, wind, and geothermal
- Iron - primarily converted to a variety of steels with addition of nickel, chromium, vanadium, molybdenum, and tungsten. Used in construction of buildings, bridges, vehicles, major appliances, and cookware
- Copper - in power transmission, telephone lines, switches, pipes, and coins
- Aluminum - buildings and aircraft.

This pattern applies to an infinite number of end produces and a similar chain relates all the mineral resources to end uses in society. An example will suffice:

- a comet explodes in the atmosphere
- a gigantic crater is formed
- rebound of the bottom brings valuable ores to the surface
- iron ore is mined and processed
- other elements, likewise obtained, are blended into the iron
- steel is formed into large sheets
- a fabrication shop fashions the sheet into complex shapes
- an automobile manufacturer assembles the parts
- the resulting vehicle is sent to a distributor
- railroads and trucks are used
- a customer purchases a new car whose frame and body are made of steel.

So are similar components of the trucks and trains that run on steel rails.

This isolated chain of events tracks only a single element that is extracted from the earth leading to related end products. A very large number of other elements have parallel tracks that lead to uncountable, essential products. Civilization is enormously complex in its recovery of mineral resources and their final form as beneficial products. Modern society could never have arisen without assistance from the cosmic bombardment. But that is not all.

The Ultimate Debt

Some of the explosions have been associated with specific interfaces of geological periods and the concurrent reduction in the number of living creatures. All extinctions must have had similar causes and alternative suggestions would be difficult to defend. Each event reset the clock of evolution. Many species disappeared while a few survived the cataclysms. Ecological niches that were previously occupied became vacant for exploitation by new creatures. Many of them moved in and adapted to the new conditions.

Something important happened about 65 My when the dinosaurs faded away. Suspecting that the cause was an asteroid crater, scientists established that *Chiczulub* was created on the correct date. With a diameter of only 200 mi or so, it seems to be an unlikely candidate for death on a global scale compared with many megacraters with diameters up to 18 times larger. Perhaps another cosmic event did the dirty work. The Deccan Traps of central India have also been dated accurately at 65 My. That lava plateau was apparently started by an explosion forming the crater that has been identified as *India*. Actually, several explosions in India tore open the seems for release of lava as described in the catalog.

Whatever the specific event, the last great explosion is uniquely significant to mankind. While large dinosaurs vanished, smaller creatures survived. Happily, a flying predator with a wingspan of 50 ft didn't make it. Flying dinosaurs about the size of chickens in the family of peradactyls became the progenitors of today's birds. Also surviving was a mouse-sized mammal. He flourished in the new environment as the hinge pin between the age of dinosaurs and the age of mammals. Forces of evolution worked with these little survivors increasing their size and branching into a vast menagerie. As brains grew, so did intelligence. A temporary peak was reached at the appearance of our own ancestors, *Homo sapiens sapiens.* Much work remains to determine the exact cause of the transition between the Cretaceous and Tertiary geological eras and the concurrent lose of dinosaurs. If the event had never occurred, evolution would have continued along a different pathway instead of diverging toward the eventual appearance of man.

Humanity owes its *very existence* to all of the colossal explosions and, in particular, the last one 65 million years ago. *Think about it!*

Appendix A. Reference Data.

"...growing recognition of the role played by cosmic impacts in earth history is changing some long-cherished ideas in both geology and biology, including how life evolves."
— David Morrison

As each new crater was discovered in this research, a name was chosen and a file was opened to hold any relevant data. Additional information was then added to those files as it became available. As the study expanded, so did the files. In parallel, other files were established by subject matter including Asteroids, Chicxulub, Comets, Earthquakes, Extinctions, Fullerenes, Geological Time, Hotspots, Inclination, Islands, Lava Plateaus, Mars, Minerals, Moon, Mountains, Multiring Structures, Nuclear Reactions, Oil, Pangea, Rifts, Sea Level, Tunguska, Volcanoes, and Waterfalls. The net result was several file drawers crammed with information. The task of boiling it down in some meaningful way was not easy.

This appendix assembles nearly all the information pertaining to the discovered megacraters. The first order of business is handled in Table A.1. Megacrater Sizes. It seems appropriate to emphasize how large they are and the increase in numbers among smaller examples. Each crater is identified by its given name in Column 1 followed by the location of its center in global coordinates of latitude and longitude. They are not as accurate as implied by the number of digits to the right of the decimal points. Generally speaking, one degree of latitude and longitude in the tropical and temperate zones amounts to about 60 mi. In most instances, that precision is expected. Centers are hard to establish in some cases where uncertainly might rise to 60 mi. On the whole, accuracy in locating centers is believed to be within the accuracy of the tenths digits, namely, ± 6 mi. Even better accuracy can be expected in some cases where the true center may lie right next door to the designated coordinates.

Column 4 provides diameters in statute miles from the largest to the smallest. The European and scientific convention of expressing distance in kilometers is essentially meaningless to Americans. But everyone in the U.S. gets a pretty good idea of a hole in the ground 200 mi across.

Column 5 provides approximate ages of the craters in million of years (My) as estimated from geological literature. Inferences are sometimes required as in equating the age of an impact with the earliest age measured from lava at the site. Numerous vacancies in this column indicate how poorly the subject is known. Since the craters have not been recognized previously, geologists have not had the opportunity to study the question. Considerable information about sequences of overlapping craters is clear from the patterns of respective damage and lava flows. While sometimes mentioned in the text, that material is not included here.

Column 6 illustrates the frequency of linear, fault system radiating from large explosions. In nearly every instance, the faults begin at the outer rims of craters and may extend for thousands of miles. Rift Angles were carefully measured on maps and photographs from space then calculated with a computer program.

Column 7 lists values of Rim Ratios that are surprisingly common. Recall that these figures are calculated from the radii of outer rims divided by those of inner rims. They must, therefore, be greater than 1.0. Values are listed only for cases where multiple rims seem to be inherent in the initial process of crater formation. Values are not included where they are so large as to imply delayed failure of internal walls.

Finally, Column 8 indicates the continents where they occurred. All that formed in salt water are collected under the heading of Oceanus based upon the present location of the continents.

The same data are rearranged in Table A.2. Megacrater Latitudes to encourage further research into possible, crater chains from cometary fragments. This information is directly applicable to chains restricted to remnants of Pangea, such as, Asia, Europe, and Africa that account for 57% of the global landmass. The information is also valuable where a chain is limited to one of the migrating continents. Some impressive suggestions arise when Pangea is reconstructed. In such cases, coordinates must be severely modified to be compatible with a vastly different geography.

Variation of earth's inclination in the past does not effect the strong tendency for comet strings to land in sequence along the same latitude. However, data in this study indicate major changes in the actual *spin axis*. That is to say, instead of the present North Pole and South Pole defining the axis today, it may have been represented by an axis from Tehran to Tahiti. Until that puzzle is solved, a comprehensive assessment of crater chains is pretty well blocked. Some hope, however, can be attributed to a Spanish philosopher, George Polya, who taught that intractable problems can sometimes be solved backwards. One assumes the answer then works backward to see how he got there. This trick may work here because some data imply latitudes inclined to the northwest from present parallels.

Because so many craters are involved, the reader needs some way of pinpointing the basic data for any crater that might be encountered in the text. For that reason, the raw data have again been rearranged in Table A.3. Megacrater Names where an alphabetical listing is provided. Any crater attracting special interest in the text or elsewhere can be found by scanning down the list and looking across the table.

Even that convenience is not enough because few people are intimately familiar with global coordinates. A verbal description is required to pinpoint crater locations in terms of everyday landmarks if any exist. That need is satisfied in the final and lengthy section of this appendix under the title of Catalog Of Megacraters. Space does not allow extensive descriptions but critical evidence can be mentioned. Mountain ranges, coast lines, river courses, and shore lines can illustrate the extent of the craters. So independent investigation of any crater can then verify its existence or reject as mere fantasy. Brevity is required to save paper so many shortcuts are used. Most grammatical sins are intentional. Dropped articles. Frequent abbreviations. Incomplete sentences. Collapsed paragraphs. Staccato language. The purpose is to transmit information succinctly. Sufficient information has been provided to demark each crater and encourage further research on any of special interest. In some instances, additional discussion addresses new concepts of geology or details of crater-controlled geography.

Table A.1. Megacrater Sizes (mi).							
Name	**Latitude**	**Longitude**	**Diam**	**Age**	**Rift Ang**	**Rm Rat**	**Con**
Joplin	37.05 N	94.31 W	3640			1.128	NA
East Pacific S	4.85 N	132.72 W	3430				OC
Indonesian	14.12 N	118.48 E	3420		- 71.5	1.111	OC
Moscow	57.42 N	45.20 E	3330		- 22.0	1.066	AS
Far East	15.15 N	144.65 E	2990			1.068	OC
Philippine Sea	17.47 N	141.05 E	2660				OC
Mid-Pacific	8.30 N	170.90 W	2580			1.161	OC
Europe	51.03 N	26.92 E	2250			1.114	EU
Wilkes Land	81.85 S	115.90 E	2200	53	19.0		AN
Atlantic	30.27 N	62.57 W	2040	30		1.061	OC
Manaus	4.82 S	62.62 W	2010				SA
Kuril	56.33 N	133.45 E	1980		- 4.7	1.065	AS
Ryukyu	35.03 N	113.06 E	1980			1.087	OC
Spitzbergen	77.09 N	18.47 E	1930				OC
Timbuktu	20.33 N	0.05 E	1930	130	-20.63		AF
Arctic Ocean	84.63 N	172.18 W	1880	95			OC
Manchuria	50.07 N	121.47 E	1880				AS
West Pacific	32.68 N	158.85 E	1880		- 82.6		OC
South China	29.03 N	110.88 E	1820		69.9		AS
Amazon	4.48 S	68.37 W	1800		5.8	1.119	SA
Central	2.00 S	170.00 W	1800				OC
Congo	2.92 S	23.40 E	1780				AF
East Pacific N	13.07 N	92.05 W	1740				OC
Ross	73.77 S	151.42 E	1680		18.2		AN
Central U.S.	37.05 N	88.36 W	1650		13.5	1.144	NA
Noril'sk - Yakutsk	61.78 N	116.55 E	1650	250	41.4		AS
Himalaya	38.45 N	87.95 E	1640	25	0.75	1.125	AS
Dar es Salaam	7.45 S	51.87 E	1630			1.038	OC
Great Western	40.77 N	106.77 W	1630		- 29.9		NA
Queen Maud	82.38 S	36.53 E	1580				AN
Alberta	54.90 N	115.48 W	1550		- 34.7		NA
Buffalo Hills	59.2 N	115.17 W	1540			1.152	NA
Chicago	40.27 N	90.40 W	1530		- 24.0	1.107	NA
Iran	35.55 N	60.53 E	1470	55	- 32.3	1.185	AS
Great Bear	59.82 N	104.82 W	1440			1.091	NA
Tunguska	66.70 N	95.00 E	1410				AS
Nanjing	32.82 N	118.57 E	1350				AS
Zambia	13.90 S	31.90 E	1350				AF
Aleutian W	60.58 N	178.37 W	1340	95	0.66		OC
Bengal	12.20 N	89.55 E	1300		49.2	1.106	OC
Sahara	19.87 N	27.87 E	1300				AF
Northern Australia	23.30 S	132.23 E	1240				AU

Brazilia	15.1 S	47.78 W	1230				SA
India Ocean	6.55S	48.18 E	1230		32.1	1.116	OC
Ho Chi Minh	9.30N	107.90 E	1210				AS
Hudson	59.50N	80.50 W	1200				NA
Mariana	17.73N	138.10 E	1170				OC
Aleutian E	62.15N	168.68 W	1140	95		1.175	OC
Weddell	70.63 S	44.68 W	1140			1.092	AN
Artesian	29.43 S	144.08 E	1130		53.2		AU
China (S)	29.03N	110.88 E	1130				AS
Coral	13.40S	152.05 E	1130			1.052	OC
Kalahari	26.93S	23.75 E	1120	250	29.1		AF
Delhi	29.10N	75.43 E	1110				AS
Chad	13.4 N	13.93 E	1100		29.7		AF
Tanzania	2.95 S	37.35 E	1100	45	10.3		AF
Great Plains	46.60N	105.42 W	1060				NA
Somolia	3.68S	44.23 E	1060			1.185	OC
Venezuela	2.30N	65.56 W	1060			1.071	SA
Burma	25.87N	134.23 E	1050				AS
Sinkiang	43.00N	82.00 E	1050				AS
Brisbane	28.77S	144.97 E	1020		32.7		AU
Resistencia	34.52S	56.80 W	1020				SA
Fiji	18.15 S	179.60 W	1000				OC
Hawaii	17.08 N	161.90 W	1000				OC
Texas	33.63 N	99.30 W	990		34.1	1.135	NA
Caspian (N)	44.00 N	52.00 E	960				AS
Montana	46.90 N	105.48 W	950			1.181	NA
Bonneville	42.52 N	112.93 W	940				NA
Alice	23.87 S	132.72 E	930				AU
Greenland	77.27 N	40.32 W	930			1.101	OC
Mariana E	13.18N	151.52 W	930				OC
Peak Hill	25.13S	119.92 E	930			1.053	AU
Libya	24.70N	17.00 E	920				AF
Sudan	11.53 N	33.09 E	900				AF
Volcano Belt	46.47N	19.17 E	890				EU
Vanuatu	17.25 S	173.03 E	880			1.145	OC
Kazakh	50.53 N	72.00 E	870				AS
Rain Forest	7.47 S	55.72 W	860				SA
Caroline W	3.18 N	139.05 E	850		- 57.8	1.114	OC
Mexico Gulf	23.60 N	93.66 W	850	200	- 76.0	1.064	OC
Philippine	8.72 N	120.00 E	840			1.211	OC
Small Bight	31.50 S	130.48 E	840				AU
Arabia	22.85 N	39.22 E	830		- 27.9	1.156	AS
Angola	16.52 S	17.98 E	820		48.0		AF
Great Basin	41.00 N	117.00 W	810		0.0		NA
Kara	71.82 N	75.22 E	810		46.5	1.161	OC
Khabarovsk	59.43 N	129.23 E	790		-10.7	1.099	AS

North Slope	72.97N	156.37 W	790				NA
North Sea	55.63 N	3.65 E	780				OC
Yellow Sea	35.58 N	120.37 E	780				AS
Territories	64.57 N	110.59 W	760	65			NA
Burma W	21.62 N	98.78 E	710			1.097	AS
Campo Grande	20.28 S	52.22 W	710				SA
Japan	39.80 N	134.02 E	710			1.138	AS
Panama	7.50 N	79.22 W	700			1.145	OC
Campeche	19.38 N	91.83 W	690		- 84.9	1.090	OC
Byrd	82.92 S	102.35 W	680				AN
Gory Putorama	70.28 N	107.40 E	680				AS
Zimbabwe	20.00 S	31.08 E	680				AF
Anchorage	60.43 N	147.30 W	670				NA
Arabian Sea	18.62 N	65.52 E	670		72.9		AS
India	21.50 N	79.10 E	670		78.6		AS
Williston	48.08 N	103.38 W	670				NA
Jalisco	20.32 N	103.02 W	660				NA
Sydney	34.41 S	144.67 E	660				AU
Klondike	61.80 N	131.97 W	650		-50.1		NA
Pelly River	61.78 N	133.57 W	650				NA
Mato Grosso	20.22 S	52.52 W	640			1.075	SA
Cordoba	26.88 S	60.43 W	630		15.1		SA
Chile	49.17 S	68.02 W	610			1.102	SA
Saudi	23.50 N	44.00 E	610				AS
Baikal	55.87 N	102.50 E	600	25	-11.9	1.183	AS
Simpson	25.50 S	139.30 E	600				AU
Alabama	32.20 N	86.20 W	590				NA
Bellingshausen	70.35 S	80.75 W	570			1.076	AN
Angkor	13.50 N	104.90 E	570		- 33.7		AS
Antilles	14.78 N	65.32 W	560	54			OC
Lena	66.90 N	135.25 E	550		- 10.7		AS
Apennine	43.40 N	6.42 E	540				OC
St. Lawrence	47.18 N	62.82 W	540		38.6		NA
Celebes	4.00 N	122.00 E	530				OC
Hungary	46.20 N	20.10 E	530		19.5		EU
Banda Sea	50.10 S	129.05E	520				OC
Black Hills	44.10 N	103.50 W	520				NA
Bothnia	65.00 N	21.30 E	510		33.4		NA
Jodhpur	26.02 N	73.13 E	510		- 0.14		AS
Sardinia	40.85 N	9.35 E	510		21.9		EU
Bohopal	24.70 N	76.52 E	500				AS
Catamarac	28.03 S	66.92 W	500				SA
Ireland	53.53 N	7.73 W	500			1.075	EU
Mourzouk	24.97 N	13.52 E	500				AF
Turkey W	38.18 N	30.78 E	500			1.122	AS
Zacatecas	22.73 N	102.80 W	490				NA

Bahia de Campeche	21.70 N	93.80 W	480				OC
Baffin Bay	73.72 N	68.07 W	470				OC
Bismark Sea	3.29 S	150.05 E	470		- 54.6		OC
Yukon	66.73 N	141.37 W	470		80.8		NA
Luknow	25.27 N	79.48 E	460				AS
Four Corners	36.00 N	109.50 W	450				NA
Pelly River A	60.37 N	126.07W	450				NA
Rio Grande	28.64 S	37.50 W	450				SA
Catalina	30.00 N	120.85 W	440			1.158	OC
Hainan	19.21 N	109.37 E	440				AS
Kimberley	16.37 S	126.68 E	440				OC
Uraguay	29.32 S	58.90 W	440				SA
Great Sandy	21.50 S	124.48 E	430				AU
Persian Gulf	26.27 N	53.13 E	430				OC
Polar Islands	70.65 N	109.52 W	430			1.138	NA
Vancouver	41.50 N	127.10 W	430				OC
Michigan	43.50 N	84.50 W	420				NA
Aurangabad	18.47 N	76.80 E	410				AS
Buenos Aires	36.03 S	60.52 W	410				SA
Chaco	20.85 S	59.77 W	410		60.5	1.153	SA
North Island	40.72 S	173.54 E	400				OC
Springfield	36.40N	92.57 W	400			1.159	NA
Sulu Sea	8.27N	119.55 E	400				OC
Ethiopia	10.30 N	39.18 E	390		35.4		AF
Turkey E	38.97 N	33.40 E	390			1.169	AS
Spokane	45.27 N	117.13 W	380				OC
Sascatchewan	57.38 N	107.35 W	370		- 35.2		NA
Cree Lake	57.47 N	106.48 W	360				NA
Scotia	58.00 S	30.00 W	360	84			OC
Aegean	37.39 N	25.27 E	350	30	- 39.2		EU
Alps	45.72 N	8.17 E	350				EU
Champaign	39.68N	88.09 W	350		22.8		NA
Guatemala	17.08 N	90.60 W	350		- 72.8		NA
Sudan Jr.	12.00 N	27.16 E	350				AF
Adriatic Sea	45.23 N	27.80 E	340			1.045	EU
Bandar Abbas	28.83 N	54.45 E	340				AS
Caspian (S)	31.43 N	52.00 E	340				AS
Trinidad	14.83 S	64.90 W	340				SA
Under Klondike	60.47 N	129.30 W	340				NA
Aldan Plateau	57.48 N	126.17 E	330				AS
Madagascar N	18.12 S	46.45 E	330		- 56.3		AF
Navajo	36.50 N	109.50 W	320			1.048	NA
Caroline E	2.00 N	149.00 E	310				OC
Gibson	23.00 S	124.00 E	310				AU
Victoria	2.06 S	33.90 E	310		- 8.6		AF
Siberia 1	59.85 N	130.37 E	300		10.0	1.085	AS

Mozambique	23.15 S	33.37 E	280				AF
Pasco	46.22 N	119.00 W	280				NA
Snake	44.47 N	114.28 W	280				NA
Venice	44.60 N	14.03 E	280		- 49.4		EU
Cuernavaca	19.27 N	99.68 W	270				NA
Mexico	18.75 N	99.68 W	270				NA
Quebec	50.03 N	70.80 W	270	220			NA
Bucharest	44.00 N	24.50 E	250				EU
Jask	25.21 N	58.20 E	250				AF
Santa Cruz	19.42 S	61.43 W	250				SA
Madagascar SW	22.18 S	44.67 E	240				AF
Bass Strait	39.55 S	146.25 E	230				AU
Madagascar S	24.03 S	45.43 E	210				AF
Chixulub	20.33 N	89.50 W	200	65			NA
Maracaibo	10.10 S	72.10 W	190				SA
Prague	49.80 N	14.70 E	190				EU
Bakersfield	35.90 N	119.50 W	170				NA
Paton Plateau	58.59 N	114.38 E	170				AS
Twin Bridges	45.80 N	112.00 W	170				NA
Pantanal	18.47 S	56.18 W	160				SA
Tyrrhenian	39.25 N	14.40 E	160				OC
Black	42.00 N	40.50 E	150				AS
Hopi	36.00 N	110.00 W	150				NA
Ungava	59.50 N	67.20 W	150				NA
Belcher	59.00 N	67.00 W	140			1.048	NA
Bighorn	43.85 N	108.05 W	120			1.127	NA
Yellowstone	44.42 N	110.38 W	90				NA

Name	Latitude	Longitude	Diam	Age	Rift Ang	Rm Rat	Cont
Arctic Ocean	84.63 N	172.18 W	1880	95			OC
Greenland	77.27 N	40.32 W	930			1.101	OC
Spitzbergen	77.09 N	18.47 E	1930				OC
Baffin Bay	73.72 N	68.07 W	470				OC
North Slope	72.97 N	156.37 W	790				NA
Kara	71.82 N	75.22 E	810		46.5	1.161	OC
Polar Islands	70.65 N	109.52 W	430			1.138	NA
Gory Putorama	70.28 N	107.40 E	680				AS
Lena	66.90 N	135.25 N	550		- 10.7		AS
Yukon	66.73 N	141.37 W	470		80.8		NA
Tunguska	66.70 N	95.00 E	1410				AS
Bothnia	65.00 N	21.30 E	510		33.4		NA
Territories	64.57 N	110.59 W	760	65			NA
Aleutian E	62.15 N	168.68 W	1140	95		1.175	OC
Klondike	61.80 N	131.97 W	650		-50.1		NA
Noril'sk - Yakutsk	61.78 N	116.55 E	1650	250	41.4		AS
Pelly River	61.78 N	133.57 W	650				NA
Aleutian W	60.58 N	178.37 W	1340	95	0.66		OC
Under Klondike	60.47 N	129.30 W	340				NA
Anchorage	60.43 N	147.30 W	670				NA
Pelly River A	60.37 N	126.07 W	450				NA
Siberia 1	59.85 N	130.37 E	300		10.0	1.085	AS
Great Bear	59.82 N	104.82 W	1440			1.091	NA
Hudson	59.50 N	80.50 W	1200				NA
Ungava	59.50 N	67.20 W	150				NA
Khabarovsk	59.43 N	129.23 E	790		-10.7	1.099	AS
Buffalo Hills	59.20 N	115.17 W	1540			1.152	NA
Belcher	59.00 N	67.00 W	140			1.048	NA
Paton Plateau	58.59 N	114.38 E	170				AS
Aldan Plateau	57.48 N	126.17 E	330				AS
Cree Lake	57.47 N	106.48 W	360				NA
Moscow	57.42 N	45.20 E	3330		- 22.0	1.066	AS
Sascatchewan	57.38 N	107.35 W	370		- 35.2		NA
Kuril	56.33 N	133.45 E	1980		- 4.7	1.065	AS
Baikal	55.87 N	102.50 E	600	25	-11.9	1.183	AS
North Sea	55.63 N	3.65 E	780				OC
Alberta	54.90 N	115.48 W	1550		- 34.7		NA
Ireland	53.53 N	7.73 W	500			1.075	EU
Europe	51.03 N	26.92 E	2250			1.114	EU
Kazakh	50.53 N	72.00 E	870				AS
Manchuria	50.07 N	121.47 E	1880				AS
Quebec	50.03 N	70.80 W	270	220			NA

Table A.2. Megacrater Latitudes.

Prague	49.80 N	14.70 E	190				EU
Williston	48.08 N	103.38 W	670				NA
St. Lawrence	47.18 N	62.82 W	540		38.6		NA
Montana	46.90 N	105.48 W	950			1.181	NA
Great Plains	46.60 N	105.42 W	1060				NA
Volcano Belt	46.47 N	19.17 E	890				EU
Pasco	46.22 N	119.00 W	280				NA
Hungary	46.20 N	20.10 E	530		19.5		EU
Twin Bridges	45.80 N	112.00 W	170				NA
Alps	45.72 N	8.17 E	350				EU
Spokane	45.27 N	117.13 W	380				OC
Twin Bridges	45.80 N	112.00 W	170				NA
Adriatic Sea	45.23 N	27.80 E	340			1.045	EU
Venice	44.60 N	14.03 E	280		- 49.4		EU
Snake	44.47 N	114.28 W	280				NA
Yellowstone	44.42 N	110.38 W	90				NA
Black Hills	44.10 N	103.50 W	520				NA
Caspian (N)	44.00 N	52.00 E	960				AS
Bucharest	44.00 N	24.50 E	250				EU
Bighorn	43.85 N	108.05 W	120			1.127	NA
Michigan	43.50 N	84.50 W	420				NA
Apennine	43.40 N	6.42 E	540				OC
Sinkiang	43.00 N	82.00 E	1050				AS
Bonneville	42.52 N	112.93 W	940				NA
Black	42.00 N	40.50 E	150				AS
Vancouver	41.50 N	127.10W	430				OC
Great Basin	41.00 N	117.00 W	810		0.0		NA
Sardinia	40.85 N	9.35 E	510		21.9		EU
Great Western	40.77 N	106.77 W	1630		- 29.9		NA
Chicago	40.27 N	90.40 W	1530		- 24.0	1.107	NA
Japan	39.80 N	134.02 E	710			1.138	AS
Champaign	39.68 N	88.09 W	350		22.8		NA
Tyrrhenian	39.25 N	14.40 E	160				OC
Turkey E	38.97 N	33.40 E	390			1.169	AS
Himalaya	38.45 N	87.95 E	1640	25	0.75	1.125	AS
Turkey W	38.18 N	30.78 E	500			1.122	AS
Aegean	37.39 N	25.27 E	350	30	- 39.2		EU
Joplin	37.05 N	94.31 W	3640			1.128	NA
Central U.S.	37.05 N	88.36 W	1650		13.5	1.144	NA
Navajo	36.50 N	109.50 W	320			1.048	NA
Springfield	36.40 N	92.57 W	400			1.159	NA
Four Corners	36.00 N	109.50 W	450				NA
Hopi	36.00 N	110.00 W	150				NA
Bakersfield	35.90 N	119.50 W	170				NA
Yellow Sea	35.58 N	120.37 E	780				AS
Iran	35.55 N	60.53 E	1470	55	- 32.3	1.185	AS

Ryukyu	35.03 N	113.06 E	1980			1.087	OC
Texas	33.63 N	99.30 W	990		34.1	1.135	NA
Nanjing	32.82 N	118.57 E	1350				AS
West Pacific	32.68 N	158.85 E	1880		- 82.6		OC
Alabama	32.20 N	86.20 W	590				NA
Caspian (S)	31.43 N	52.00 E	340				AS
Atlantic	30.27 N	62.57 W	2040	30		1.061	OC
Catalina	30.00 N	120.85 W	440			1.158	OC
Delhi	29.10 N	75.43 E	1110				AS
South China	29.03 N	110.88 E	1820		69.9		AS
China (S)	29.03 N	110.88 E	1130				AS
Bandar Abbas	28.83 N	54.45 E	340				AS
Persian Gulf	26.27 N	53.13 E	430				OC
Jodhpur	26.02 N	73.13 E	510		- 0.14		AS
Burma	25.87 N	134.23 E	1050				AS
Luknow	25.27 N	79.48 E	460				AS
Jask	25.21 N	58.20 E	250				AF
Mourzouk	24.97 N	13.52 E	500				AF
Libya	24.70 N	17.00 E	920				AF
Bohopal	24.70 N	76.52 E	500				AS
Mexico Gulf	23.60 N	93.66 W	850	200	- 76.0	1.064	OC
Saudi	23.50 N	44.00 E	610				AS
Arabia	22.85 N	39.22 E	830		- 27.9	1.156	AS
Zacatecas	22.73 N	102.80 W	490				NA
Bahia de Campeche	21.70 N	93.80 W	480				OC
Burma W	21.62 N	98.78 E	710			1.097	AS
India	21.50 N	79.10 E	670		78.6		AS
Timbuktu	20.33 N	0.05 E	1930	130	-20.63		AF
Chixulub	20.33 N	89.50 W	200	65			NA
Jalisco	20.32 N	103.02 W	660				NA
Sahara	19.87 N	27.87 E	1300				AF
Campeche	19.38 N	91.83 W	690		- 84.9	1.090	OC
Cuernavaca	19.27 N	99.68 W	270				NA
Hainan	19.21 N	109.37 E	440				AS
Mexico	18.75 N	99.68 W	270				NA
Arabian Sea	18.62 N	65.52 E	670		72.9		AS
Aurangabad	18.47 N	76.80 E	410				AS
Mariana	17.73 N	138.10 E	1170				OC
Philippine Sea	17.47 N	141.05 E	2660				OC
Hawaii	17.08 N	161.90 W	1000				OC
Guatemala	17.08 N	90.60 W	350		- 72.8		NA
Far East	15.15 N	144.65 E	2990			1.068	OC
Antilles	14.78 N	65.32 W	560	54			OC
Indonesian	14.12 N	118.48 E	3420		- 71.5	1.111	OC
Angkor	13.50 N	104.90 E	570		- 33.7		AS
Chad	13.47 N	13.93 E	1100		29.7		AF

Mariana E	13.18 N	151.52 W	930				OC
East Pacific N	13.07 N	92.05 W	1740				OC
Bengal	12.20 N	89.55 E	1300		49.2	1.106	OC
Sudan Jr.	12.00 N	27.16 E	350				AF
Sudan	11.53 N	33.09 E	900				AF
Ethiopia	10.30 N	39.18 E	390		35.4		AF
Ho Chi Minh	9.30 N	107.90 E	1210				AS
Philippine	8.72 N	120.00 E	840			1.211	OC
Central Pacific	8.30 N	170.90 W	2580		18.2	1.161	OC
Mid-Pacific	8.30 N	170.90 W	2580			1.161	OC
Sulu Sea	8.27 N	119.55 E	400				OC
Panama	7.50 N	79.22 W	700			1.145	OC
East Pacific S	4.85 N	132.72 W	3430				OC
Celebes	4.00 N	122.00 E	530				OC
Caroline W	3.18 N	139.05 E	850		- 57.8	1.114	OC
Venezuela	2.30 N	65.56 W	1060			1.071	SA
Caroline E	2.00 N	149.00 E	310				OC
Central	2.00 S	170.00 W	1800				OC
Victoria	2.06 S	33.90 E	310		- 8.6		AF
Congo	2.92 S	23.40 E	1780				AF
Tanzania	2.95 S	37.35 E	1100	45	10.3		AF
Bismark Sea	3.29 S	150.05 E	470		- 54.6		OC
Somolia	3.68 S	44.23 E	1060			1.185	OC
Amazon	4.48 S	68.37 W	1800		5.8	1.119	SA
Manaus	4.82 S	62.62 W	2010				SA
India Ocean	6.55 S	48.18 E	1230		32.1	1.116	OC
Dar es Salaam	7.45 S	51.87 E	1630			1.038	OC
Rain Forest	7.47 S	55.72 W	860				SA
Maracaibo	10.10 S	72.10 W	190				SA
Coral	13.40 S	152.05 E	1130			1.052	OC
Zambia	13.90 S	31.90 E	1350				AF
Trinidad	14.83 S	64.90 W	340				SA
Brazilia	15.15 S	47.78 W	1230				SA
Kimberley	16.37 S	126.68 E	440				OC
Angola	16.52 S	17.98 E	820		48.0		AF
Vanuatu	17.25 S	173.03 E	880			1.145	OC
Madagascar N	18.12 S	46.45 E	330		- 56.3		AF
Fiji	18.15 S	179.60 W	1000				OC
Pantanal	18.47 S	56.18 W	160				SA
Santa Cruz	19.42 S	61.43 W	250				SA
Zimbabwe	20.00 S	31.08 E	680				AF
Mato Grosso	20.22 S	52.52 W	640			1.075	SA
Campo Grande	20.28 S	52.22 W	710				SA
Chaco	20.85 S	59.77 W	410		60.5	1.153	SA
Great Sandy	21.50 S	124.48 E	430				AU
Madagascar SW	22.18 S	44.67 E	240				AF

Gibson	23.00 S	124.00 E	310				AU
Mozambique	23.15 S	33.37 E	280				AF
Northern Australia	23.30 S	132.23 E	1240				AU
Alice	23.87 S	132.72 E	930				AU
Madagascar S	24.03 S	45.43 E	210				AF
Peak Hill	25.13 S	119.92 E	930			1.053	AU
Simpson	25.50 S	139.30 E	600				AU
Cordoba	26.88 S	60.43 W	630		15.1		SA
Kalahari	26.93 S	23.75 E	1120	250	29.1		AF
Catamarac	28.03 S	66.92 W	500				SA
Rio Grande	28.64 S	37.50 W	450				SA
Brisbane	28.77 S	144.97 E	1020		32.7		AU
Uraguay	29.32 S	58.90 W	440				SA
Artesian	29.43 S	144.08 E	1130		53.2		AU
Small Bight	31.50 S	130.48 E	840				AU
Sydney	34.41 S	144.67 E	660				AU
Resistencia	34.52 S	56.80 W	1020				SA
Buenos Aires	36.03 S	60.52 W	410				SA
Bass Strait	39.55 S	146.25 E	230				AU
North Island	40.72 S	173.54 E	400				OC
Chile	49.17 S	68.02 W	610			1.102	SA
Banda Sea	50.10 S	129.05 E	520				OC
Scotia	58.00 S	30.00 W	360	84			OC
Yukon	66.73 N	141.37 W	470		80.8		NA
Gory Putorama	70.28N	107.40 E	680				AS
Bellingshausen	70.35 S	80.75 W	570			1.076	AN
Weddell	70.63 S	44.68 W	1140			1.092	AN
Ross	73.77 S	151.42 E	1680		18.2		AN
Wilkes Land	81.85 S	115.90 E	2200	53	19.0		AN
Queen Maud	82.38 S	36.53 E	1580				AN
Byrd	82.92 S	102.35 W	680				AN

Name	Latitude	Longitude	Diam	Age	Rift Ang	Rm Rat	Con
Adriatic Sea	45.23 N	27.80 E	340			1.045	EU
Aegean	37.39 N	25.27 E	350	30	- 39.2		EU
Alabama	32.20 N	86.20 W	590				NA
Alberta	54.90 N	115.48 W	1550		- 34.7		NA
Aldan Plateau	57.48 N	126.17 E	330				AS
Aleutian E	62.15 N	168.68 W	1140	95		1.175	OC
Aleutian W	60.58 N	178.37 W	1340	95	0.66		OC
Alice	23.87 S	132.72 E	930				AU
Alps	45.72 N	8.17 E	350				EU
Amazon	4.48 S	68.37 W	1800		5.8	1.119	SA
Anchorage	60.43 N	147.30 W	670				NA
Angkor	13.50 N	104.90 E	570		- 33.7		AS
Angola	16.52 S	17.98 E	820		48.0		AF
Antilles	14.78 N	65.32 W	560	54			OC
Apennine	43.40 N	6.42 E	540				OC
Arabia	22.85 N	39.22 E	830		- 27.9	1.156	AS
Arabian Sea	18.62N	65.52 E	670		72.9		AS
Arctic Ocean	84.63 N	172.18 W	1880	95			OC
Artesian	29.43 S	144.08 E	1130		53.2		AU
Atlantic	30.27 N	62.57 W	2040	30		1.061	OC
Aurangabad	18.47 N	76.80 E	410				AS
Baffin Bay	73.72 N	68.07 W	470				OC
Bahia de Campeche	21.70 N	93.80 W	480				OC
Baikal	55.87 N	102.50 E	600	25	-11.9	1.183	AS
Bakersfield	35.90 N	119.50 W	170				NA
Banda Sea	50.10 S	129.05E	520				OC
Bandar Abbas	28.83 N	54.45 E	340				AS
Bass Strait	39.55 S	146.25 E	230				AU
Belcher	59.00 N	67.00 W	140			1.048	NA
Bellingshausen	70.35 S	80.75 W	570			1.076	AN
Bengal	12.20 N	89.55 E	1300		49.2	1.106	OC
Bighorn	43.85 N	108.05 W	120			1.127	NA
Bismark Sea	3.29 S	150.05 E	470		- 54.6		OC
Black	42.00 N	40.50 E	150				AS
Black Hills	44.10 N	103.50 W	520				NA
Bohopal	24.70 N	76.52 E	500				AS
Bonneville	42.52 N	112.93 W	940				NA
Bothnia	65.00 N	21.30 E	510		33.4		NA
Brazilia	15.15 S	47.78 W	1230				SA
Brisbane	28.77 S	144.97 E	1020		32.7		AU
Bucharest	44.00 N	24.50 E	250				EU
Buenos Aires	36.03 S	60.52 W	410				SA

Name	Lat	Lon					
Buffalo Hills	59.20 N	115.17 W	1540			1.152	NA
Burma	25.87 N	134.23 E	1050				AS
Burma W	21.62 N	98.78 E	710			1.097	AS
Byrd	82.92 S	102.35 W	680				AN
Campeche	19.38 N	91.83 W	690		- 84.9	1.090	OC
Campo Grande	20.28 S	52.22 W	710				SA
Caroline E	2.00 N	149.00 E	310				OC
Caroline W	3.18 N	139.05 E	850		- 57.8	1.114	OC
Caspian (N)	44.00 N	52.00 E	960				AS
Caspian (S)	31.43 N	52.00 E	340				AS
Catalina	30.00 N	120.85 W	440			1.158	OC
Catamarac	28.03 S	66.92 W	500				SA
Celebes	4.00 N	122.00 E	530				OC
Central	2.00 S	170.00 W	1800				OC
Central Pacific	8.30 N	170.90 W	2580		18.2	1.161	OC
Central U.S.	37.05 N	88.36 W	1650		13.5	1.144	NA
Chaco	20.85 S	59.77 W	410		60.5	1.153	SA
Chad	13.47 N	13.93 E	1100		29.7		AF
Champaign	39.68 N	88.09 W	350		22.8		NA
Chicago	40.27 N	90.40 W	1530		- 24.0	1.107	NA
Chixulub	20.33 N	89.50 W	200	65			NA
Chile	49.17 S	68.02 W	610			1.102	SA
China (S)	29.03 N	110.88 E	1130				AS
Congo	2.92 S	23.40 E	1780				AF
Coral	13.40 S	152.05 E	1130			1.052	OC
Cordoba	26.88 S	60.43 W	630		15.1		SA
Cree Lake	57.47 N	106.48 W	360				NA
Cuernavaca	19.27 N	99.68 W	270				NA
Dar es Salaam	7.45 S	51.87 E	1630			1.038	OC
Delhi	29.10 N	75.43 E	1110				AS
East Pacific N	13.07 N	92.05 W	1740				OC
East Pacific S	4.85 N	132.72 W	3430				OC
Ethiopia	10.30 N	39.18 E	390		35.4		AF
Europe	51.03 N	26.92 E	2250			1.114	EU
Far East	15.15 N	144.65 E	2990			1.068	OC
Fiji	18.15 S	179.60 W	1000				OC
Four Corners	36.00 N	109.50 W	450				NA
Gibson	23.00 S	124.00 E	310				AU
Gory Putorama	70.28 N	107.40 E	680				AS
Great Basin	41.00 N	117.00 W	810		0.0		NA
Great Bear	59.82 N	104.82 W	1440			1.091	NA
Great Plains	46.60 N	105.42 W	1060				NA
Great Sandy	21.50 S	124.48 E	430				AU
Great Western	40.77 N	106.77 W	1630		- 29.9		NA
Greenland	77.27 N	40.32 W	930			1.101	OC
Guatemala	17.08 N	90.60 W	350		- 72.8		NA

Hainan	19.215N	109.37 E	440				AS
Hawaii	17.08 N	161.90 W	1000				OC
Himalaya	38.45 N	87.95 E	1640	25	0.75	1.125	AS
Ho Chi Minh	9.30 N	107.90 E	1210				AS
Hopi	36.00 N	110.00 W	150				NA
Hudson	59.50 N	80.50 W	1200				NA
Hungary	46.20 N	20.10 E	530		19.5		EU
India	21.50 N	79.10 E	670		78.6		AS
India Ocean	6.55 S	48.18 E	1230		32.1	1.116	OC
Indonesian	14.12 N	118.48 E	3420		- 71.5	1.111	OC
Iran	35.55 N	60.53 E	1470	55	- 32.3	1.185	AS
Ireland	53.53 N	7.73 W	500			1.075	EU
Jalisco	20.32 N	103.02 W	660				NA
Japan	39.80 N	134.02 E	710			1.138	AS
Jask	25.21 N	58.20 E	250				AF
Jodhpur	26.02 N	73.13 E	510		- 0.14		AS
Joplin	37.05 N	94.31W	3640			1.128	NA
Kalahari	26.93 S	23.75 E	1120	250	29.1		AF
Kara	71.82 N	75.22 E	810		46.5	1.161	OC
Kazakh	50.53 N	72.00 E	870				AS
Khabarovsk	59.43 N	129.23 E	790		-10.7	1.099	AS
Kimberley	16.37 S	126.68 E	440				OC
Klondike	61.80 N	131.97 W	650		-50.1		NA
Kuril	56.33 N	133.45 E	1980		- 4.7	1.065	AS
Lena	66.90 N	135.25 E	550		- 10.7		AS
Libya	24.70 N	17.00 E	920				AF
Luknow	25.27 N	79.48 E	460				AS
Madagascar N	18.12 S	46.45 E	330		- 56.3		AF
Madagascar S	24.03 S	45.43 E	210				AF
Madagascar SW	22.18 S	44.67 E	240				AF
Manaus	4.82 S	62.62 W	2010				SA
Manchuria	50.07 N	121.47 E	1880				AS
Maracaibo	10.10 S	72.10 W	190				SA
Mariana	17.73 N	138.10 E	1170				OC
Mariana E	13.18 N	151.52 W	930				OC
Mato Grosso	20.22S	52.52 W	640			1.075	SA
Mexico	18.75 N	99.68 W	270				NA
Mexico Gulf	23.60 N	93.66 W	850	200	- 76.0	1.064	OC
Michigan	43.50 N	84.50 W	420				NA
Mid-Pacific	8.30 N	170.90 W	2580			1.161	OC
Montana	46.90 N	105.48 W	950			1.181	NA
Moscow	57.42 N	45.20 E	3330		- 22.0	1.066	AS
Mourzouk	24.97 N	13.52 E	500				AF
Mozambique	23.15 S	33.37 E	280				AF
Nanjing	32.82 N	118.57 E	1350				AS
Navajo	36.50 N	109.50 W	320			1.048	NA

Noril'sk - Yakutsk	61.78 N	116.55 E	1650	250	41.4		AS
North Island	40.72 S	173.54 E	400				OC
North Sea	55.63 N	3.65 E	780				OC
North Slope	72.97 N	156.37 W	790				NA
Northern Australia	23.30 S	132.23 E	1240				AU
Panama	7.50 N	79.22 W	700			1.145	OC
Pantanal	18.47 S	56.18 W	160				SA
Pasco	46.22 N	119.00 W	280				NA
Paton Plateau	58.59 N	114.38 E	170				AS
Peak Hill	25.13 S	119.92 E	930			1.053	AU
Pelly River	61.78 N	133.57 W	650				NA
Pelly River A	60.37 N	126.07W	450				NA
Persian Gulf	26.27 N	53.13 E	430				OC
Philippine	8.72 N	120.00 E	840			1.211	OC
Philippine Sea	17.47 N	141.05 E	2660				OC
Polar Islands	70.65 N	109.52 W	430			1.138	NA
Prague	49.80 N	14.70 E	190				EU
Quebec	50.03 N	70.80 W	270	220			NA
Queen Maud	82.38 S	36.53 E	1580				AN
Rain Forest	7.47 S	55.72 W	860				SA
Resistencia	34.52 S	56.80 W	1020				SA
Rio Grande	28.64 S	37.50 W	450				SA
Ross	73.77 S	151.42 E	1680		18.2		AN
Ryukyu	35.03 N	113.06 E	1980			1.087	OC
Sahara	19.87 N	27.87 E	1300				AF
Santa Cruz	19.42 S	61.43 W	250				SA
Sardinia	40.85 N	9.35 E	510		21.9		EU
Sascatchewan	57.38 N	107.35 W	370		- 35.2		NA
Saudi	23.50 N	44.00 E	610				AS
Scotia	58.00 S	30.00 W	360	84			OC
Siberia 1	59.85 N	130.37 E	300		10.0	1.085	AS
Simpson	25.50 S	139.30 E	600				AU
Sinkiang	43.00 N	82.00 E	1050				AS
Small Bight	31.50 S	130.48 E	840				AU
Snake	44.47 N	114.28 W	280				NA
Somolia	3.68 S	44.23 E	1060			1.185	OC
South China	29.03 N	110.88 E	1820		69.9		AS
Spitzbergen	77.092N	18.47 E	1930				OC
Spokane	45.27 N	117.13 W	380				OC
Springfield	36.40 N	92.57 W	400			1.159	NA
St. Lawrence	47.18 N	62.82 W	540		38.6		NA
Sudan	11.53 N	33.09 E	900				AF
Sudan Jr.	12.00 N	27.16 E	350				AF
Sulu Sea	8.27 N	119.55 E	400				OC
Sydney	34.41 S	144.67 E	660				AU
Tanzania	2.95 S	37.35 E	1100	45	10.3		AF

Name	Lat	Long					Cont
Territories	64.57 N	110.59 W	760	65			NA
Texas	33.63 N	99.30 W	990		34.1	1.135	NA
Timbuktu	20.33 N	0.05 E	1930	130	-20.63		AF
Trinidad	14.83 S	64.90 W	340				SA
Tunguska	66.70 N	95.00 E	1410				AS
Turkey E	38.97 N	33.40 E	390			1.169	AS
Turkey W	38.18 N	30.78 E	500			1.122	AS
Twin Bridges	45.80 N	112.00 W	170				NA
Tyrrhenian	39.25 N	14.40 E	160				OC
Under Klondike	60.47 N	129.30 W	340				NA
Ungava	59.50 N	67.20 W	150				NA
Uraguay	29.32 S	58.90 W	440				SA
Vancouver	41.50 N	127.10 W	430				OC
Vanuatu	17.25 S	173.03 E	880			1.145	OC
Venezuela	2.30 N	65.56 W	1060			1.071	SA
Venice	44.60 N	14.03 E	280		- 49.4		EU
Victoria	2.06 S	33.90 E	310		- 8.6		AF
Volcano Belt	46.47 N	19.17 E	890				EU
Weddell	70.63 S	44.68 W	1140			1.092	AN
West Pacific	32.68 N	158.85 E	1880		- 82.6		OC
Wilkes Land	81.85 S	115.90 E	2200	53	19.0		AN
Williston	48.08 N	103.38 W	670				NA
Yellow Sea	35.58 N	120.37E	780				AS
Yellowstone	44.42 N	110.38 W	90				NA
Yukon	66.73 N	141.37 W	470		80.8		NA
Zacatecas	22.73 N	102.80 W	490				NA
Zambia	13.90 S	31.90 E	1350				AF
Zimbabwe	20.00 S	31.08 E	680				AF

Catalog Of Megacraters.

> *"A complete assessment of the role of large body impacts in terrestrial evolution will no doubt require new minds and fresh handles on the many problems."*
> - Leon T. Silver

Adriatic* Sea.** Near Posegy, Slavonia with islands left on the rim in Adriatic Sea. Associated with five others in sculpting the Italian Peninsula. Descriptions have been assembled under ***Sardinia. Must see.

Aegean. The most obvious clue to this crater is the southern arc of islands from the multi-pronged Peloponnesus peninsula in southern Greece to the southwestern corner of Turkey. Crete and Rhodes, along with many lesser islands, are the exposed crests of an underwater ridge. Numerous mountain peaks in Greece and Turkey complete the circle of the crater. The geology of the region is extremely complex with sedimentary rocks upturned as on a rim. Some areas are dominated by volcanic intrusions. Some rocks revealed a temperature-pressure history accounting for relief of 6,000 atmospheres. Such data suggests an original, crater depth of about 18 mi. Rebound of the bowl provided access to marble and valuable ores. Earthquakes follow slippage between an overthrusted slab and the one beneath. The Aegean is circumscribed by a dense record of earthquake epicenters as would be expected around a crater. Concentric fault systems surround the Aegean Sea and, in fact, the bottom is subducting to the northeast down the Hellenic Trench although the movement has been slight. Among the uplifted islands, two are volcanic, a common feature of large craters. In a massive explosion, Thera (Santorini) blew its top about 1500 B.C. An underwater ridge on the rim in the Northern Sporades contains an extinct volcano that can be seen from excursion boats. A formula developed by scientists yields an approximate value for the impact energy being equivalent to 140 million H-bombs of nominal size. In deep antiquity, somebody knew about this crater. Delos Island at the center of the sea became a prominent shrine for Apollo. Astronomical observations made from its peak, Mt. Kynthus, determined the sites of temples. Reconstruction of the heavens in the past establishes the dates with startling accuracy. A second shrine of Apollo, Delphi, was located for no apparent reason on the flank of Mt. Parnassus. It is on the crater rim. These matters were thoroughly explored in a previous book by the author (McCampbell, unpublished). Until recently, geologists believed that the Red Sea, separating Africa and Arabia, unzipped slowly from south to north as a continuation of The Great Rift Valley. Recent research, however, has shown that the original rift that widened into the sea was created suddenly. The reason can now be explained. The massive explosion of ***Aegean*** sent out a linear fault system instantly that, upon widening and flooding, became the present Red Sea. The original rift system goes down the center of this remarkably linear sea where hot water escapes from deep sources. High cliffs on both sides of the sea attest to massive fracture.

Alabama. A crater of intermediate size centered in northern Alabama can be traced in detail for about 180 ° of arc. While originating in Kentucky and passing through Tennessee, a convenient starting point is where Interstate 40 spans the Tennessee River. It then crosses the border into Mississippi where it cuts a narrow bite out of the northeastern corner 150 mi south of the town of Cochrane. On the Mississippi side, it passes through the gorge of

Columbus Lake, the Tennessee-Tombibee waterway, and Aliceville Lake. The arc enters Georgia near Columbus and continues to Macon. Satellite photos show concentric rings of low hills separated by farmland. Close to the outer rim is a zone known as the Black Belt. Economical development of the state before the space age was based upon the iron and steel industry of Birmingham. Convenient deposits of iron ore, coal, and limestone were close by. Oil production became significant in 1944 and subsequent prospecting for oil and gas has been encouraging. Mineral deposits of every kind have been shown to be determined by geological processes following a massive explosion. A major rift system radiates from the center of this crater in the form of parallel ridges of linear, mountain ranges extending 1000 mi to New York. Guntersville Lake lies in the trough where the Tennessee River was dammed. Elongated, geological regions are Valley and Range, and Blue Ridge. Another clue to the configuration of **Alabama** is the drainage pattern. Streams from an interior plateau encounter a sudden change in elevation producing rapids and small waterfalls. Hence the formal name of Fall Line. It describes an arc around the rim and continues in a straight line to the northeast beyond the Potomac River at Washington, D.C. It marks the original edge of North America when it separated from Eurasia. Land between the rift and the ocean is sedimentary. The entire Appalachian Range is among the oldest mountains on earth, being the remains of much larger mountains after eons of erosion. The linear rift to New York continues into the Hudson River Valley and the linear valley holding Lake Champlain. Beyond that it intersects the rift system of the St. Lawrence River with an impressive record. Eight, prominent features called Montenegrin Hills in Quebec rise as cones of extinct volcanoes. They had accumulated thick layers of sediment then, when elevated, suffered nearly complete erosion. The hills are popular areas for recreation, sightseeing, and residence. As an example near Montreal, Mount Royal provides a top of 2 sq mi raised to 769 ft. The highest and largest of these volcanic remains consists of three intrusions raised to 1755 ft with a top covering 20 sq mi. Especially severe damage to earth's crust has been recognized at the intersections of rift systems from megacraters. Montenegrin Hills are a fine example. When North America separated from Eurasia, Alabama was directly offshore from Western Morocco so some clues might have been left there. A shallow indentation of the Atlantic coast between the countries of Morocco and Western Sahara is highly suggestive. But some complications are encountered. Further evidence is provided by the 350-mi arc of the Canary Islands. Shoals at the center of that arc imply the rebound of a crater 250 mi in diameter centered at about 30 N 16 W. Another crater is suggested with the central uplift of the Madeira Archipelago at 32.41 N 16.15 W with a diameter of 960 mi. These candidate craters may be called **Canary** and **Madeira**, respectively. It should be pointed out that the size of **Madeira** is close to that of **Alabama**. The details along this line of inquiry, however, are too scrambled to include them in this Catalog.

Alberta. Canadian islands in the Pacific Ocean indicate the outer rim of a large crater whose center is in the province of Alberta near Swan Hills. This chain runs about 1000 mi from Branof Island (site of Sitka) to Vancouver Island at Seattle. About midway, the west coast of Queen Charlotte Islands emphasizes the outer fault. Careful map work established the center at 375 mi north of the provincial capital of Edmonton. Concentric hills in the region provide an arcing course for the Athabasca River flowing northward into Lesser Slave Lake. This basin is noted for its mineral deposits. Oil from a major field in northwestern Alberta and northeastern British Columbia is delivered by a pipeline to Edmonton. Enormous reserves have been found. Also mined in the region are gold, lead, zinc, uranium, and radium. Along the northern border with Northwest Territories lies a unique, geological formation known as

Caribou Hills. It is a nearly round plateau rising more than 1000 ft above the plains just east of the Wood Buffalo National Park. It appears to be a volcanic intrusion. Arcs of mountains surround the formation resembling the configuration at the well known asteroid crater in Eastern Canada, Manicouagan. Caribou Hills must record an independent and later strike of an asteroid or a volcanic intrusion associated with *Victoria*. A linear fault system at - 34.7 ° clearly set the course of the mighty Mackensie River flowing to the Arctic Ocean.

Aldan Plateau. A small crater south of Tommot in southern Siberia. Its rim is the mountain range of Stanovoy Khrebet. There appears to be an inconsequential crater inside of Aldan at 58.30N 125.75E with a diameter of only 120 mi. These features resulted from fracture of the thin, oceanic crust.

Aleutian Islands. It is convenient to treat twin impacts in the Bering Sea together and limit the information in this section. *Aleutian* **E** and *Aleutian* **W** were described in considerable detail in Chapter 1 (see Figs. 1.19 and 1.20). The first explosion went off close to Alaska producing the Alaskan Peninsula and connecting islands out to Umnak. An outer rim is indicated by the Kenai Peninsula and the large Kodiak Island. The trough between the rims is the Cook Inlet leading to Anchorage in which oil is produced. No subduction was created because all the geological movement was on the North American continent. No evidence of a linear rift is found for the same reason. The second explosion was near the intersection of the International Date line and parallel of 60 N. Islands in the chain were raised all the way to Asia and overthrusting created the Aleutian Trench. All of these islands are intensely volcanic as magma finds channels to the surface between the upper and lower slabs of an overthrust. A major rift system was created southward initiating lava flows that became the Emperor Seamounts. All of these features were created because bedrock of the ocean floor is thin. The sequential birth and death of the Emperor Seamounts took millions of years but their pathway had to be marked in a very short period. Coordinates of each one were measured to a hundredth of a degree. Their calculated azimuths from Detroit showed maximum deviations within ± 7°. The azimuth from Detroit to the next seamount, Tenji, is 169.25° whereas the value for the last one before the knee, Toba Guyot, is 169.46°. This chain is remarkably straight with a Rift Angle of 169.35° - 180.00° = - 10.65°. The initial crack progressing at any speed would display a westward drift due to earth's rotation toward the east. None was observed while a westward displacement of Toba Guyot of 1° could easily be detected. So the earth would rotate that amount in 1° / 15 °/hr = 0.06677 hr, the maximum amount of time available for the crack to proceed from Detroit, a distance of 1280 mi. The propagation speed must have been at least 1280 mi / 0.0667 hr = 19,200 mi/hr! It could have been 2 or 3 times faster.

Alice. Dominates central Australia including Great Sandy Desert, Gibson Desert, and Great Victorian Desert that must be independent, smaller craters of more recent vintage. Center is west of Alice at Mt. Kiel, the highest peak on the continent as part of a central uplift of volcanic origin. Also are included are two, massive, lava flows in the vicinity of MacDonnel Range and Musgrave Range. A dramatic, reddish-brown rock in the outback is one of Autralia's most popular destinations for tourists. It has become a symbol of the country on par with the kangaroo, kiwi, and koala. Ayers Rock rises 1143 ft above the plains 200 mi SW of Alice. The size is impressive at 1.5 mi long by 1 mi wide. It has been called a terrestrial iceberg because only the tip is exposed above the surface. Some scientists believe that the bottom may be 10,000 ft deep. This stone, called Uluru in the native tongue, is sacred to the aborigines and contains numerous carvings of great age. The Uluru National Park of 487 sq mi was established around the rock in 1958. Fourteen miles to the west is a cluster of 61 more domes above the desert known as The Olgas. Both Ayers Rock and The Olgas are sedimentary in origin but the layers have been tilted to near vertical. The sandstone of Ayers has become fluted by erosion of the softer layers. The Olgas are composed of stones and boulders that are cemented into a mass by sand and silt. The geological origin of these rare formations has baffled scientist but a new one is proposed here. Near the park to SW is the Peterman Range that seems to reappear to the east. A circle though these features encounters Davenport to the NW of Alice Springs. These factors suggest that the anomalies in the park define the rim of a very ancient crater that would be 550 mi in diameter with the center at Mount Wedge. It is surprising how many national parks around the world, based upon unusual geological formations, are found to be related to megacraters.

Alps. A relatively weak explosion near Livorno created circular faults in northern Italy. Tectonic forces from the south then drove the land to high elevations of the Alps. All the other mountain ranges in that country seem to have been created by additional explosions as illustrated in a sketch under *Sardinia*. A somewhat larger crater whose center is in northern Sardinia raised the arcing Apennines for 350 mi. This range stretches from the vicinity of La Pezia down almost as far south as Naples. These two craters, *Alps* and *Sardinia,* clearly established the valley receiving all southern drainage from the Alps that flows into the Po River. An impact in the Tyrrhenian Sea carved the *instep of the boot* and the north shore of Sicily along with the corresponding mountains. Etna, the famous volcano, is on the rim of the crater, *Tyrrhenian*. A shallow indentation on the eastern shore of Italy from Ancona to the *spur of the boot*, Cape Gargona, is probably related to another crater. Its effect went inland far enough to reverse the curvature of the Apennines. Called *Adriatic* for easy identity. A fifth crater named *Venice* for the famous city nearby carved the headwaters of the Adriatic Sea with a center at the island of Cres in Yugoslavia. The Volcano Belt in western Italy also forms an arc for nearly 300 mi parallel to the coast. Constituents are Larderello, Amata, Vulsini, Cimini, Sabatini, Alban Hills at Rome, Roccamonfina, Vesusius at Naples, and Vulture at the intersection with the Apennine range. This crater, called *Rome*, is defined by the diameter of 680 mi at 44.97 N 17.20 E in Yugoslavia. Most, if not all, volcanoes in Italy are thus seen to be associated with the craters. They were considered to be of little interest in the initial studies of truly large craters of global significance. But they are included here to illustrate how exquisitely they mold local geography and dominate volcanism and earthquakes. Similar results will be found throughout the world when the study of megacraters expands into neglected territory between the smallest of their kind to the largest, known asteroid craters. Not all of these craters are included in the Catalog Of Megacraters because some are too small to monkey with. See the sketch under *Sardinia* for details.

Amazon. Fig. 1.14 Amazon And Andes showed how this huge crater dominates the northern half of South America. First of all, the impact was west of Manaus near Sao Paulo de Olivenca where all the tributaries in the basin converge into the mighty Amazon River. Significantly, it carved the circular west coast of the continent. That zone is under compression so the Pacific floor dives under the continent. In doing so, the circular portion of the Andes Mountains was lifted in a concentric belt inland. A double rim is conspicuous for at least 1500 mi from Medellin, Colombia in the north to at least Lima, Peru in the south.

The rim ratio of 1.119 is typical. Petroleum is produced all along the Andes Arc as far east as Guyana. Another aspect in this context is a very small crater called *Maracaibo.* It is bracketed by the range of Sierra de Perlia on the west and Cordellera de Merida on the east. A nearly circular Lake Maracaibo fills the basin to sea level from the Gulf of Venezuela. A large fraction of South American petroleum is produced in this basin. This explosion struck on the rim of *Amazon* compounding the massive damage to the crust that had already been done. Because the Amazon Basin is largely covered by forest, many more craters are expected to be found there. A startling discovery is revealed by examining the complete Pacific coast. The southern half runs essentially in a straight line for 2400 mi from about Lake Titicaca to the village of Wellington in southern Chile. Again, compression in the crust and subduction of the Pacific Plate raise the entire length of the Andes. This geology indicates a linear rift system emanating from the center of *Amazon* along an angle of 5.8 °. South America is almost a carbon copy of Africa and both were formed by megacraters. Africa's bulge was carved by *Timbuktu* and the coast south of Benin City, Nigeria followed the linear rift. Crater diameters are **Timbuktu** (1930 mi) compared to *Amazon* (1800 mi); Linear rifts are *Timbuktu* (1780 mi) and *Amazon* (2400 mi). And, of course, they have the same orientation. Additional megacraters in South America will be discussed under their own headings.

Anchorage. Center in Prince William Sound 80 mi east of Anchorage. Small size defined by the perfect ring of the Alaska Range containing Mt. McKinley, the highest peak in North America and another national park. Ties into Wrangel Mountains with some distortion. May extend out to part of the Kuskokwim range that has been used to record the diameter. The crater could be as small as 400 mi. The explosion was directly in the trough of the double-rimmed **Aleutian E** and is apparently younger.

Angkor. A startling surprise accompanies the study of Vietnam's rounded shore on the South China Sea. It is a true arc with diameter of 800 mi from Dong Hoi in the north to the Ca Mau Peninsula in the south. Distortions have been produced by the Mekong Delta and the swampy peninsula that is presumed to be composed of recent sediments. A coastal mountain range, Annamese Cordillera or Central Highlands with peaks above 7000 ft, appears to be the inner rim. To the west, it may be represented at the root of the Malay Peninsula and the Bilauk Tsun Range whose ridge is the international border between Thailand and Burma. (Many international borders have been identified as crater rims.) Stay on your side of the mountain. A linear fault system can be traced through the elongated Great Lake (Tonle Sap), Tonle River to it's confluence with the Mekong and the latter's final run to the sea. With a diameter of 920 mi, the center in Cambodia is within 50 mi of the famous ruins of Angkor Wat! Perhaps the temple complex is actually the correct center. Construction began in the 9th century A. D. By 1000 A. D. it had become the largest city in the world with a population of 15,000,000. Under rule of King Suryavarman II in the 12th century, the largest temple was built to honor Vishnu. Knowledge of geography at the founding six centuries before Columbus, was hardly adequate for selecting that site but it happened. Throughout history, locations that became holy shrines continued their status even through wars. Old structures were destroyed but the sites were retained and rebuilt in the architectural style of the conquerors. The question here is the whether or not somebody had a refined knowledge of geography very early.

Angola. Well inland close to the border with Namibia in western Africa. A slight bulge into the Atlantic from Lobito down to Swakopmund is the key. This crater was created in Pangea prior to the separation of South America. The rupture starting at ***Timbuktu*** came down the coast along a straight line. It then detoured around the rim of the crater already in place that had created a circumferential fault system. Upon completing the detour, it resumed the original course. Further evidence is the Namib Desert. Inward drainage ends in a salt flat 60 mi across but it is offset from the center. Inland rims have been lost. A NE rift system toward the center of Lake Tanganzania opened a linear valley 140 mi long by 35 mi wide. Dozens of lakes fill this depression, most prominently Lac Upemba. The rift passes through the mineral-rich Katanga Province of Zaire. Diamonds and iron are the principal products. Exact location of the diamond mines has not been established and they could be independent pipes of volcanic Kimberlite.

Antilles. Named for the island arc at eastern end of the Caribbean Sea. Primary islands are Antiqua, Dominica, Grenadines, and Grenada. A straight chain of much larger islands connects them to Cuba including Puerto Rico and Hispaniola. A fainter but similar pattern is repeated along the north coast of Venezuela composed of Margarita, Tortuga, Willemstad, and Aruba. The western boundary of Nicaragua defines the sea as 1500 mi long and 400 mi wide, a substantial body of water covering 600,000 sq mi. The floor is an independent, tectonic plate with no permanent connections to the South American, North American, or Atlantic Plates. An entrenched theorem of geology explains this formation as a consequence of eastward drifting of the Caribbean Plate starting with the island arc somewhere west of Central America. The island arc was supposedly created in some unexplained manner related to the tectonic drift. This theory has been broadly applied to all other island arcs but it seriously flawed. An alternative theory is proposed that reverses the existing cart before the horse. A massive explosion over the thin, oceanic crust of the Atlantic raised a double-rimmed crater whose islands are merely exposed peaks along the ridges. Subduction of the Atlantic floor began when the interior of the crater was lifted and driven over the floor. The Americas continued their westward drift while the crater remained anchored in place relative a fixed magma. Something had to give. It was slippage along two faults forming the N-S boundaries of the sea. Lava oozing out of cracks built the corresponding islands. It should be mentioned that the east coasts of Nicaragua, Costa Rico, and Panama form a perfect arc that is backed up by mountian ranges. Its radius of 300 mi is the same as for the Antilles Arc. This geometry suggests that the western half of ***Antilles*** is actually the east coast of Central America where the original center was about 12.5 N 80.0 W compared to the present arc at 14.78 N 65.32 W. The meaning is clear. The impact cut the large bay in Central America at the same time it raised the island arc. A relatively narrow ribbon of the Atlantic floor dove under the islands while both the American Plates continued a westward drift. The extent of this differential drift provides intelligence on the time since the explosion. A major fault at Central America established the western boundary of the Caribbean Plate corresponding to the eastern bountry at the subduction trench. The Plate, therefore, never moved anywhere. It only elongated in place. The northern and southern boundaries were not primeval but slowly grew from nothing at the rate of slippage along the two, transform faults. From the longitude, one obtains the total movement of (80.0 - 65.32) deg x 68.74 mi/deg = 1009 mi. A detailed analysis of age of the ocean floor north of Cuba shows that this distance was accomplished in 87 mil yr. The long-term average rate of drifting was 1009 mi / 87 mil yr = 11.6 mi per million years. Over the same period, the changes in latitude of the original and final centers was (1.50 - 14.78) = 0.72 deg. At that latitude, one degree equals 68.75 mi. Hence, the total movement is 0.72 deg x 68.75 deg/mi = 49.5 mi. The long-term movement of South America due to rotation counter clock wise becomes 49.5 mi / 81 mil yr = 0.611 mi/ mil yr. Rates of plate movements today are very accurately determined by the Global Positioning System. They are useless, however, in dealing with significant periods of geological time. The above method of studying plate movement can improve accuracy of long-term drift wherever the geography is favorable. Two other factors confirm this crater. A great basin sits inside the Antilles arc. And the floor of the entire Caribbean Sea is covered by a world-class, lava flow.

Apennine. This crater created the Apennine Mountains that form the backbone of Italy. It is so closely associated with other craters that their write ups are assembled under the single heading, *Alps* in this catalog. And illustrated by sketch in *Sardinia*.

Arabia. To see this one, the geological clock must be turned back prior to creation of the Red Sea. Recall that the original rift came from *Aegean* about 30 My. By bringing the Africa and Arabia in their beginning positions, one can see the whole crater. With its center at Jiddah, the crater extends from approximately Riyadh to the Nile River. Just west of the capital, two mountain ranges link together like inner and outer rims. They are Ad Dahna and Jabal Tuwayq. The latter has recently been found to hold the largest oil reserves in the world and the government is searching for a corporation to develop and exploit it. This huge oil field could have been predicted by careful study of the crater enhanced by knowledge of the local geology. Creation of the Red Sea may have an alternative explanation. Being considerably larger than *Aegean* and right on the coast of the Red Sea, this explosion could have been responsible, possibly functioning in conjunction with the European explosion. The rift would then have been at the angle of - 27.9 °. Clues in Egypt are more subtle. Most of the Nile River flows in an arcing valley between highlands of the rims. When dammed in the gorge at Aswan, the waters backed up 270 mi forming Lake Nasser. The stream was unaffected up to Ad Dabbah. After a suspicious offset down to Khartoum, the original stream is seen as the channel of the Blue Nile. As for the offset, a small explosion raised the land in the face of the Nile requiring it to detour 175 mi to the east. Faint evidence of the original course between Khartoum and Ad Dabbah is a dry streambed that is used by camel caravans. Previous concepts of Red Sea formation involved a slow unzipping from the great rift in Ethiopia. Recent research, however, has shown that the sea was opened all at once, thus supporting the role of megacraters.

Arabian Sea. Of intermediate size in the Arabian Sea between Oman and India about 300 mi south of Karachi, Pakistan. Locally, the only clue is a circular basin revealed by deep water. At some distance, however, the linear rift has profound consequences. It projected SW into Pangea with force enough to crack it along the south coast of Yemen and the north coast of Somalia. A major fault separated the land masses to the root of the Gulf of Aden. Land on both sides of the fault drifted apart to the present distance of 200 mi, more or less. Hence the strait shores of the Gulf of Aden. The little known but interesting island of Socotra, terminating the Horn of Africa, stayed with that plate and is a continental fragment. Altogether, six megacraters drew the outline of Arabia and accounted for its associated islands. The northern coast on the Persian Gulf was the rim of *Iran.* It was refined on the Oman Peninsula by the combined effects of two weak events, *Persian Gulf* and *Jask*. The Red Sea coast by the rift from *Aegean* and the Gulf of Aden coast by the linear rift from this crater. Refer to the sketch under *Persian Gulf* and Fig. 1.4 Europe And Middle East.

Arctic Ocean. Many atlases do not show the arctic region as seen from above. Great distortions are introduced in all projections onto a flat map representing the temperate and tropical zones. Everyone knows that the Arctic Ocean lies above Alaska but there is more. The northern shore of North America describes a smooth arc in Alaska and mainland Canada. A similar curvature is found in the old world from Norway, through Russia, to eastern Siberia that is half a world away. These curves describe a large, circular region called the North Sea. It is understood here to be a megacrater called *Arctic Ocean*. Everyone remembers the enormous effort and expense of building the Alaskan Pipeline to bring oil from Prudoe Bay down to Valdez at Prince William Sound. Additional resources are in the adjacent U.S. Naval Reserve and National Wildlife Preserve that are being debated. All these deposits are on the rim of the Arctic Ocean. Prospecting in northern Siberia has hardly begun but indications for petroleum and other commodities are reportedly very promising. A major crater near the North Pole is as intriguing as those found in the Antarctic.

Artesian. Now to Australia. This crater covers the eastern half of the continent known as the Artesian Basin. The southeastern coast was formed from Townsville to Melbourne, a distance of 1500 mi. The inner rim in the north is the Connors Range tying into the Great Dividing Range with the peak of Mt. Koscinsko at 7310 ft. Inland rivers drain toward the center of the basin near the corner of New South Wales and South Australia. In the northwest, the rim is probably revealed by the MacDonnell Range near Alice Springs plus the Musgrave and Everard Ranges. The rim appears to enter the Bright Belt at Point Fowler about 800 mi west of Melbourne. About 600 mi of the rim are missing between Point Fowler and Cape Jaffa. A linear fault system radiates from the center raising the short Flinders Range. A great chain of swamps drain into Spencer Gulf. This wedge-shaped gulf, 200 mi by 80 mi at the ocean, was shown elsewhere to have been created by an Antarctic explosion, *Ross.* The entire, mountainous zone along the eastern shore produces coal and some oil. It is perfectly clear that *Artesian* raised the Great Dividing Range along Australia's east coast. An outboard fault cut off a strip, 930 mi long and 110 mi wide on average that became New Zealand. Drill cores from the ocean floor indicate that this continental fragment separated about 33 My. Approximate reconstruction of the original continent is simply cut-and-paste mapping. The narrow peninsula of North Island wraps tightly around the coastal bend at Brisbane reaching almost as far north as Rockhampton. The stubby peninsula of North Island fits slightly north of Brisbane. But that relocation is rather crude compared to results from the new paradigm. Great Dividing Range takes a sharp turn at Brisbane adding 240 mi of mountains. They form a neat arc with horns pointing east. Reversals of curvature along a mountain range indicate the junction of two craters. Sometime after *Artesian* completed its work, a small object struck north of Brisbane. Current reckoning places it at roughly 26S 154E with a radius of 200 mi. About half of that crater rode away on the back of New Zealand. Fortunately, the eastern rim is the long narrow peninsula of North Island at Auckland whose radius also amounts to 200 mi. There is no trick in matching the two halves to find exactly where New Zealand began its journey. Similar but more complicated circumstances apply to the creation of another island of continental origin. See the description in *Madagascar.*

Atlantic. A crater in the Atlantic Ocean carved the east coast of North America. A steep fall to the Hatteras Abyssal Plain descends to a depth of 17,400 ft. A similar structure near

Newfoundland descends the same depth to the Sohn Abyssal Plain. Rising about 150 mi from the center is a scattered uplift including Bermuda. Strong indications of a double rim are found in the Caribbean Islands lying between Florida and Puerto Rico that are entirely unrelated to the Cuban chain. They begin with Grand Bahama passing through Nassau, to Great Onagua. Very narrow strips of land are correctly oriented following the rim that is always perpendicular to a radius. The Bahamas and Cat Islands provide the sharpest indication of a double rim with a spacing of 58.9 mi. The resulting Rim Ratio is thus found to be 1.061, well within the expected range. A probable clue to the eastern rim in the Sargasso Sea is an isolated cluster of volcanic seamounts at 35.45N 50.03W including Justus Corner, Castle Rock, and Rockaway.

Aurangabad. One of several craters in India about 500 mi in diameter related to the Deccan Traps. Treated along with others under *India*.

Baffin Bay. A piece of the globe in the far north of Canada, west of Greenland, is so remote that it is known only to a small population of Eskimos, Air Force personnel at Thule, and their bosses in Washington. Perhaps pilots on direct, polar flights see it once in a while. Separating Greenland from the vast area of large islands in Arctic Canada is bowl-shaped bay with increasing depths toward the center. It has enough earmarks to be included in this study. Little more is known about it.

Bahia de Campeche. The Gulf of Mexico is a large body of water between Texas and the Yucatan Peninsula. A few curious people have suggested its origin by a cosmic impact but the idea has not penetrated into professional circles that should be interested. Supporting evidence can be found under the heading of *Mexico Gulf.* Its west coast along the country of Mexico shows a spiral effect of shorter radii from Jalapa along the coast near Tabsco and Campeche to Merida. In fact, the southern third of the Gulf of Mexico exhibits properties of a megacrater. Coastal regions in the state of Vera Cruz are sediments brought down from a mountain chain. In this general region, scientists found on the shore near Merida, an asteroid crater named *Chiczulub*. It is widely believed to be the largest crater on earth with an uncertain diameter of about 200 mi. But it is dwarfed right next door by *Bay de Campeche* at 480 mi. Oddly enough, *Chiczulub* lies on the rim of its larger neighbor. Lacking a double rim suggests that the Bay is a clean bowl made by an asteroid instead of a comet. Respective latitudes of both craters are of interest. The difference amounts to less than one degree, in particular, 20.33 - 19.38 = 0.95 °. So it is quite possible that they were created by fragments of the same asteroid. In addition, several other craters are found close to their latitudes as shown in Table A.2. Megacraters By Latitude. Among them are two giants in Africa, namely, *Sahara* (1400 mi) and *Timbuktu* (1930 mi). But their ages are unknown. It should be obvious that *Chicxulub* might have been flying through space as a bead on a string. Other fragments may have snuffed out the dinosaurs with *Chicxulub* being a minor or negligible contributor. It will be exciting to discover what really killed the beasts so this avenue of research should be traveled.

Baikal. An impressive lake in southern Siberia close to Mongolia contains one-fifth of the world's fresh water excluding the ice caps. About 30 mi wide, its axis is an arc 395 mi long. It is very deep and flanked on both sides by mountain ranges reaching 6600 ft. These features suggest a double-rimmed crater. Vodokhranilishche is earth-filled 410 ft high and 3.2 mi long. It is a major source of hydroelectric power. A central uplift near Brofsk is nearly surrounded by the river. This crater of 600 mi is no baby. Complex faults in lockstep with the lake's arc extend beyond its ends for a hundred miles. Most critical are the Primor inboard and the Bargusin outboard. A cross section of the lake shows about a dozen faults in bedrock below sediment that has essentially filled the mile deep gorge. Analysis of that sediment indicates the age of the lake at 25 My. Vents at the bottom, still emitting steam, were discovered in 1990. Interesting conclusions were pronounced by the research team of Americans and Russians. The rift system is opening up at about 7 inches per year. The lake's origin was explained upon the shopworn concept of India crashing into Asia and transferring the affect about 1500 mi. And the geography is all wrong. Believe it or not. The present perception of Lake Baikal makes it exactly analogous to its African twin, Lake Tanzania. Both were formed between rims of their own craters becoming, respectively, the deepest and second deepest lakes in the world. Further information is presented in *Tanzania*.

Bakersfield. About 100 mi north of Los Angeles is a crater so small that it hardly belongs in the catalog. Inclusion is based upon its unique relations to major, mountain ranges and local oil fields. Two mountain ranges dominate California. They run nearly parallel along the axis of the state with the Sierra Nevadas on the east and the Coastal Range on the west. Both have been traced to the major explosions, *Bonneville* and *St. Louis*. The annulus between them was originally a deep gorge that slowly filled with alluvium from mountains much larger than at present. The resulting valley set the courses of the Sacramento and San Joaquin Rivers. In addition, they became the agricultural treasures of today when irrigation systems were built. Vegetables of all kinds, fruits, nuts, and fibers furnish much of the U.S. market. The valley also supports a lively export market. The minor explosion in southern California raised the mountainous rim that tied the major ranges together and created its own farmlands. The eastern rim is defined by the Greenhouse Mountains of the southern Sierras where the arc is used as part of the California aqueduct from China Lake. In addition to the abundant produce, the crater was responsible for many oil fields. While the southern Sierras are barren, many fields are located at the edge of the Coast Range across the valley. Major pools appear to be linear with lengths typically five times greater than the width. Dozens of smaller fields dot the area with the same lineation. All of these are clearly associated with the Coastal Range. The fields are very small portions of the originating crater. Other nearby fields are distributed along a curve that is concentric with the rim of *Bakersfield* and others are scattered throughout its basin. A major rift system the full length of the Great Valley has isolated patches of oil and gas wells. They must be assigned to the major craters with very little contribution from *Bakersfield*.

Banda Sea. Bounded by island chains of Malaya Archipelago and Lesser Sundra Islands. Outer rim includes major islands of Obi, Misool, and Irian Java with numerous small islands along the ridge. Inner rim contains Propinsi Maluku and Keplauan along with many lesser islands. By size, this crater belongs in the catalog but there is a problem. The Rim Ratio, outer radius divided by the inner radius, is much too large. The value of 1.414 is far out of the range for typical megacraters of about 1.100. It probably is not a true megacrater in terms of its creation. Better explained as an unusually large basin whose inner walls failed. During rebound, the inner ridge was elevated to be exposed as islands.

Bandar Abbas. One of Iran's ports on the Persian Gulf. Would not have been included except for the author's personal knowledge of the Southern Shores and Islands Province. Outer rim is established as Bushehr-Bandar Abbas- Kerman-Yadz plus the islands of

Hormus, Queshm, and the extended chain from Foriar to Sheykh She'eyb. A spectacular, geological formation along the coast to the town of Jask was eventually understood by the author in the context of a huge crater centered near Masad, Iran. It is described under a separate entry of *Iran*. *Bandar Abbas* probably contributed to the rise of the Zagros Mountains that were critically important in ancient history. Inscriptions in three languages high upon a cliff lead to the deciphering of Egyptian hieroglyphics.

Bass Strait. A waterway separating Australia from Tasmania at Melbourne. Outer rim in Australia is the peculiar kink in the Great Dividing Range below the smooth arc of 850 mi north. In Tasmania, the northern mountains are on the outer rim. The inner ring is revealed by the islands of Hunter and King on the west and the Furneau Group on the east. The latter is a arc containing Clark Banks and Flinder Island. It will be recalled that both Australia and Tasmania ride on the same tectonic plate. Their southern and western shores, respectively, are remnants of **Ross** that separated them from Antarctica. A very small object subsequently pushed a depression in the plate that did little damage. It left the shallow strait separating two pieces of the same cake.

Belcher. A bay along the eastern shore of Hudson Bay in Canada. The latter has been widely discussed as a possible impact crater and a few scientists have made the same suggestion regarding Belcher Bay. Although one of the smallest in the catalog, it offers many important details supporting the cosmic interpretation. The eastern shore completes nearly a half of a perfect circle running 160 mi. A narrow band of elongated islands lie close to shore for most of its length. In the south, a prominent island 22 mi long is aptly named Long Island. The Nastapoka Island arc continues the pattern northward involving about 20 islands named the Hopewell Islands. Their distance off shore is very uniform and the sum of their lengths amount to about 50% of the total distance. This pattern clearly shows a continuous ridge with flooding of the lower levels. This double rim has a ratio of 1.048 that is an acceptable value for a megacrater. Offset from the center of this bay is a large cluster of islands that suggest a central uplift or typical mountain range inside a crater. It covers about 3600 sq mi. But their shapes and arrangement are unusual. Five parallel arcs extend about 60 mi with notable gaps. Close study of satellite photographs clarify the dilemma. Belcher Islands form part of the rim of Hudson Bay! See *Hudson* in this catalog. Other small islands in Belcher Bay may be uplifted peaks. Further evidence of an impact crater extends far inland that has been depressed into a shallow dish. Rivers and streams, therefore, converge toward the center of the crater.

Bellingshausen. Named for a sea in Antarctica. Evidence of a crater is the Alexander Group of islands west of the Antarctic Peninsula. It is dominated by the large Alexander Island that has a very strange shape. A mountain range along its east coast suggests a rim. But the island roughly resembles a capital **J** with several radiating valleys filled with glaciers. Islands to the north have been badly displaced by the tectonic drag from South Sandwich Islands as described in Chapter 1. See *Antilles* for a more detailed treatment of a similar process in geomorphology. Because the evidence is weak and not supported by additional clues, this crater must be considered suspect. Alexander might be the bones of an extinct volcano of prodigious size. Terrain hidden by the ice cap might clarify the question.

Bengal. A notably circular sea separates India and Burma by 1300 mi. The eastern rim is the Malay Peninsula. Parallel to the shore is a complex arc of islands that would be expected as the remnants of an inner rim. The Myanman and Merghi Archepelagos are long and narrow with the prominent example of Sabikyun. Using that island as typical of the group yields an inter-rim distance of 54.9 mi that translates into a Rim Ratio 1.076. Such a typical value supports the concept of a megacrater and there are plenty of other clues. Additional islands are found within the crater, namely the Adamans and Nicobars. The island chains form a perfect arc for a distance 810 mi from the Myanmar Peninsula of Burma to the northern tip of Sumatra. Such a ridge must be related to a separate crater and so it is. The neighboring *Indonesia* at 3420 mi has an outer rim corresponding to these islands. Their existence determines a time sequence of the two craters. Bengal had to strike first followed by *Indonesia* that formed its rim in the sea. The latter overthrusted the floor of the Indian Ocean along 2000-mi of the Sundra Trench. So the Indo-Australian Plate is plunging under the Eurasian Plate. The extent of this motion has been very uncertain. Sophisticated technology has been used to measure the drift rate of tectonic plates. The systems in use are a) Global Positioning System, b) Mark II, and c) Very Long Baseline Array. A few dozen sites are scattered around the globe whose movement relative to a fixed magma is determined. Some large regions have no sites, so information is missing. However, it was shown under *Antilles* that present rates of drift do not represent the average over many millions of years. The large-scale geology of *Bengal* and its huge neighbor is especially interesting because it introduces some new ideas. The Indo-Australian Plate carrying the crater is sliding under the Eurasian Plate. This motion can be visualized as a library-style, magnifying glass sliding, left to right, under a book. The extent of overlap indicates the displacement at least since formation of the crater. That distance, measured as 440 mi, is less than half of the diameter. At a reported rate of subduction of 6 cm/yr, the corresponding time can be found. Without displaying the calculations, the answer is 11.8 mil yr. A complicated analysis of this question yields an age of 49.1 mil yr. The concept of subduction at the world's trenches must be re-examined without prejudice. For two objects in relative motion, either one may be considered at rest. Motion at the trenches is now conceived as movement of the ocean floor relative to a fixed, overlying slab, usually continents. An alternate and proper point of view is that the ocean floors are fixed while continents or portions thereof, slide in the opposite directions over the ocean floors. While this idea may not apply world-wide, it has some merit in the present case. The configuration of the *Bengal-Indonesia* complex permits the revised concept. Recovery from the latter's crater drove the crust out and over the ocean floor. Such processes can be rapid at the beginning but take millions of years to complete, especially toward the end. The final return to grade level is known to be extremely slow. Witness the islands of *Aegean* that continue to rise after 26 mil yr. The very large crater of *Indonesia* is mostly covered by oceans and has a long way to go. So the internal disk of the crater is still expanding. Earthquakes inside the arcs, relying upon this relative motion, will be produced in any case. Same for volcanoes. In other words, the so-called rates of subduction may be attributed to expansion of the crater linings that have not recovered completely. Therefore, some fundamental ideas about plate tectonics and continental drift should be reviewed with care. An especially significant feature in the Indian Ocean is part of the scenery. The Ninety East Ridge is a straight line of lava beds running north-south for 2450 mi. The volume of lava qualifies it as among the greatest of the world's *large igneous provinces.* Like an arrow, this ridge aims straight at the heart of *Bengal.* It must, therefore, be a linear rift system created by the explosion. Furthermore, relative positions of the crater and 90E have not changed one iota because they are riding double on the same donkey. Knowledge about 90E can thus be applied to the crater.

160

Bighorn. Diminutive example in northwestern Wyoming. Such obvious features should have attracted earlier attention. Boundaries are ranges of Bighorn, Wind River, and Absaroka Mountains that have been distorted. Highest mountain is Francs Peak at 13,153ft. S. Fork Shoshone River arcs between rims flowing into Buffalo Bill Reservoir near Cody. May have contributed to the scrambled geology at Yellowstone National Park.

Bismarck Sea. Located immediately above the tip of Papua New Guinea encircled by the Bismark Archipelago. Appears to have been compressed into an oval that would form a nearly complete string of islands. Then traceable through New Handover, Admiralty Islands, Karkar Island, and Long Island. If so, the original crater would have been larger with the center farther west. Of no particular significance. Included mainly for statistical purposes.

Black. The eastern shore is a circular arc of more than 180 degrees from Trabzom, Turkey past Sukhum, Georgia. Very small and unimportant.

Black Hills. A large crater with impressive, central peak rising well above the plains. Forests of Ponderosa pine were recognized in the adopted name. A massive intrusion of molten rock covered 6000 sq mi where most has been eroded down to 3000 ft. Still standing boldly is Mt. Harvey at 7242 ft as the tallest in North Dakota. Vertical movement from unknown depths brought up rich deposits of various minerals, primarily gold. The hills contain the Precambrian granite of Mt. Rushmore where former presidents gaze at their prosperous legacies. Expert's descriptions of the Black Hills as part of the Rocky Mountains must be in error. The crater size is established by the curved range of Missouri Coteau. As faded rims, they would have guided the Missouri River to take its 500-mi journey along an arc from Ft. Peck Reservoir in Montana through western North Dakota. No evidence has been found in the central plains where rims could lie buried under a few miles of sediment. Uplifting of the Rocky Mountains apparently destroyed the western half of this old crater. A cogent piece of missing information would be the age of the Black Hills.

Bohopal. One of several craters in India about 500 mi in diameter related to the Deccan Traps. Treated along with others under ***India***.

Bonneville. The Great Salt Lake in Utah is a well-known landmark and the epicenter of this explosion is the northern tip of the lake. The resulting crater, not yet fully recovered, has left many mountain ranges and a large basin. Rains created a huge lake called Bonneville covering 19,000 sq mi in western Utah and portions of Idaho and Nevada. Weather changes over long periods flooded or dried the basin. Salts in solution precipitated out creating an absolutely level surface. Eons of such climate changes left vast areas of salt flats that may be familiar to automobile racing fans. The last cycle from wet to dry is about half completed. The Great Salt Lake now fills only the deepest part of the basin. The present area is only 1700 sq mi. Average depth is 15 ft with deepest spot at 35 ft. Shallow, indeed. Salt from the flats and lake provides commercial quantities of sodium and potassium chlorides. Farther a field, the explosion cut a giant cookie out of bedrock that, to the west, overshot the floor of an ancient ocean. Normal results were volcanic origin disposed in a perfect arc. Changes in elevation of land and sea have left snow-capped peaks in Oregon and northern California. They are mountains with famous names of Adams, Hood, Jefferson, Washington, Thielsen, Scott (Crater Lake), McLoughlin, Shasta, and Lassen. On

May 18, 1980 (not so long ago), Mt. St. Helens erupted only 32 mi west of Mt. Scott. This ridge of mountains delineates the Sierra Nevada Mountains that continue south to Mt. McKinley, the highest peak in the contiguous United States. Quite a show for a megacrater only 940 mi in diameter.

Bothnia. The Bothnian Sea in northern Europe separates the Scandinavian Peninsula from Finland on the continent of Asia. Its northern end is a suspicious circle only 90 mi in diameter. The terrain shows that the water fills the deepest hole at the center of a much larger basin 500 mi across. An explosion apparently formed the Peninsula with a sharp ridge along the border between Norway and Sweden. Numrous streams in Sweden converge toward the center. To the southeast, shorelines of large bodies of water imply the location of a rim. They include the Gulf of Finland, Lake Ladozhskoye Oiere near St. Petersburg, Russia. Also along that sweep is another Russian lake then the White Sea entering the harbor at Murmansk. Parallel shores of the Bothnian Sea suggest a linear rift directly aligned with the small basin.

Brazilia. About 40 mi north of the new capital of Brazil. The primary clue to this crater is the smoothly curved coast along the Atlantic Ocean from Rio de Janeiro to Salvador, Bahia. The coast is backed up by steeply rising mountains, sometimes even escarpments. Four diamond producing regions form their own arc with a similar radius having distances inland varying from 360 mi to 80 mi. A strong relationship has been established between megacraters, volcanoes, ancient pipes of Kimberlite containing diamonds, and diamonds found in alluvial deposits downstream encapsulated in sandstone. A major industry in the region is the manufacture of jewelry from a great variety of precious stones. Some portions of the crater have hardly been explored. Primary value of the basin is ranching. Drainage to the north by parallel rivers Auguaia and Tocanlins that are suspected channels of a linear rift. These two rivers along with the Amazon control drainage of the northern half of the continent. The crater existed prior to the breakup of Pangea, fitting into the shallow indentation at Luanda, Angola. Inland from the coastal plain of Angola is a major escarpment that is a challenge to climb over in a small airplane. The linear rift from **Timbuktu** encountered the weakened rim of **Brazil** raced along its path and released the entire crater to its future in South America. Diamonds were discovered in 1725 in the state of Minas Geraes immediately north of Rio de Janeiro. Other sources were found near Sao Paulo in Parana plus Goyaz, Matto Grosso, and Bahia. Diamantino is the present name of the four most productive mines at the former town of Tejunco where diamonds are distributed over 9000 sq mi.

Brisbane. One of many megacraters in Australia situated northwest of the city of Brisbane. Contributed to forming the wide, desert region from Alice Springs to the Great Dividing Range. Established coast south of Townsville and the Clarke Range nearby. May have had a linear rift at 32.7 ° contributing to the course of the Darling River. But primary cause of that river was most likely **Ross** in Antarctica as in Fig. 1.9. Birth Of Australia.

Bucharest. Small crater on Balkan Peninsula west of the Black Sea. Rimmed by Transylvanian Alps and Balkan Mountains. Insignificant except that it destroyed a long stretch of the Parathians that was on the SE rim of **Hungary**.

Buenos Aires. Small, immediately south of the Argentine capital. Hemispheric coast to Bahia. Numerous streams converge toward the city draining a shallow basin. Together with ***Uraguay*** carved the wedge-shaped bay at the mouth of the Rio de la Plata, 150 mi wide at the ocean running 200 mi inland.

Buffalo Hills. Named for Buffalo Head Hills in northern Alberta, Canada. Outer fault set western coast of large islands of British Columbia. Included are Baranof (city of Sitka), Queen Charlotte Islands, and Vancouver Island. If the axis of inland sea is treated as an inner fault, the rim ration is 1.152. Probably contribiuted to the location and shape of Great Bear Lake. Provided more than 250,000 sq mi of plains for wheat farming on both sides of the border with U.S. North of the center is a large outcropping called Caribou Mountains that is often found inside large craters.

Burma. Near eastern border of Burma at same latitude as Mandalay. Defined double arc of Arakan Yoma (mountains). Inner fault drained by Irrawady River outside Pegu Yoma range. Course controlled from bend at eastern end of Himalayas for 800 mi to the apex of the delta above Rangoon. Clearly distinguished, double rims with radii of 357 mi and 326 mi sets Rim Ration of 1.097. No evidence of a linear rift from this crater.

Burma W. Located 200 mi east of Mandalay, Burma containing a central uplift 240 mi in diameter. Raised the range of Arakan Yoma that governs course of Chindura River.

Byrd. Inland along the axis of Antarctic Peninsula about 480 mi from South Pole. Contains various mountain ranges. Cut coast of Maria Byrd Land from Coats Land to Edward VII Land. See Fig. 1.9. Birth Of Australia.

Campeche. The southern most part of the Gulf of Mexico is identified separately as Bahia de Campeche whose circular coast is a dead giveaway. Radius is 340 mi. Oddly enough, its eastern rim nearly coincides with the site of ***Chicxulub***. The centers of ***Campeche*** and ***Chicxulub*** differ in latitude by only 65 mi. Because asteroid fragments fall close to the same latitude, this crater could have been part of ***Chiczulub***'s parent. With a crater diameter of 690 mi, this event could have been a major factor in the dinosaur problem. Also, it could have generated tidal waves in Cuba for which evidence has been reported. The date would have to be as close to 65 My as modern techniques allow. A most peculiar, geological arrangement lies nearby. A chain of nineteen volcanoes, crossing southern Mexico, form a straight line from the Pacific Ocean to the Gulf of Mexico aimed generally toward this crater. They are out of place because volcanoes are normally found along continental edges whereas these are all on the North American Plate. They could have grown along a linear fault from ***Campeche*** but that is denied by other evidence. Linear rifts have been shown to emanate from the *centers* of craters. But, in this case, a large mismatch is noted. The responsible crater has to lie somewhere along the line of volcanoes. Projection into the Pacific was fruitless. Projection into Bahia de Campeche was successful. It crossed dead center over a crater that is practically invisible on maps but abundantly clear in satellite photos. Supportive evidence confirms this identity that is described under ***Trans Mexico*** below.

Campo Grande. Moderate size in South America 160 mi ENE from Campo Grande in Motto Grosso State of Brazil. Overlaps about one-third of ***Brazilia***. Center of large basin is at confluence of three rivers into Rio Parana. Drains Motto Grosso Do Sul. Gorge at Iguasu Falls may be involved.

Caroline E. Coincides with East Caroline Basin. Landmarks on the rim include Admiralty Islands (40), Manus, New Hanover, Kapingamerangi Atoll, Nukkuoro Atoll, Mortlock Island, Puluwat Atoll, and Lamotrek Atoll. Excellent definition of rim around 300,000 sq mi of ocean.

Caroline W. Ocean basin of same name 850 mi in diameter. Double rim defined to south as Yapen and Biack in the Malay Archipelago. These two established rim ration of 1.114. Also clockwise are Mervi, Peleliu, Babelthuap, and Yap. A linear rift to the southeast runs for 1860 mi including the Solomon Ridge, New Hebrides, New Caledonia, and New Georgia. Beyond the last island of Oroa, the ridge has been distorted by later impacts. Frequently, elongated islands occur in a double row with variable spacing up to about 70 mi. It is important to consider the shapes and orientations of the islands. Most are elongated and curve with the rims. Others are elongated and parallel to linear rifts. Some deviation of this pattern must be expected from vagaries of geology.

Caspian N. Strikingly different from *Caspian S* as it is a large, shallow basin. No physical impact to provide excavation. Diameter of depression is 550 mi but surrounding lowlands are 960 mi across. Very ancient with badly eroded evidence. Revealed for short distances along ridges of international borders: Iran-Turkmen, Kazakh-Russia, and Armenia-Iran. Volga River south of Kuybyshe follows the arc inside the rim to Volgograd. Right angle turn into Radial fault. Outer rim may be indicated by Zhiguli Hills west of the river. Also, Volga Plateau extending south as the Yergini Plateaus.

Caspian S. Major sea between Europe and Asia. Bottom topography and hydrology indicate two distinct parts. This portion is generally circular and would be more so if lowlands on eastern shore were flooded. Physical impact dredged to an unknown depth leaving a present bottom at 3360 ft on top of the accumulated silt. Raised Elburz Mountains to peak of Mount Qolleh-ye-Damavan at 18,386 ft, visible from Tehran. Details suggest two separate events created the entire sea. An asteroid blow for the south and an atmospheric explosion of a comet for the north. Major oil and gas discoveries are located on both sides of the sea where the two craters overlapped.

Catalina. Unusual geography is found off the coast of southern California that has been named after the famous, vacation spot of Catalina Island. The center is about 200 mi offshore from Los Angeles. First to be noticed is the smooth arc of the shoreline from Point Conception to San Diego that represents only the edge of sediment washed down from the mountains. This structure consists of arcs as follows: a) outer rim is Santa Ynez Mountains east of Point Conception past Santa Barbara to coastal ranges behind Los Angeles, b) middle arc includes Catalina, the Santa Rosa group, and Rodriguez Seamount 30 mi west of San Miguel Island, and c) inner rim is San Clemente and San Nicolas islands. Radii from the epicenter are respectively 220 mi, 190 mi, and 160 mi. Islands on this inner arc are few, small, and widely spaced suggesting the crest of material sloughed off the crater wall. Oil and gas fields abound in the region. Several are located along the shore near Santa Barbara including offshore wells and major production at Goleta and Rincan. Dispersal of lesser fields throughout the Los Angeles basin implies that it is another crater about 40 mi in diameter, too small to be of much interest. Long lines of parallel fields are found in the vicinity of Long Beach also with prominent, offshore production. These southern fields are distributed for up to 50 mi in narrow bands parallel to the coast and each other. Similar results were obtained for the coastal fields of southern Texas in *Mexico Gulf.*

Catamarac. East of the Andes Mountains in northern Argentina. West rim is bent in lower section of Andes. Roughly follows towns of Presidentia Rogue Saeng Pena, Santiago dol Estero, Sina Miguel de Tucuman, and San Salvador de Jujuy. Overlaps another crater to the north where destruction of adjoining rims opens a wide, connecting valley. Geological significance is limited.

Celebes. A nearly circular sea between Borneo and Philippines. Rim defined by ridge along NE Borneo plus extended peninsula Propinsi Sulawesi Utara, Greater Sundra Islands Sangihe and Keuplauan, and southern Philippine coast. NW quadrant has been destroyed and replaced by the crater of **Sulu** Sea. Reversed arc of the new craters consists of Isabella, Jollo, and Tareakan in the Sulu Archipelago.

Central. Southwest of Hawaii overlapping *Central Pacific.* Bottom topography shows numerous seamounts scattered randomly within a boundary of submerged ridges and island chains. While less than obvious, the boundary of this basin is decidedly circular. A tremendous amount of work by oceanographers was required to describe this distinctive feature but its origin was not considered. The size of this basin, 1800-mi diameter, is typical of megacraters. In the central region, one finds the widely scattered Phoenix Islands that are uninhabited, coral atolls with a combined surface area of only 11 sq mi. They were at one time large, volcanic islands whose loss indicates great age. The rim can be followed from the southwest by the whole chain of Gilbert Islands, Samoa Islands, Manihiki Atoll, Starbuck, Christmas-Fanning-Washington-Palmyra and Cape Johnson Guyot. The author landed on Johnson Island for refueling upon returning from the Pacific Theatre in WWII. A mere speck of coral with an area measured in acres, barely room for a runway with elevation less than 10 ft.

Central Pacific. Not to be confused with the above *Central* that it overlaps. A huge tract of the Pacific Ocean is devoid of islands except the miniscule and isolated Johnson Island described above. This basin covers 5,200,000 sq mi. Farther south, thousands of islands poke up out the sea that are directly related to the basin. A great deal of discussion on this crater is not required because it was treated at some length in Chapter 1 and illustrated in Fig. 1.10. South Pacific Islands. Very different scales are required in a single map to show the shapes and orientations of the islands and also indicate their positions relative to the crater rim. Groups of islands on the southern rim are the Marshal Islands, Cook Islands, and French Polynesia. Finding the desired maps is not easy. The most valuable one turned out to be an Australian map of the Pacific Ocean at Equatorial Scale 1:34,531,200 by Robinson (undated). Surrounding the primary map are many inserts at different scales displaying individual islands, archipelagos, and island nations. Let it suffice to consider a little detail for selected subjects. Take Hawaii, for example. The array of atolls and reefs out to Midway are approximately linear whose creation has been explained. Present-day islands of volcanic origin, on the other hand, are notably nonlinear. In fact, they form a short arc on the rim of this crater. Well-know ages show a temporal progression from the oldest Kawaii to the youngest and still active Hawaii. About 20 mi southeast of the Big Island, a new one being born, the Loihi Seamount. In a million years or so, it will take its place among the older islands. A map of Hawaii shows that it is not situated at the end of straight line. It is marching ahead of a volcanic column along the rim, actually the circular fault. Next consider the Marshal Islands. They are disposed along two parallel chains with axes correctly aligned for the crater rim. The western chain includes Kwajalein and Bikini, both familiar to the author from military days and nuclear weapons testing. The eastern chain runs from Majuro, capital of the Federated States of Micronesia, to Taka Island. The distance of 200 mi between these chains provides a basis for determining the Rim Ratio. With an outer radius of 1290 mi, it becomes 1290 / (1290 - 200) = 1.183. Naturally, some variation from that figure is to be expected. A complete review of this crater would burn up too many hours and pages. So the reader is left with a lot of homework to reach a private opinion. The magnitude of this task if fully appreciated whether the reader confirms the present findings, wishes to modify them, or rejects them completely.

Central U.S. Close to Paducah, Kentucky where Ohio River joins Mississippi. This detail is extremely significant. Primary rivers draining large basins on continents have been noted in ***Congo*** and ***Amazon***. Major rivers converge toward the epicenters, in this case, Missouri, upper reaches of Mississippi, and Ohio. Along a major rift to the sea, secondary rivers join the main flow to the sea. In the U.S., they include the Tennessee, Cimeron, Arkansas, and Red Rivers. The major fault terminates at the SW corner of Mississippi State while everything to the south is sedimentary fill. This very old basin predated the Appalachian Mountains that truncated it in the east. Remaining area of drainage is about 1,700,000 sq mi. The main channel for the Mississippi River is a graben, that is, a region where land sinks between faults, sometimes escarpments. Between Natchez, Mississippi and Harrisburg, Louisiana, steep banks are 130 ft high on the west and 190 ft high in the east, defining a valley 29 mi wide. Being half filled with gravel and silt forces the river to meander on the surface. The values above are compatible with most linear rifts. Close to the epicenter, 300 residents of New Madrid, Missouri were subjected to the greatest release of seismic energy ever in the United States except Alaska. Three shocks during the next three months have been estimated as greater than 7.6 on the Richter Scale. After shocks included one about as intense along with 104 stronger than 4.3 R. Over 2000 shocks were felt for five months. They continue to this day in the region considered the most hazardous in the U.S. These disturbances come from movement along the very deep, New Madrid Fault System. While epicenters have been recorded throughout a sizable region, they are strongly concentrated along fault lines. Basically, the fault runs SW from Cairo, Illinois 150 mi to Carouthers, Missouri. Instead of a long chain of epicenters in a straight line, the pattern resembles a capital **T**. All this seismic activity is completely out of place because earthquakes are usually associated with continental margins and rumbling volcanoes. The combination of two straight lines in the **T** suggests an intersection of linear faults from two, independent impacts. Whether linear or circular, these intersections of major faults create the greatest damage to the crust. The site can remain unstable for many millions of years. Low populations and small towns in 1811 limited the injuries, deaths, and property damage. Another Richter 8 or above on the New Madrid fault would destroy virtually everything out to about 50 mi with serious damage out to 150 mi. Casualties could be in the millions. In recognition of this potential, seven states and the Federal Emergency Management Agency (FEMA) have developed response systems coordinated by the Central U.S. Earthquake Consortium with headquarters in Memphis.

Chaco. Small crater in northern Paraguay 310 mi south of Santa Cruz. At southern tip of Defensures del Chaco National Park. Displays a drainage line between two rims in the southwest.

Chad. In the land locked nation of that name in northern Africa. Center at Lake Chad along southern edge of Sahara Desert where borders are shared between Niger, Nigeria, Cameroon, and Central African Republic. The lake and associated swamps are 140 mi long with an average width of 60 mi. Size changes according to rain fall. These conditions are common within megacraters. Perimeter is defined by volcanic mountains pushed up through the desert. Included peaks are Jabal Marrh (10,131 ft), Emi Koussi (11,204 ft), and Mt Greboune (6,562 ft). Emi Koussi is the highest peak in the extensive Tibesti Range of great wealth in tungsten, tin, niobium, tantalum, and uranium. Early exploration for oil was disappointing but reserves 500 million barrels have been established. Extending out to sea the rift released lava for creating the islands of Santa Antonio, Principe, Sao Tome, and possibly St. Helena 1500 mi away.

Champaign. Close examination of a U.S. map shows a remarkable pattern of rivers in Illinois and Indiana. They form a jagged but true circle with center 30 mi south of Champaign, Illinois. Starting near Chicago, the Illinois swings along the western side and joins the Mississippi River continuing the arc past St. Louis. The Ohio River on the east passes Cincinnati and Louisville on an arc. They converge at Cairo, Missouri. A few miles to the north, an arcing range of hills runs from Carbondale to Shawneetown in the Shawnee National Forest. A linear rift runs from the crater to Helena, Arkansas, starting at the crater rim as seen in every other case. This rift dictated the course of the Mississippi River whose meanders and oxbow lakes are confined within a wide gorge. The typical width is about 19 mi. However in the sector near Reelfoot Lake it broadens to 27 mi. The reason is surprising. The river wanders around the center of a small crater at 30.25N 89.52W. With a diameter of 140 mi, it is only noted and ignored. A major fault system that poses a serious threat at that location is the New Madrid Fault. A remarkably detailed map of that area was prepared at the University of Memphis showing epicenters of 4,387 earthquakes from 1974 to 1998. While they are scattered lightly over 13,000 sq mi, a very dense concentration is found along the Mississippi River. It tilts toward the NE. This cluster is 86 mi long averaging 9 mi wide. Like a capital **T**, it has a cross stroke with the same density that is slightly askew. This pattern is close to what would be expected at the intersection of a linear rift with the corresponding peripheral fault. The location for the record is 35.33N 89.54W. As the referenced map covers only four square degrees, it affords pinpoint accuracy in measurements down to a dozen ft or so. The **T** leans toward the NE at 44.43° and should point to the center of Champaign. But it does not. The latter azimuth is 23.98°. The difference of 20.70° is far too great to attribute New Madrid Fault to Champaign. The pattern of earthquakes, however, must be linked to another crater that has not been found.

Chicago. Among the larger craters in the catalog at 1530 mi, the center is at Macomb, Illinois 200 mi SW of Chicago. Dominated eastern North America from upstate New York to the Sangre de Christo Mountains of New Mexico. North-South spans from Baton Rouge, Louisiana almost to Hudson Bay, Canada. Prominent landmarks on the ring are the great valley along the border of Tennessee and North Carolina between the Cumberland Plateau and Great Smoky Mountains. Its average width of 65 mi gives a rim ration of 1.107. Between Lakes Erie and Ontario at the Canadian border, the rim generated the gorge of the

Niagra River although the escarpment at the falls has not been linked to any crater. Profound effects are traced through western Canada by a straight chain of major lakes, namely, Woods, Winnepeg, Athabasca, Great Slave, and Great Bear. Special significance is attached to the bay in Great Bear at Yellow Knife and the straight, narrow alignment of Winnepeg. Other craters contributed to shaping these lakes but their locations are correct for a linear rift from *Chicago.*

Chixulub. Small crater at northern tip of Yucatan, Mexico. Subject of intense study and endless publication in articles, conference proceedings, and books. Proof of an asteroid impact at the correct time came from geological samples at the Cretaceous-Tertiary boundary that everybody knows was 65 My. Samples from around the globe showed abnormally high concentrations of iridium. That element is much more common in asteroids than on earth. The argument was that dust from the asteroid impact drifted around the world being deposited in the narrow layer of rocks marking the transition between geological eras. Since Chiczulub was the largest crater known at that time, it was naturally judged to be guilty. Within the present context, however, that seems to be unlikely. It has never been shown that Chiczulub was the only crater of the right age. Other impacts within the accuracy of time measurement could have caused the disaster. A strong possibility is the crater herein called *India* that is 3 to 4 times larger than *Chixulub.* The great lava plateau known as Deccan Traps covers all of central India and has been dated at 65 My with remarkable accuracy. There is no reason to disqualify this alternative. Furthermore, evidence suggests otherwise. The mistaken concept of a sprinting sub-continent would require India to have been thousands of miles farther south at that time. Since India never moved independently of Asia, a new question must be explored. Is it possible for Chiczulub and India to be apples off the same tree? The database in Appendix A shows their respective latitudes as 20.33N and 21.50N, a difference of 1.17°. In the current configuration of continents, that difference amounts to 80.5 mi. In the geological past, continental drift may or may not have maintained that alignment. The point is that Chiczulub may be only a fragment among larger ones that pelted earth at the same time. Such combined damage could have easily ended a geological period and severely disturbed the global biota. An intensive study is required to determine which megacraters were created at that time. Then the truth about the K-T boundary will come forth.

Chile. In South Atlantic at San Julian, Argentina close to Cape Horn. Outer rim indicated by innumerable islands off the west coast of Chile south to Darwin. Then farther east through Isla de los Estados and East Falkland. Inner rim is the Chilean Alps to West Falkland. Sunken, annular fault system is primary waterway on the Pacific Coast, Strait of Magellan, and Falkland Sound separating the two main islands. Caution is advised here because the topography has been subjected to enormous forces from drag on the northern boundary of the Scotia Plate. See *Scotia*.

China S. More than 1000 mi diameter with center south of Yichang. Contains two more that are only noted here

27.87 N 110.57 E 450 mi
29.63 N 106.82 E 300 mi

Simply ignoring such worthy craters is required to keep the book within reasonable bounds.

Congo. Encompasses all of Zaire. Much larger than expected by virtually spanning the entire African continent. Center near Yolona east of Salonga National Park. Extends north to Ndele, Central African Republic. Inner rim noted as Adamawa Range in Nigeria including Mt. Cameroon (13,435 ft). Extends to Bioko Island whose creation is partly attributed to ***Chad.*** Western rim is lost at sea down to Luanda, Angola. A tight cluster of negligible craters lie east of Luanda with diameters of 150, 120, and 100 mi. Southern rim is ill defined. Annulus between rims in east is East Rift Valley including Lakes Turkana, Natron, and Manyara. Basin displays typical pattern of many tributaries joining the Congo River to the ocean.

Coral. A sea between Australia and Solomon Islands that is seriously distorted. Follows Great Dividing Range along NE coast of Queensland for 800 mi passing Cairnes and Townsville. An arc of islands is found parallel to the shore. Of special interest are those in Halifax Bay. Their arc is in correct position for an inner rim relative to the coastal mountains. Includes Great Palm Island. Invaded from NW nearly to center by peninsula of New Guinea. Base of peninsula corresponding to rim is Mt. Wilhelm (14,793 ft). Northern rim is probably New Britian but the region has been badly chopped up. Must be very old.

Cordoba. Identified as Grand Chaco in Argentina. Drainage includes tributaries to Parana River near center. Physical evidence includes mountain arcs on east and west. Other portions of rims are missing.

Cree Lake. In central Saskachewan, Canada, almost entirely defined by other lakes. They are Raindeer, Weleston, Athabasca, Peter Ford, and Ile-a-la Grosse. This crater is closely aligned with others in the Klondike Cluster described under ***Klondike.*** They are designated as *A* through *D* with increasing longitudes. The craters are so large that their separation is slight notwithstanding considerable spacing of their centers. They are close together. Because *Cree* is larger than any of the Klondike Cluster, it may well have been the first fragment of a chain. If that were the case, not proven, the sequence of explosions would be almost evenly spaced as

$$\textbf{Cree} + 0.34 \text{ hr} + \textbf{A} + 0.37 \text{ hr} + \textbf{B} + 0.39 \text{ hr} + \textbf{C} + 0.21 \text{ hr} + \textbf{D}$$

for a total time of 1.31 hr or 1 hr 19 min. Such cosmic events can roll along pretty fast.

Cuernavaca. About 70 mi west of Mexico City near Cuernavaca with an uplifted center 80 mi in diameter. Volcanoes abound. Volcan Iztaccihuati at 17, 340 ft, Volcan Popocatepetl at 17,887 ft, Cerro La Malinche at 14,636 ft, and Pico de Orizaba at 18,701 ft. Standing so tall gives away their youth. In 5 mil yr, they will be worn away to stubs. Many other factors in this study have suggested that the present concepts of geological ages may be exaggerated.

Dar es Salaam. In the Indian Ocean 840 mi east to Dar es Salaam. Carved coasts of southern Somalia, Kenya, and Tanzania. Raised the arc of Comoros Islands although they may have drifted from their original positions but not much. Islands of Pemba and Zanzibar south of Mombasa may be all that is visible of an inner rim providing a tentative Rim Ratio of 1.038.

Delphi. Named for familiar city to indicate northern India. Center, however, is 300 mi south near Kota. Faint record with northern arc buried under sediment of the Ganges Valley. Most revealing features are a short, southern arc from Bhopal to Vadodara. Narmada River flows between inner Vindhy Range and outer Satpura Range.

East Pacific N. Center at Pacific Coast opposite Guatemala City consisting of the Cocos Plate. Also known as Guatemala Plate. Eastern half has been lost down the American Trench. Westward drift of Central America stretched out the Caribbean Basin providing an everlasting supply of lava that paved the floor of the Caribbean Sea. During the same period, Central America rode over the Cocos Plate. The extent thus far amounts to 870 mi of the oceanic crater. The interplate boundaries have produced a world-class, lava plateau, supporting the idea that the plate was created by an explosion. A mental movie of these processes in fast forward is a fascinating example of tectonic motion. In other words, the impact carved the east coast of Central America. Both Americas drifted west as a ribbon of ocean floor dove under the Antilles Arc. Lava pored into the wound gradually paving the entire Caribbean Basin. Westward creep of Central America covered nearly half of Cocos Plate. These details warrant serious study.

East Pacific S. Off the west coast of South America as shown in Fig. 1.11. Hidden Giants. Evidence is barren so little more can be said in support of this crater.

Ethiopia. East African nation with intriguing geology that was covered briefly in Chapter 1 and Fig. 1.3. Afar And Environment. Key points are a) the entire country is an igneous province known as the Ethiopia Plateau whose volcanic skeletons near the epicenter are honored in a national park, and b) before creation of the Red Sea by the rift from ***Aegean***, lava spread into the heel of Arabia, and that portion of the Plateau became separated from the main body, c) the linear fault from this explosion opened the Gulf of Aden, d) the Afar Triangle is the delta of an extinct river that flowed along the East African Rift Valley during wetter periods when people fished in lakes under the present sands of the Sahara Desert, and e) water for the small country of Djibouti and its capital is supplied by underground branches of the defunct river. The Arabian Peninsula was formed by crustal damage from several explosions. The Red Sea coast was formed by ***Aegean***. The rift from ***Ethiopia*** cut the coast on the Gulf of Aden. The coast along the Persian Gulf was part of the circular faults from **Iran**. Geography in the Near East is dominated by an extension of the East African Rift Valley via the Gulf of Aqaba, Dead Sea, and Jordan River to Mt. Hermon.

Europe. Very large and old crater covering most of Europe with center 60 mi NWW of Kiev in western Russia. Created the lower Ural Mountains and the arc of major islands in the Mediterranean Sea including Sardinia, Sicily, Crete, and Cyprus. Outer and inner diameters of 2250 and 2020 mi give a rim ratio of 1.114.

Far East. A huge expanse of the Pacific Ocean south of Japan. Rim is traceable along major islands and groups starting with the southern Japanese islands and peninsulas west of Nagoya. They are Kyoto, Shikoku, and Kyushu. Inland Sea is fault system between ridges separated by 95 mi. Resulting rim ratio is 1.068. Farther south, the crater includes Okinawa, Taiwan, Philippines, Halmahera, and New Guinea.

Fiji. Consists of a large cluster of islands in the South Pacific. In addition to the principal islands of Viti Levu and Vanua Levu, many others are distributed along a neat circle surrounding the Koro Sea 160 mi across. All of them are situated atop a huge mountain whose slopes plunge into the deep sea. The main islands will eventually be worn away leaving a circle of coral atolls and sea mounts. Thus the Koro Sea will become a central lagoon of 20,000 sq mi. To the east, the Tonga chain runs from the northern most Tafahi 480 mi to Ata. All members of this chain are exposed peaks of an underwater ridge. A gentle curvature, concave toward Fiji, suggests a crater rim. The final clue is an overthrust of the ocean floor creating the Tonga Trench. The Tonga chain must be the inner rim while the outer rim has been lost to subduction. The latter's original radius can be estimated by the product of the inner radius time a typical rim ration of, say 1.144. The resulting radius thus becomes 440 mi x 1.144 = 500 mi or the corresponding diameter of 1000 mi.

Four Corners. Modest-sized crater with center in NE Arizona. Close to the famous Four Corners of Colorado, New Mexico, Arizona, and Utah. The Colorado Plateau surrounds that point with a radius of 220 mi. Very old volcanoes have been eliminated by weather with vertical pipes remaining and collars of rubble surround their bases. Crater rim to NE is defined by the inner San Juan Mountains and the outer Sangre de Christo Range. Rim crosses Grand Canyon between Zion and Bryce Canyon National Parks. Abundantly clear in satellite photos.

Gibson. In southern portion of Western Australia. So far outback that no convenient landmarks exist.

Gory Putorama. A crater of intermediate size east of the city of Noril'sk, Siberia and inside *Tunguska*. Reaches east to the Lena River and possibly beyond where a portion was erased by a later event. Western mountains are the Central Siberian Plateau. A double rim sets the valley of the Lower Tunguska River to Turukhansk and the Yenisey River to Dudima. Central coordinates are considered to be especially accurate because it is surrounded by a perfectly circular structure at a distance of 85 mi. The mechanics of formation of such small bowls is not understood. Perhaps, it is part of the rebound process as the same feature has been encountered elsewhere.

Great Basin. A major geographical feature spanning several western states. Roughly speaking, this basin consists of deserts and mountains ringed by the cities of Boise, Reno, Las Vegas, and Salt Lake City. It is by no means an isolated crater as it contains many smaller ones, shares some boundaries with others, and has been substantially damaged in the north. Geographic clues on the perimeter include the Wasatch Mountains south of Provo and the Warner Range in northern California. High probability should be assigned to a circular rift system establishing the Grand Canyon and the upstream course of the Colorado River. Outfall south from Hoover Dam would correspond to a linear fault radiating from the basin's center. Along with its tributaries, this mighty river drains the west slop of the Rocky Mountains. Some thirty dams prevent flooding, generate electricity, and provide invaluable water for general use and irrigation.

Great Bear. Mammoth crater with center on the northern border of Saskachewan, Canada. Created the vast, wheat belt of Saskatchewan and Alberta west to the Canadian Rockies. Plains of grain extend into Montana and South Dakota. Established flat, lake country east to

Hudson Bay. Shaped the peninsulas and elongated bays of Great Bear Lake whose spacing of 60 mi corresponds to a Rim Ratio of 1.091. Linear rift to the southeast is disclosed by Lake Winnepeg that is 270 mi long by 60 mi wide. Probably extends an additional 300 mi along a chain of lakes, namely, Lower Red, Cass, Leech, Plicon, and Mille Lacs west of Dutluth, Minnesota. The bearing gives Rift Angle of - 31.0 °.

Great Plains. An impressive geological oddity is found west of Rapid City, South Dakota. Named Black Hills, it is an intrusion of molten rock reaching the present height of 7242 ft at Harney Peak. Dimensions of these lava mountains are about 110 mi long by 40 mi wide. Included is the Wind Cave National Park at Hot Springs. The region is noted for mineral production, especially gold that was discovered in the 1870s. At the north end of the Black Hills at Lead is one of the largest gold mines in the western world that is owned and operated by the Homestake Mining Company. Mining is one of the primary, economic foundations of the state along with ranching and more recent tourism. This plateau is apparently associated with an enormous crater centered 190 mi NNW. The outline passes through Saskatoon, Winnepeg, Sioux Falls, Salt Lake City, Boise, and Calgary. The great Rocky Mountain Trench with dimensions appropriate for linear fault systems was probably form by this event along a rift angle of - 41.7 °. This large number indicates a great age.

Great Sandy. In northern portion of Western Australia without any landmarks for pinpointing the center. Several uplifts occurred near the center. Within the western half, a ring of hills suggests another crater that is not treated separately.

Great Western. Located 110 mi west of Cheyenne, Wyoming and older than the Rocky Mountains that destroyed part of the crater. Notable aspect is the Rocky Mountain Trench along the bearing of 330°, 1800 mi long with width varying from 70 to 120 mi. Very appropriate geography for a linear rift. Width is comparable to other crater-generated examples such as the East Africa Rift at Addis Ababa averaging 70 mi across.

Greenland. Crater is the entire, ice-filled basin of Northern Greenland. Most evident feature is the waterway Nares Strait separating Ellsmere Island, Canada from Greenland by connecting the Arctic Ocean and Baffin Bay. Is a water-filled graben between double rims with ration of 1.101. Established the northern coast of scalloped glaciers. Probable rim to east is Greenland Sea and submerged ridge 25 mi offshore. No tectonic evidence and the island is part of the North American Plate.

Guatemala. Center in northern Guatemala at the base of Mexico's Yucatan Peninsula. A triple-rim crater indicated by Sierra de Chama at 140 mi, Sierra de Las Minas at 170 mi, and Cordillera del Meredom at 190 mi. Intervening water courses are Rio Polochic and Lake Isabal flowing into Bahia Amatique in the Caribbean Sea. Next outboard is Rio Molagua. The eastern extremity is a host of islands from Gladden Cay north to Amergris Cay. An oval-shaped plateau has intruded within the crater with an area estimated at 14,000 sq mi. The eastern quadrant contains a 170-mi crater at 17.02N 83.37W that is otherwise ignored.

Hainan. A major island of 13,000 sq mi lies off the coast of Vietnam is probably a central uplift of this 440-mi crater rising to 6207 ft. Not known to be volcanic. The rim appears as mountain ranges on the mainland that lie behind the alluvial deposits of the Asian coast and seem to be distorted. Valuable primarily as a data point.

Hawaii. Crater is in the central Pacific Ocean about 750 mi SW of Honolulu. The state of Hawaii including its 1500-mi chain of islands, seamounts, and guyots has been discussed at some length. Their creation by drifting of the ocean floor over a hotspot was found to be untenable. Their origin must be understood in the context of marching volcanoes. But something strange about the vacation isles is noticed. Needle-point dividers with a large-scale map reveal the startling fact that they lie on an arc. From Kawaii eastward, the fault crosses the back side of Oahu, across Molokai, between Lanai and Maui then on to the big island. The Kilauea vent on Mauna Loa gushes lava into the sea creating a plume of steam that widens in the wind. The next island on the arc has been born. A transition took place at Tenet. To that point, the volcanoes followed the linear fault from another crater far to the west where they were deflected along the rim of *Hawaii*. The dormant fault was awakened by weight of the growing volcano there. Secondary clues to the rim are La Perous, Necker, Nihoa, and Nu'hua where tourism is prohibited to preserve Polynesian culture. The eastern half of this crater can be followed along seamounts. A linear fault from *Hawaii* is indicated by Johnson Island, Christmas Ridge, Malden, Starbuck, and Carolines. The azimuth is 147.4°.

Himalaya. The most spectacular mountain range in the world includes many peaks nearly 5 mi in elevation. Examples are Manashi, Xixabangma Feng, Cho Oyn, Everest, Makalu, and Kanchanjunga. Traditional geomorphology adheres to a simple and direct explanation that wayward India crashed into Asia and drove the terrain upward while it dove under the continent. But facts and perceptions developed herein require abandoning that theory. Current concepts utterly fail to explain the perfect arc of the southern face. Besides, the basic idea of India's migration was based upon contrived and false evidence. Experts have shown the basis of migration was "...the biggest paleontological fraud of all time." The newly introduced concept of continental drift over hotspots was stretched too far. It was used to explain an illusory relationship between India, Maldive Islands, Chagos Archipelago, and Kerguelen Island at an incredible 49.50 S. Such migration of India simply does not square up with well-known, continental drift in that part of the world. Next, studies of fossilized sea shells supporting the migration were based upon inaccurate claims of their recovery. Full appreciation of these matters and the misconception about India will take some independent study by the reader. Earthquake epicenters disposed along an arc have, in all instances, been seen as overthrusted slabs of rock along the rims of megacraters. That requires the driving force to be inside the arc. The same geology applies to the Himalayas whose frontal arc is noted for a history of earthquakes that has not subsided. Earthquakes are caused by differential motion in a subduction zone so some of India under Asia is implied. Can any clues provide estimates of the extent? Yes. Recall the fault systems of *Antilles* and *South Sandwich*. Both ends of their subduction trenches blended into linear faults. As ocean floor is lost down the trench, continued drift of adjacent floor slipped past the island arcs. That is to say, relative motion between the impact basin and surrounding ocean floor was accommodated along straight, strike-slip faults. The extent of subduction could be determined from ages of the ocean floor. Similar geology is observed at the Himalayan arc. Linear faults project to the north from both ends of the Himalayas. In the west, the Himalaya Frontal Thrust joins the extension of the Owens Fracture Zone near Chitral, Pakistan. This fault reaches the Persian Gulf about 225 mi west of Karachi, Pakistan. At the east end of the mountains, the second, southward trending fault runs 1800 mi before blending into the arc of *Indonesia*. The western fault extends to a point below the Arabian

173

Peninsula on the Carlsburg Ridge that is the mid-oceanic ridge of the Indian Ocean in that region. It was created by the linear fault system from **Iran** that sliced the straight coastline of western India. As land masses on both sides drifted away, the granitic Seychelles Islands and submerged mountain chain were moved 650 mi away from India. Could that process account for portions of India being lost under the Eurasian Plate? *Au contraire!* That possibility is denied by the geological record. In floating away from the spreading ridge, both India and Seychelles would normally be removed an equal distance. And, indeed, they did within accuracy of map measurements. India moved 850 mi NE while Seychelles moved 800 mi SW. Any loss of India under the Himalayas would show up as a deficit on the Indian side. None exists. Therefore, India never subducted a measurable distance down the Himalayan Frontal Thrust. In the island arcs of **Antilles** and **South Sandwich,** both strike-slip faults projected away on the concave side. Such faults from both ends of the Himalayan arc, however, project in the direction of the convex side. In the case of island arcs, the strike-slip faults accommodated relative motion between the crater and undisturbed ocean floor. The absence of such a pattern at the Himalayas utterly prohibits extensive subduction! Prominence of earthquakes along the front indicates continued instability from a geologically recent event. Measurements using the Global Positioning System have indicated relative motion between India and the Eurasian Plate of about 15 cm/yr. That value converts to a more convenient number of 122.43 mi per million years. The age of the ocean floor at Seychelles and off the Indian coast was found by the Deep Sea Drilling Program to be about 28 mil yr. That would be about the age of the great explosion named **Iran** and the time available for spreading at the Carlsburg Ridge. So the extent of India's subduction, implied by this data, is about 350 mi. But such a figure would easily be reflected in geology of the Indian Ocean. Because it is not, the result must be wrong. The most likely source of error is the unstated assumption that the present motion correctly represents the average value over tens of millions of years. It clearly does not. The absence of extensive subduction by India does not challenge these considerations. Earthquakes along the mountain front indicate some motion between the lower and upper slabs but it can not be much.

Ho Chi Minh. Center in South China Sea 60 mi offshore SSE of Ho Chi Minh city, Vietnam. Carved the entire Malay Peninsula down to Singapore. Then Malay Archipelago and Sundra Island. Distance of 27 mi between Rempangand Biau Islands sets rim ration of 1.047. Continues as mountainous backbone of Borneo. Not traceable in the north. This crater occupies the western quadrant of **Indonesia**. Perhaps the greatest rift system on earth radiates from the epicenter with a bearing of 116.22°. This finding may be spurious because an enormous chain of islands may be related to **Indonesia** that should be consulted along with descriptions in Fig. 1.6. Vietnam Cluster and Fig. 1.7. Global Fracture.

Hopi. The practiced eye is more likely to recognize craters that have previously been overlooked. This small crater has clues that are too weak to inspire confidence. Center is on Hopi Indian Reservation. The rim appears to follow the San Juan River, Glen Canyon on the Colorado River, and Painted Desert continuing through Winslow, Holbrook, Sanders, and Gallup.

Hudson. An inland sea of 196,000 sq mi was reportedly discovered in 1610 by Henry Hudson although Native Americans knew about it for countless generations. Little outside attention was paid to Hudson Bay in Canada for 200 yr until thoroughly mapped in 1929-

1931. The sea has an average depth of 330 ft with a smooth bottom. For at least two decades, Hudson Bay has been suspected to be an impact crater. Considerable discussion of this idea has appeared in the literature but the author has not found any published conclusions. Perhaps the question will be settled here. The region now flooded is only the central part of a much larger area of flat lake-lands that drain toward the Bay befitting a large basin from an explosion. But many features common among megacraters are missing. No central peak is observed. Elevated rims are not found. James Bay on the SE may indicate a linear rift but that is problematic. Belcher Bay on the eastern shore is a much stronger candidate whose details are described under **Belcher**. A small difference in latitudes of Hudson, Ungava, and Alberta suggests a possible relationship. Instead of a rarity, chains of craters from fragments could be the norm.

Hungary. Almost the whole country of Hungary occupies the floor of an ancient crater professionally known as the Hungarian Basin that was once a lake. The center is a short distance south of Budapest. Multi-ring mountain chains form the northern half of the rim in an arc from Vluna, along the Carpathian Mountains through border regions of Czechoslovakia, Ukraine, and Russia. The southern rim is missing. Flowing through the crater is the largest river in Europe based upon water volume and length of 1750 mi. The famous Danube has always been a route for exploration, conquest, and transportation. Fishermen, riverboat captains, and students living on its banks know exactly where it is. But the rest of the world is uncertain about its course. While generally flowing west-to-east, it suddenly takes an offset of 180 mi north-to-south passing through the basin. Entry is through the Gates Gorge. The river crosses Hungary in a linear fault but meanders hinder accurate measurement of its azimuth.

India. The size of this crater is based upon the distribution of the Deccan Traps that cover nearly all of central India. A normal assumption is that the lava flowing from a single impact filled the crater. Actually, that country was pummeled by four objects that created separate craters from 400 to 500 mi in diameter. Assigned names of the largest four are based upon cities near their center as below.

Jodhpur	26.02N	73.13E	510 mi
Bohopal	24.70N	76.52E	500
Luknow	25.27N	79.48E	460
Aurangabad	18.47N	76.80E	410

and there are few smaller ones. Several aspects of these craters bear examination. Jodhpur, in the far NW created the Thar Desert and influenced the course of the Indus River. Differences in latitudes and sizes of the first three are remarkably small. The maximum variation in latitude of 0.75° amounts to separation of only 51.6 mi. Crater diameters vary around the average of 490 mi by only 4%. These figures strongly imply a string of fragments striking close together. Longitudes are also meaningful. If these craters are found of the same age, the sequence of impacts would have been

$$\textbf{\textit{Luknow}} + 11.8 \text{ min} + \textbf{\textit{Bohopal}} + 13.6 \text{ min} + \textbf{\textit{Jodhpur}}.$$

Such rapid-fire bombardment of intermediate power could do as much damage as a lone monster. The entire event over 25.4 min would be comparable to the time interval between the Aleutian impacts of 38.8 min. Certainly, the combined effects of these explosions could account for the Deccan Traps as similar correlations have been found at many locations. Off India's west coast, a string of islands runs straight south for thousands of miles very close to 73° east longitude. It bears earmarks of a linear fault from an impact crater corresponding closely with Ninety East Ridge from **Bengal.** Most of the 2000 islands are too small for habitation. Separate groups are Lakshadwee (formerly Laccadive), Maldive Islands, and Chago Archipelago. The last group consists of a few coral islands on the rim of an extinct volcano with a lagoon of 6500 sq mi. Off by itself in the archipelago is an irregular atoll, Diego Garcia, home for natives and a military base for British and American bombers so effectively employed in Kuwait and Afganistan. The linear rift from **Jodhpur** split open the wedge-shaped Gulf of Khambhat along its eastern shore then opened the crack for growth of the island chain. Westward curvature of this rift down to the equator is too slight to measure. So the speed of propagation must have been extraordinary, perhaps as high as 100,000 mi/hr. Apparently, the growth from there to Diego Garcia was much slower. The average speed from **Jodhpur** to Chagos of 23,600 mi/hr is comparable to that for Ninety East Ridge and Emperor Seamounts. The most southerly crater in the Indian group, **Aurangabad**, could well have been part of the bombardment. Its latitude of 18.47N is suspiciously different from the 25.33N average of the northern trio. This crater is well positioned for creating the northern portion of the Deccan Traps. If included in the bombardment, the sequence would become

Luknow + 10.7 min + **Aurangabad** + 1.1 min + **Bohopal** + 13.6 min + **Jodhpur**.

Quite a show!

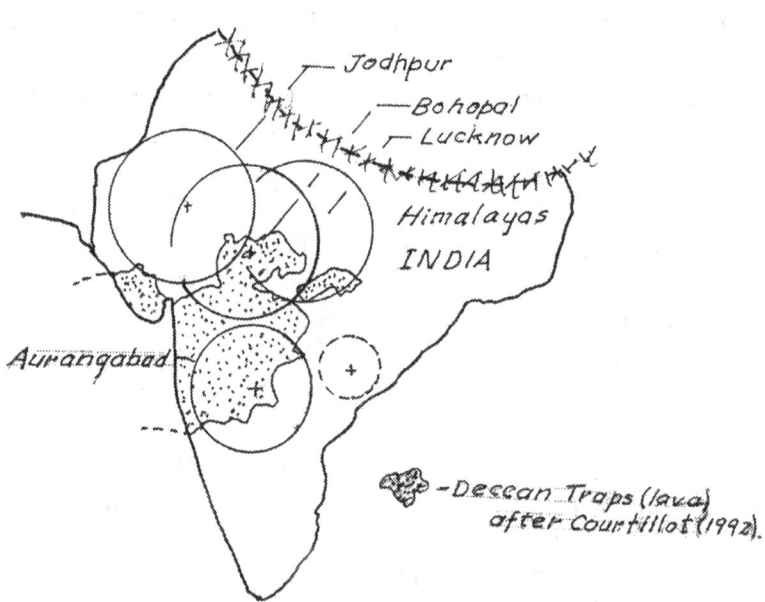

Indian Ocean. A large crater at 1230 mi whose existence is discovered by assembling diverse hints. Situated 620 mi off the African coast at Dar es Salaam, Tanzania. Inner rim noted as islands of Pemba, Zanzibar, and Mafia close offshore plus the Comoros Archipelago. Resulting rim ratio comes to 1.116, about mid-range for that index. This explosion preceded **Zambia** that pried Madagascar loose because the evidence pertains to its original position. The linear rift proceeded along the azimuth of 212.12° creating the straight east coast of Madagascar. (This crater should dominate a conflict of interpretation on this point.) At the same time, the sloping northern coast was carved by the rim. So the northern tip of Madagascar becomes another region of rim-rift intersection. That point is sustained by the rising of Maromokoto at Massif de Tsaratana including four other peaks still soaring to 9470 ft. In addition, the prospect of petroleum riches on the Tanzanian coast is reinforced by oil seeps on shore and discovery of natural gas. These resources are exactly where they are expected after percolating through tilted layers of porous rock and blocked by vertical offsets on the crater rim. Other details on the formation of Madagascar are presented in *Zambia* and *Madagascar N*.

Indonesia. Located about 250 mi off the northern tip of Borneo. A double rim, including Samatra and Java, was discussed in Chapter 1 under the heading of South East Asia and illustrated in Fig. 1.6. Vietnam Cluster. It must date back to the time when earth's axis was tilted away from the orbital plane about 44°, much greater than the present 23.5°. This event drove one of the world's greatest fault systems 4040 mi to the southeast. This aspect was addressed in Chapter 1 under South Pacific Islands and illustrated in Fig. 1.7. Global Fracture. As a refresher, rebound from the explosion drove the outer portion of the original bowl over the ocean floor to form the Java Trench where subduction is still in progress. Primary land areas on the inner rim are Sumatra and Java. The outer rim is composed of the Adaman Islands, Nicobar Islands, Kepluan, and Mentawi. The region is plagued by earthquakes and volcanoes including the famous explosion of Krakatoa and the greater but lesser known Toba. The great rift to the southeast set the stage for growth of New Guinea, Solomon Islands, Vanuatu, Fiji, French Polynesia, Pitcairn, and Sala y Gomez.

Iran. Located 510 mi east of Tehran near Mashad, Iran, this large event exerted a profound influence in central Asia and beyond. A linear rift system along a bearing of 147.7° took a slice out of western India, creating its straight, western coast. This crack eventually became the Indian Ocean Ridge as India and its shard drifted apart. The fragment is now the underwater ridge supporting the Seychelles Islands. It is a critical point that the islands are of continental granite rather than basaltic material as in most other islands. The crater rim cut the wedge-shaped Gulf of Kutch in the state of Gujarat west of Ahmadabad. The intersection of the rim with its own rift suffers the greatest damage. Such an unstable region should be prone to earthquakes. After a long history of seismic shocks, a 7.7 R struck on January 26, 2001 along an east-west trending fault where a portion of the crust is pushed over an adjacent slab. Early reports indicate more than 20,000 death, 50,000 injured, and 500,000 homeless. But that is not all. The region also lies on the rim of *Jodhpur*, a more modest crater only 510 mi in diameter. The great arc of *Iran* set the courses of the Tigris-Euphrates River whose spacing reflects a double rim. Also sculpted was the main axis of the Persian Gulf from Abadan, Iran through the Gulf of Oman. Irregular geography from Bahrain to the Oman Peninsula gave shape to the United Arab Emirates. These details have been treated under *Persian Gulf* and *Jask*. These rather small events contributed to *Iran* in trapping all oil and gas in region. World dependence upon these resources from Iraq, Kuwait, Saudi Arabia, Iran, Bahrain, and Qatar hardly needs emphasis. Oil is also produced in India where the rim of *Iran* enters the state of Gujarat. The eastward trending fault through that state extends far inland along the Vindhya Range and the parallel Narmada River.

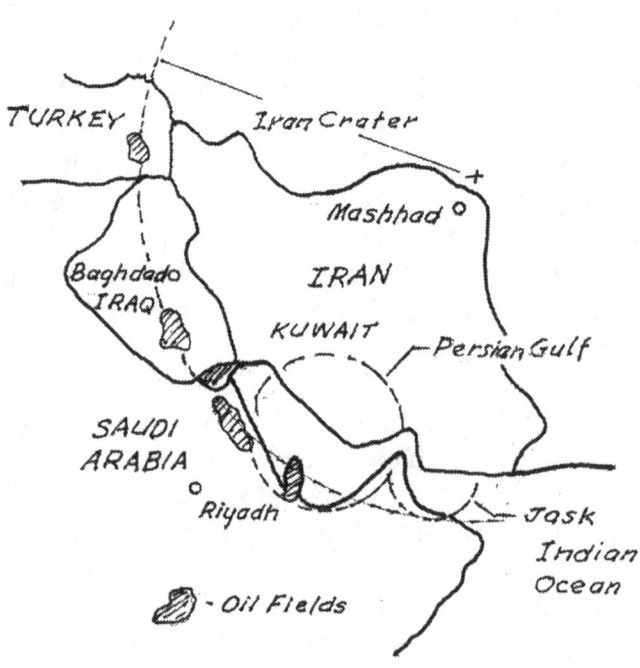

Ireland. At Lake Shannan in Ireland's plains surrounded by perfect circle of hills, such as, Sliene Gamph (NW) and Wicklow Mountains (SE). A tiny diameter of 125 mi suggests that it is an independent event but other data deny it. The most likely explanation is a central feature of a much larger crater. Primarily responsible for the island of England with an average radius to her backbone mountains of 250 mi. This arc goes along the southern peninsula at Lands End and the Isles of Scilly. Highlands flanking the British Channel imply a double rim with a ratio of 1.075. The English Channel between England and France must be a graben on the arc. Local indications are the white cliffs of Dover and the French escarpments scaled by Allied troops on D-Day. To the north, that arc passes through the center of the **North Sea** that is itself a crater. It is famous as a source of petroleum. Almost without exception, one finds oil within a small crater that rides on the rim of a larger one.

Jalisco. Struck about 20 mi north of Guadalahara in Jalisco State of Mexico. Carved coastline of Colima and Sante de Cuabe states. Manzanillo Bay is paralled to the coast but is too narrow to indicate an overthrust. No evidence of a linear fault is found.

Japan. Created the Sea of Japan between that country and Asia. Also formed the northern section of the Kyukyu Trench up to Mt. Fuji. From there a fault links into the Japan Trench. Subduction of Pacific Ocean floor is now substantial at 150 mm/yr or 93.2 mi per million years (mi/mil yr). But it has not always been so rapid. Analysis of the extended Hawaiian chain demanded rejection of the drift-over-hotspot theory. So the extent of Pacific floor loss under Japan must be much less than prevailing opinion. Subduction is normally lead by the oldest floor but ages vary from 11 My to 34 My, south to north, along the Trench. It appears that the long-term movement of the Pacific Plate must have been nearly due north instead of west, perpendicular to the trench. The extent of subduction could, therefore, be quite limited. The circular shape of Sea of Japan is obvious when attention is called to it. From the north, the crater is defined by the short peninsula of Hokkaido terminating at Hakodate. It continues into northern Honshu where the Bay of Matsu where a double rim is indicated.

178

The spacing of 43 mi leads to a Rim Ratio of 1.138. If the sea level were high enough, mountain peaks of Japan would be a magnificent, concentric arc of islands. The entire island of Honshu is a pronounced arc down to Hiroshima. The Inland Sea out of Osaka strongly suggests a double rim although separation of mountain ranges is abnormally large. Further evidence of the rim is the Asiatic coast from Pusan, Korea to Vladivostok, Russia but tectonic distortion is apparent.

Jask. In headwaters of Gulf of Oman 40 mi southeast of Jask, Pakistan. Carved the east coast of Oman Peninsula where United Arab Emirates pokes into Iran at Bandar Abbas. The Strait of Hormuz is seen as the point where two craters were tangent, namely, *Persian Gulf* and *Jask*. Primary evidence of the rim in Oman is provided by the mountain ranges of Al Jabal and Al Akhdar. See *Bandar Abbas*.

Jodhpur. One of several craters in India about 500 mi in diameter related to the Deccan Traps. Treated along with others under *India*.

Joplin. Well within *Springfield* near Joplin, Missouri, a very ancient explosion left all local evidence under miles of sedimentary rock. The center location must be inferred from a rim 1320 mi away that formed the west coast of North America from San Diego, California to Guadalahara, Mexico. The outer rim begins in the mountain range inland from San Diego and continues down the back bone of Baja California to the tip at San Jose del Cabo and beyond to the volcanic island of Barcena. The inner rim along the coast of mainland Mexico is separated by the Gulf of California (Sea of Cortez). The ratio becomes 1.128. Farther south, the inner rim is exposed by Tres Marias Islands. Lying between the rims along the coast is Bahia de Bandera at Puerto Vallarta flanked on the coast by Punta Mita and Cabo Corrientes. The proximate valley between the rims holds Lake Guadalahara and others past Mexico City to Cordoba on the Gulf of Mexico. The entire rim across Mexico is sprinkled with volcanoes. Rims at Guadalahara are the Sierra de las Minas and the aptly named Neovolcano Range. Paricutin Volcanoe was observed at birth in a field on February 20, 1948. Experts diligently observed a massive explosion and lava flows until activity ceased after nine years. Other volcanoes along that rift system are dormant except for Popocatapetl near Mexico City. It continues to threaten 30 million citizens with fumes and rumbles. In summary, *Joplin* created the Baja Peninsula, a few obscure islands, Acapulco Bay, Lake Chapala, and the string of volcanoes from the Pacific Ocean to the Gulf of Mexico. They are the only such volcanoes in Mexico and are allegedly out of place on a continent. Such a large crater must have left some evidence of a linear fault system of continental scale. A strong candidate is the string of major lakes across western Canada. They would be the narrow portion of Lake Winnipeg, a contributor to Lake Athabasca, and the northwestern bay of Great Slave Lake. Also the final run of the MacKinsey River from Ft. Good Hope to Ft. McPherson at the apex of a large delta. These items are not as straight as expected so are problematic. If the lake lay between two rims, the northern side would correspond with the outer rim. Computer calculations give that distance as 1910 mi compared with the best radius of the crater of 1820 mi. This 4.9% difference may reflect the uncertainty in locating the epicenter. Who wishes to quibble? At any rate, the shape and location of Great Slave Lake are appropriate for the intersection of a linear rift and the rim of its own crater.

Kalahari. Almost at the center of the great Kalhari Desert in Boswanna, Africa. Carved coast of South Africa from Maputo to Cape Town and raised the concentric arc of

Drakensburg Mountains for 600 mi. This crater was present when the linear rift from *Timbuktu* sliced through the western portion that drifted several thousand miles away with South America. In this breakup, most of an igneous province, Karoo Traps, was also hauled off. The missing portions of *Kalahari* were shown in Fig. 1.6. Continental Trimming and described under *Montevideo*. Dry climates produce deserts in depressions that turn into swamps and shallow lakes during wet seasons. A unique illustration is the Okavango Delta in the Kalahari Desert. Seasonal rains in the Angolan plateau drain through the Caprivi Strip, enter Boswana near Shakawe sixty miles downstream, then split into several streams forming the largest inland delta. Almost all the water disappears into the sand during dry seasons. At full flow, the river creates a lake of 5000 sq mi. Southeast of Okavango Delta are flats of an ancient lake. Radiating to the northeast, a major rift system 50 mi wide runs NE through Zambia along the border with Zimbabwe and into the center of the Katanga Province of Zaire. A gentle bend begins there tying into the 50-mi wide fault filled with the waters of Lake Tanganyika. Mineral wealth of South Africa has been noted elsewhere.

Kara. The Kara Sea north of Russia is part of the Arctic Ocean. Distinguishing feature is a peninsula at Khabarovo and a 600-mi arc of the major islands of Ostrov Vaygach and Novay Zemlya. This western half of the rim has been distorted from a true arc and the eastern portion can not be found. In the vicinity of the center on the Gydamsky Tolustrov Peninsula, the land is too convoluted to interpret. The northern 300 mi of Ural Mountains are virtually straight and aligned to the center of the crater. South from a kink in the Urals, the mountains have been raised along the rim of another huge crater in Western Siberia. The Urals are by far the largest concentration of valuable ores in all of Asia. Production and reserves include petroleum, zinc, uranium, bauxite, copper, phosphate, gold, lead, and nickel. However, the short stem south of *Kara* is barren. Crater rims produce geological changes favorable to mining whereas linear faults do not. The elongated Novaya Zemlya has mountain ridges along both sides of a valley filled with snow. Their spacing of 49.0 mi corresponds with a rim ratio of 1.137°.

Kazakh. Faint evidence indicates a crater covering most of Kazakh S.S.R. The southern rim is marked by the curving Alay Range between Kirghiz S.S.R. and Tajik S.S.R. that contain two of the highest peaks in U.S.S.R. The highest point was thought to be Lenin Peak until 1928 when another nearby was found to be a few feet higher. It was named Stalin Peak (24,599 ft) but shifting political climates required a change to Communist Peak as it is known today. Very often the highest peaks anywhere on earth are found along rims of megacraters. The Irtysh River to the east follows the boundary of Kazakh-Siberia that continues to the southern tip of the Urals where a pair of short ridges is found. Completion of the circle is the western shore of Aral Sea and the Amu Darva River along the Uzbek-Turman border. The Aral Sea is itself another shallow depression that filled with water to become the fourth largest lake in the world. Many of the major seas and lakes have been created by impacts. The name derives from "Sea of Islands" because of their large number, namely 1130 larger than 2.5 acres. This geological oddity is not known to the author to have been previously explained. The sea supported extensive fishing and shipping industries. But diversion of input streams and progressive reduction of rainfall have virtually dried it up. Small freighters that used to ply the lake can now be seen lying in the middle of a desert. Aral Sea must be suspected as a crater although no evidence of a rim can be found. As this event was relatively modest, thick continental crust probably withstood the onslaught without failure of the crust. Ground zero is 45.17N 60.02E but a diameter of 125 mi disqualifies it from the fraternity of megacraters.

Khabarosk. Situated in eastern Siberia fronting on the Sea of Okhotsk. The coastal arc of 430 mi runs from the city of Okhotsk south to Antykan backed up by the concentric mountain range of Khrebet Dzhughur. The small island of Feklistova is properly located and oriented to be part of an outer rim giving a ratio of 1.099. Geographical features inland consist of valleys: NE - the upper reaches of the Lena River. S- the valley holding lake Zeyskoye Vodokkhran. This is a young crater as satellite photos show a complete circle of distinct changes in vegetation. See Fig. l.21. Siberian Cluster.

Kimberley. This crater is the volcanic Kimberley Plateau in northwestern Australia, a very remote region with few people. The center lies in a circular depression 160 mi in diameter inside the 440-mi crater. A double rim to the east is indicated by the Duiac Range and the Carr Boyd Range. A faint double rim is seen in satellite photos to the south west 27 mi apart. Lower reaches of Fitzroy River arc around the outer rim to empty into King Bay. About one third is flooded by the Indian Ocean south of Timor island. Gold discovered in 1885 launched minor production. Modern exploration was delayed until the 1950s and continued into the 1980s. Striking diversity of mineral discoveries include nickel, niobium, tantalum, flourite, tin, beryllium, and tungsten. A world class source of diamonds was found at Argyle. Even craters of moderate size can rearrange the crust, trigger massive volcanism, and expose mineral wealth.

Klondike. In the southeast corner of Canada's Yukon Province west of Lake Frances. Not especially significant except for modifying the landscape. Raised the Franklin Mountains in Northwest Territories and established the course of the middle third of the MacKensie River. Created a narrow linear fault as the course of Tanana River. An interesting aspect is the company it keeps as illustrated in the accompanying sketch. The comet landed nearly on top of an earlier crater, obliterating about one third. This nearly buried crater of 340- mi diameter is centered at 60.47N 129.330W. It will be listed as ***Under Klondike***. Even more interesting is a string of small craters labeled as A through D with the following parameters. They are included in the accompanying sketch.

A	59.15N	114.27W	100 mi
B	60.38N	120.26W	210
C	60.47N	125.88W	140
D	61.23N	130.15W	60

Several factors suggest they may be a chain of fragment craters. Close proximity to the average latitude of 60.41° seems correct. A slight northerly trend could come from a low angle of attack. Notice that the sizes are about the same and the last three decline in diameter. Also the time intervals between impacts would be about the same. These factors reflect many photographs of strings of fragments in the night sky. If they were a true chain, they have to be younger than ***Klondike*** because C and D landed inside the larger crater. Otherwise, their minor record would have been destroyed. These will be called ***Klondike Cluster*** but they are too small to treat individually and overlooking them will have no influence upon the statistics. Crater chains on earth are expected to be close together because the rate of rotation is slow. Tectonic drift need not be considered for craters traveling on the same platform. Even so, the identity of these craters must remain open.

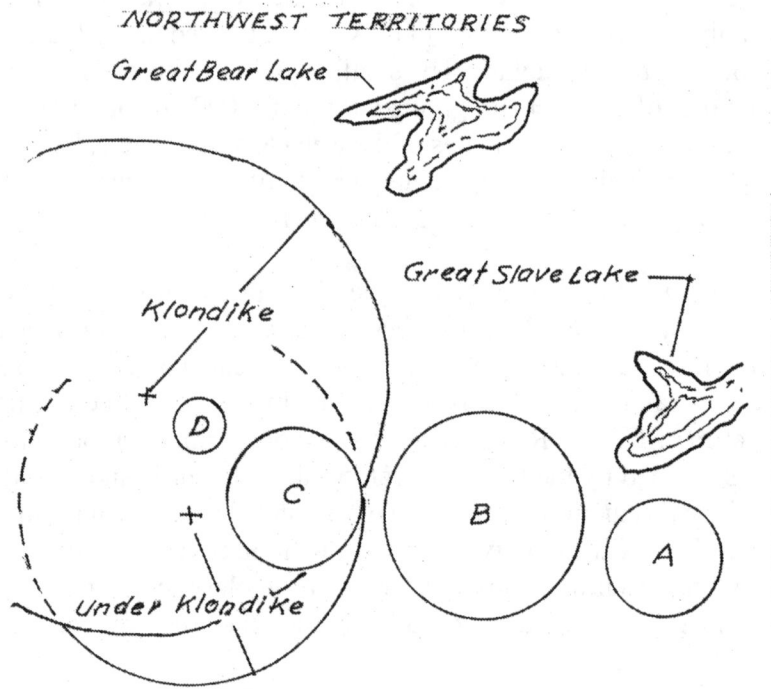

Kuril. Center is close to the town of Tomtokan in Siberia 150 mi inland from the Sea of Okhotsk. This very powerful explosion of 1980 mi diameter had profound effects in the Pacific and Arctic Oceans. Absence of any clues in Asia is attributed to the steady and massive bombardment of the continent. It created the Kamchatka Peninsula and 56 islands in the Kuril arc running for a total of 1500 mi. Overthrust in the Pacific created the Kuril Trench (34,596 ft deep) where subduction triggers endless earthquakes. The entire arc contains 227 volcanoes with 25% remaining active. Guyots and hot springs abound. The line of volcanoes in eastern Kamchatka include Klychevskaya Sopa at the terminus of the Aleutian Islands. See Fig. 1.19. Bering Twins. Parallel mountain ranges down the spine of the islands, Svedinning and Vostochny, are on the rim. Rim ration of 1.065 is established from their separation of 51.0 mi. The double rim terminates at the southern tip of Kamchatka and only a single arc of islands are present. An exception exits off of Hokkaido, Japan where elongated islands are parallel. Kunashu and Shitotan along with numerous, satellite islands. At 45.0 mi apart, compared to 51.0 mi between the Kamchatka ranges, resets the double rim. A major fault runs north along an uncertain path to the mouth of the Lena River. From that point it extends into and across the Arctic Ocean to the NE tip of Greenland. Tectonic forces have widened this 1500-mi crack in the Arctic Ocean to an average width of 280 mi making the dominant, geological feature. Southward from Greenland the fault becomes the Mid-Atlantic Ridge.

Lena. Near Batagay, Siberia where the typical pattern of interior rivers coverge. Tributaries are Adyche Dulgalakh, Nel'gese, and Boralukh. Final run to Arctic Ocean by the Lena River picks up the flow from Sedennyakh River for total area drainage of 280,000 sq mi.

182

One inner rim with radius of 300 mi in the Verkhoyansk Range appears to be failure of the crater wall. Absence of overthrusting and attendant, double rim is appropriate for a continental impact. The mountains control the 750-mi length of the Lena River that forms a perfect arc approaching 180° from Khandyga to Bulun. The western slope of theVerkhoyansk Range provides radial feeder streams of about 100 mi that follow radii from the crater center. This crater corresponds with one of the great lava flows of the world, Siberian Traps. The area within the crater is about 300,000 sq mi. Recent discoveries found the lava under deep sedimentary rock to extend to the Ural Mountains pushing the area to about 1,000,000 sq mi. The corresponding volume of lava is about 600,000 cu mi. Several different methods of analysis agree on the age of the Siberian Traps as 500 My. That figure is in agreement with the boundary of the Permian - Triassic geological periods marking the greatest extinction of life in earth's history. A link between the extinction, Siberian Traps, and a megacrater is compelling. Wide-spread deaths have been attributed to release of noxious gases, a theory that has garnered much support. Several aspects of the Siberian Traps were covered in Chapter 1, Fig. 1.22. Russian Mammoths and Chapter 2, Table 2.1. Exterminating Craters. All of the consequences of megaexplosions must be considered in concert to explain extinctions, including searing heat, shock waves, raging winds, poison gas, seismic surges, tidal waves, nuclear winter, starvation etc.

Libya. Situated out in the Sahara Desert, far from any place with a recognizable name, about 25 mi south of the Gulf of Sidra in the Mediterranean Sea. An embryonic industry was born by the discovery of oil n 1956 in the western region close to Algeria. By 1974 when production peaked, Libya had been rated among the second tier of nations in the Organization of Petroleum Exporting Countries (OPEC). This oil field is clearly related to the crater, in fact, inside the inner rim. Rain is very scarce in Libya. The coastal zone on the Mediterranean recieves a nominal amount while the vast desert is practically nil. Widely scattered, however, two major oases provide for irrigation of up to 25,000 acres of farmland. Ancient layers of sandstone in their basins slope inward toward the fresh water springs and lakes. Huge reserves are stored in underground aquifers that have been in place for thousands of years or even much longer. Libya has begun a $30 billion project to mine this fossil water and transport it for use in cities and fields to the north. Known as the Great Manmade River, the project was planned in several stages. Phase 1, completed in 1996, obtained water from 500 wells at the oases of Kufra (24N 23E) and Sabha (27N 14E). Phase 2 will pump water in the sand-filled Mourzouk Basin. The completed project will carry 6 million cubic yards per day. A crescent of the Tassili in Adje Mountains defines the western boundary with escarpments rising to 1250 ft. A double rim can be discerned nearby providing a Rim Ratio of 1.118, a typical value. This detail implies an overthrust in a former sea with thin crust. Large reserves of petroleum have also been found in the eastern Murzuq Basin. As pointed out in other contexts, layers of sandstone slope inward to the open water of oases. That pattern represents the final configuration of sedimentary rock from a rebounding crater. Moisture from the entire crater can be collected in the sandstone and drained to the oases under an imperious layer. The reverse process drives oil upward and outward to the rim where pools accumulate against impervious walls. At 260-mi diameter, Mourzouk is borderline as a megacrater but it will be treated as such because of its economic importance. It is curious to learn that a megacrater can set the stage for concentrating both potable water and petroleum. Desert craters are not easily spotted in satellite photos because of the uniformly light, tan color. However, faint indications of two small craters in southern Libya deserve some attention. Because oases develop at the center of depressions, their presumed origin is a cosmic event. So all oases should be studied with that in mind, particularly, the historically important Siwa in the Egyptian desert.

Luknow. One of several craters in India about 500 mi in diameter related to the Deccan Traps. Treated along with others under *India*.

Madagascar N. The large island east of Africa in the Indian Ocean. Known to be of continental material, popular fantasy says that it is a fragment of a lost continent. This hypothetical source has been called Lemurian in recognition of the small arboreal animal found there but no where else. Of course, it is not. More knowledgeable observers fully recognize it as a fragment from Africa but the exact site of its birth is somewhate uncertain. A new perspective offers the solution. While the island is peppered with tiny craters, three significant ones participated in the severance as shown in the sketch. Reviewing that map now will simplify the following description. Statistics for each one are included in the master tabulation. Let it suffice to point out their diameters as *Madagascar N* (330 mi), *Madagascar SW* (240 mi), and *Madagascar S* (210 mi). Restoring the island to its original position shows a snaking rift system established between two rims on the island and one on Africa that is *Mozambique 1*. The axis of Lake Nyasa is well aligned with *Madagascar N* and may be its linear rift. A dichotomy thus prevails between *Victoria* and *Mozambique* as the source of the lake with evidence favoring the latter. Pushing Madagascar back into is old slot is not easy because the shores do not always reflect the bedrock reality. Sediment has unevenly extended the island and continental shores. This knotty problem is unraveled by examining impact structures whose fragments are found at both sites. For more details, see the write ups on *Zimbabwe* and *Mozambique*. The former is rather large at 680-mi diameter. Its rim formed the narrow and steep-sided gorge of the Zambesi River that has been dammed to create a chain of large lakes. With the island still attached, the Zambezi gorge continued along the rim creating a N-S rift valley at the north end. It appears to be straight but is a short section of long radius. That feature can be accurately restored by placing the north end of the island at the apex of the Zambesi delta. The southern portion can also be reset by recognizing the geography caused by *Mozamique*. This smaller crater of 280-mi also penetrated Madagascaar leaving its own arc. It raised the Massif du Makay and the valley of the Manandrea River. These details are illustrated in the sketch by dotted extensions of the African craters. Completing the precise relocation of Madagascar exceeds the present intent. The crater, *Madagascar SW*, nearly overlaps *Madagascar S* so it is omitted from the sketch for simplicity. Further comments are provided under *Zambia*.

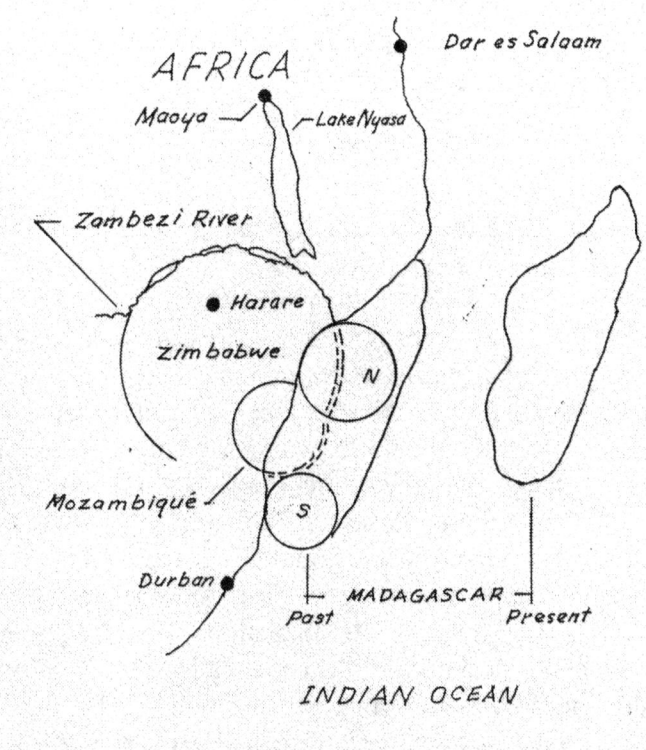

AFRICA

Dar es Salaam

Maoya — — Lake Nyasa

— Zambezi River

• Harare

Zimbabwe

N

Mozambique

S

Durban

MADAGASCAR

Past

Present

INDIAN OCEAN

Madagascar S. One of three sizable craters that are best described together in *Madagascar N.*

Madagascar SW. One of three sizable craters that are best described together in *Madagascar N.*

Manaus. In northwestern Brazil about 270 mi west of Manaus on the Amazon River. This crater is the entire Amazon Basin that is extremely well defined. In the north by the coast of French Guiana on the Atlantic to Venezuela on the Caribbean. The western rim is the entire AndesRange from Colombia to Arequipa, Peru. In the typical pattern for megacraters, six major rivers converge at the epicenter near Manaus and continue to the ocean in the major river while picking up the discharge from minor rivers. The total area of the basin drained by the Amazon is about 11,000,000 sq mi that is, of course the area of the crater. A double rim is found along the entire crescent of the Andes whose spacing varies considerably. Perhaps most representative is the separation of Cordillera Oriental in Bolivia and the southern section of Cordillera Occidental, namely, 200 mi. The resulting Rim Ratio is 1.249 although in Ecuador it is only 1.096. The latter value seems more realistic but there is no reason to think that the value would remain constant over such an extensive rim. The rim annulus in southern Peru contains Lake Titicaca, the remaining reservoir of a much larger, ancient lake of Bolivia that previously included the present Lake Poopo about 48 mi SE. This region, known as Alto Plano, lies between the Coastal and Eastern Andes whose origin has become clear.

Manchuria. Close to the northwestern border between Manchuria and Soviet Union at the same latitude as Sakhalin Island in the Sea of Okhotsk. Inner rim is the arcing Khrebet Sikhote range concentric with the coast. This 800-mi arc runs from Nakhoda close to

Vladivostok northward to the end of Tartar Strait. The northern end of Sakalin is the original outer rim although distortion or shifting has influenced the southern part. Very little remains of this crater inland because the land has been torn up so badly. In all probability, the explosion opened the Sea of Japan forcing the southern part of Sakhalin away from the Chinese coast.

Maracaibo. The largest lake in South America located in western Venezuela immediately south of the Caribbean Sea. Massive explosions have shallow basins in which layers of porous and watertight rock slope upward to the periphery. Any oil in the porous zones percolates over millions of years outward until blocked by an imperious wall in the mountainous rim. So it was with *Manaus*. Petroleum has been found at the expected places from the mouth of the Orinoco River on the Caribbean around to Ecuador on the Pacific. A unique event modified this pattern at the north end of the Andes by landing squarely on the inner fault of the Andes. It blasted out the sedimentary rock under a lake 190 mi across to a depth of 4 mi. The rim of this crater is preserved by the lofty Sierra de Perija on the west and Cordillera de Merida on the east. These mountains completed the circle to the north as presently shown by two peninsulas separating the Gulf of Venezuela from the Caribbean. The resulting accumulation of oil under Lake Maracaibo is disposed along a zone 65 mi long and 20 mi wide exactly where it had been collected eons before. Removal of overburden by this secondary explosion placed the oil within reach of modern, drilling rigs. Compare these details with similar circumstances concerning the coastal oil fields in eastern *Venezuela*. Lake Maracaibo had a turbulent history involving pirates and civil wars. Until 1917 it was only a small coffee port. Then discovery of petroleum stimulated phenomenal growth. This basin is the most concentrated source of oil in the world and a huge petrochemical industry has been built at the city of Maracaibo. Products from these installations go into an enormous variety of consumer goods from soap to rope. Two lessons are learned from this case. Vast wealth is found where a small crater is formed on the rim of a giant. Benefits to mankind are endless.

Mariana. An island arc in the western Pacific including Guam and Saipan. Both are major military bases but quite different from the time of the author's visit during WWII. The islands consist of limestone that was dredged up then punctuated with volcanic peaks. Outboard is the famous Mariannas Trench with the deepest place on earth at 36,201 ft below sea level. No outer rim is noted. Typical earthquakes associated with subduction for a zone about 220 mi wide indicating the extent of subduction before melting of the slab.

Mariana E. Situated 110 mi east of the Mariana Islands consisting of a ring of sea mounts. Naturally, would be islands at lower sea level. Of statistical interest only with out geoligical significance.

Mato Grosso. Crater in southern Brazil, slightly distorted into an oval of about 620,000 sq mi. Double rim at Atlantic Ocean is the Great Escarpment remaining undistorted for 1200 mi between the cities of Salvador and Cuaitiba. Steep slopes are especially impressive inland from Sao Paulo where a masterpiece of civil engineering provides sweeping freeways with lengthy bridges spanning precipitous canyons. Farther inland is the Brazilian Highlands at average elevation of 2, 500 ft and peaks over 9,000 ft. The entire region is a world-class, igneous plateau. It constitutes a major part of the Kara Traps formed before the divorce from Africa. A small portion remains on that continent. The western rim is Sierra de Amambo

Mountains in Paraguay. Too messy to see anything up north. Rich in gold, diamonds, and semi-precious stones providing material for a thriving, jewelry industry in Rio de Janeiro. The Rio Grande and Sao Jose dos Douzados become the Parana River. While collecting other rivers in the basin, it follows a straight fault to Posadas. Downstream, the course follows a ragged arc around the Kara Highlands to Rio de Janeiro-Montevideo. A 210-mi crater sits inside this one at 18.47S 56.60W. It seems to be a plain atop a former lava lake. A recently discovered and ancient civilization there supported a large population from many rivers and hundreds of miles of irrigation canals.

Mexico. Center is a small circular structure 60 mi SW of Mexico City, former capital of the Aztecs, Teno Chtitlan. The latter, built on an island in Lake Texcoco, was overrun by Cortez with a small force and an army of sympathetic Indians. Growth of the city under Spanish rule reduced the lake's size and excessive pumping from wells for the large population has essentially dried it up. Furthermore, extraction of water has caused severe subsidence in the modern city. Distrito Federal contains extensive lava beds and Popocatapetl contines to smoke and rumble. It is young compared to the crater. Along with others nearby, the mountain grew because of damage compounded by a linear rift from another source. See ***Trans Mexico*** and ***Springfield***. The rim consists of Sierra Madre and Sierra Madre Oriental with pronounced curvature. Balsas River flows inside the rim.

Mexico Gulf. The circularity of the Gulf of Mexico has been noted by several authors as indicating a possible impact crater. That suggestion was too outrageous for mainstream geologists. It is logged in here with a diameter of 850 mi that is not overly large in the present company. At some time when the sea was higher than now relative to North America, it may have been much larger. Large portions of southern U.S. would have been flooded. The rim may have reached a) the Sierra Madre Oriental in eastern Mexico, b) Oachita Mountains in southern Oklahoma and Arkansas, and c) the entire Florida peninsula. In that case, the diameter would have been 1120 mi. Considerable detail has been presented concerning the oil fields along the coasts of Mexico, Texas, Louisiana, and the projected arc to the SE in deep water of the Gulf. A gap in production is noted along the Mexican coast from about Cuidad Victoria to Veracruz. But oil fields there are known to experts who predict them to be largest reserve in the world. This continuous arc of potential fields is 1400 mi long. Their location and that of others around the Gulf confirm the originally suggested diameter of 850 mi. A central peak is Sigsbee Knoll. Spacing between the center and the Tropic of Cancer, only 0.50° may have some hidden importance. A linear rift struck off on the azimuth of 140° setting the stage for a straight line of islands. They are, of course, Cuba, Hispanola, and Puerto Rico. Islands of the Antilles were created by a separate explosion dubbed ***Antilles***. A strong similarity exists between ***Mexico Gulf*** and ***Indonesia*** because each created a straight chain of islands to the southeast. Respective rift angles are - 76.0° and - 71.5°. They are far outside the earth's present inclination and must be very old.

Michigan. Almost dead center on the Michigan Peninsula in the U.S. This one is a tough call because evidence suggests the site of two craters of greatly different size. Basement rock under the peninsula has retained a circular basin in the Precambrian floor plunging to 13,000 ft below present grade. This is unmistakable proof that something very powerful dented the earth. The rim is easily followed. The outer rims are ***Wisconsin*** shore of Greenbay. Inner rim is the Door Peninsula and Washington Island along with similar structures to the north. Southward, the west coast of Lake Michigan lies on the inner fault.

Northern portions are defined by islands of Drummond Cockburn and Manctoulin separated from the mainland by North Channel. In the NE, a chain of small islands to the tip of Bruce Peninsula is on the rim. A curious accident of nature landed an asteroid on the outer rim at Sudbury, Canada. That tiny crater of 125-mi diameter surrounds the city of the same name. So the previous damage from *Michigan* was augmented at Sudbury. All theories of its origin except asteroid impact have been abandoned. The originally circular crater has been distorted by tectonic forces. The site is a metallic cornucopia spilling out gold, silver, platinum, palladium, nickel, copper, and iron plus a host of others. It is the world's largest source of nickel and copper. The value of one outcropping of iron is estimated at $30 billion.

Mid-Pacific. The idea that the Kermedec-Tonga Trench may have extended north beyond Somoa is reinforced the existence of the North Tokelau Trough, a non-subducting zone with the correct alignment. Beyond the cluster of the Phoenix Islands lies a vast ocean to Hawaii that oceanographer call Mid-Pacific Basin. Little suspicion has been raised that it might hold the record of cosmic origin. One is appropriately astonished upon noting the closest island groups, prominent seamounts, and isolated islands lie upon a circle 2580 mi in diameter. This geography is illustrated in Fig. 1.10. Vast Ocean. Major Hawaiian Islands are found on the northern rim excluding the extended chain to Midway. Starting at Hawaii, the rim can be traced counter clockwise by Resolution Guyot, Ewing Seamount, Marshal Island, Kiribati (Gilbert), Tuvalu, Tokelau, Samoa and, farther east, the isolated Maniki and Pernrhyn. To be part of rim, they must all have a common center, give or take 50 mi or so. Elongated islands must also expose ridges perpendicular to their radii from the center. Finally, elongated islands should show a double rim. Layouts of island groups in Fig. 1.10. are shown by inserts of arbitrary scale. Marshal Islands are dominated by two lineations, appropriately E-W of Ratak Chain and Ralik Chain. Their separation would correspond to an acceptable, rim ration of 1.161. The large number of islands prevents displaying names but three most likely to be recognized have been visited by the author; Eniwetok, Bikini, and Kwajalein. While individual coral islands may be small, they are frequently mounted on the rims of huge volcanoes. The former island of Kwajalein was larger than Oahu but smaller than Hawaii. It had beaches that now enclose a lagoon covering 840 sq mi! Coral atolls now mark the lost beaches. Such islands were built by massive volcanism. The chain of Kiribati Islands is seen on the inner rim while Ellis Islands are on the outer rim. Also on the smooth arc of the outer rim are Savii and Uplu in Western Samoa, Tutila and Manua Islands in American Samoa. The last exposed land to the southeast is Maniki and Penrhyn. As in other great explosions, the crust was ripped open in the weak ocean bottom along a linear fault. In this case to the southeast as indicated by the Line Islands, Christmas Ridge, Tuamotu Ridge, Archipelago of French Polynesia thence to Pitcairn at a total distance of 3050 mi. The azimuth from Palmyra to Pitcairn of 136.9° gives a Rift Angle of - 43.1°.

Montana. Central Montana in Great Plains 180 mi NE of Lake Manato NE of Billings. Typical pattern of rivers with Yellowstone and Little Missouri flowing into the Missouri. Rim clockwise from high noon is the lake filling an annulus between rims, Devils Tower, and Little Belt Mountains. This basin is a strong candidate for sediment spread over a former, lava lake. If drilling records show lava at some depth, the age of impact could be estimated. Present techniques could accurately determine the ages of recovered samples of lava. Measurements for bottom and top layers of lava would give the time of impact and the duration of lava flow.

Moscow. One of the largest megacraters about 80 mi NE of Moscow, Russia that stretches from the Atlantic Ocean to Ural Mountains. Rim includes north coast of Finland-Sweden-Norway. Created Baltic Sea between Yugoslavia and Italy then the islands of Crete and Cyprus. Concentric mountain ranges run for 920 mi from Tashkent to Urumqi where spacing gives an expected Rim Ratio of 1.066. Eastern part was penetrated by Ural Mountains. Large, igneous outcrops are near Moscow.

Mourzouk. A 500 mi crater in southwestern Libya. Recognized as a basin called Edeyin al-Mourzouk. Photographed from a U.S. spacecraft released as NASA-Photo:66-HC 1757 (S55-54525) Gemini. Extremely clear image of the crater accompanied by a geological interpretation. This basin is only the central part of a double-rim crater of greater size. A rim ratio of 1.522 is not registered because this one surely collapsed internally. A great deal of research is required to isolate this type, determine the minimum size for failure, describe and compare all known examples, and establish the physics of the process. These structures must be confined to continents because large explosions at sea create island arcs. Complex combinations are found for impacts on land whose rims reach out into the ocean. Not at all surprising that many island arcs lie close offshore.

Mozambique. A 280-mi crater in southern part of the country that created the bulging coast north of Maputo. Hardly deserving attention except for its participation in the withdrawal of Madagascar from Africa. Refer to the discussion and sketch under *Madagascar N*.

Nanjing. Mid-sized crater 70 mi north of Nanjing, China. Carved east coast of Korean Peninsula and west coast of Kyushu Island in Japan. Overthrust in the Phillipine Sea lifted arc of 600 mi from Ryukyu Island down to, but excluding, Taiwan. Depression of the sea floor by weight of the overthrust slab initiated the Ryukyu Trench. Island positions at Gunto and Okinawa indicate a double rim. Separation by 50 mi gives a Rim Ratio of 1.047. Inland terrain has been so chopped up in China that remains can not be found on the continent.

Navajo. Named for the Navajo Indian Reservation 32 mi west of Gallup, New Mexico. The crater is actually the Colorado Plateau that is obvious from orbit. Southwestern boundary is Moggollose Ridge bordering on Coconino Plateau. Northern rim indicated at Abajo Peak. Rio Grande River flows outside of Sandia Mountains west of Albequerque. Near the center is the Painted Desert so famous for brightly colored minerals and a petrified forest. A forest was leveled then covered by debris falling back into the crater. Thus protected from decay, minerals replaced wood. Now exposed by erosion, petrified tree trunks and fragments are scattered across the desert. The extent of the original forest is not known. But details of the explosion may be deciphered by studying the layout of ancient trees. This small crater apparently did not bring up mineral deposits of commercial interest.

Noril'sk-Yakutsk. A powerful event in Siberia that was adequately described in Chapter 1, Fig. 1.22. Russian Mammoths. Critical implications for the flood basalt held back by ridges at Lake Baikal and Paton Plateau. Also blocked by a ridge at the Lena River. Recent discoveries indicate its probable extension lay hidden under sedimentary rock an additional 1100 mi west to the Ural Mountains. The Great Dieing dated at 250 My corresponds with the age of the lava with remarkable accuracy. Even the size of this crater (1650 mi) seems inadequate to account for the worst loss of life in earth's history. Could it have had some help? Unfortunately, an answer is hard to find. Over extremely long periods, previous continents aglomerated into a single super-continent. North America, South America, and Antarctia had not started their journeys at that time. Consequently, present latitudes of craters are of doubtful value in reconstructing a crater chain linked to Noril'sk-Yakutsk. Some fragments, however, may have landed in Eurasia. Restoring a world map to show the original Godwanaland would be helpful. Even so, changes in earth's inclination and spin axis introduce annoying complications. Any search should consider the crater shapes. For small asteroids whose energy and momentum are carried by their velocities can produce non-symmetrical craters. On the other hand, if falling vertically, they produce circular craters. These, however, are too small to contribute to mass extinctions. Aerial explosion of comets always produce circular craters. The kinetic energy and momentum are inconsequential. Forces from explosions radiate equally in all directions. The question of their approach angle is mute but most are thought to have descended vertically. Some possible candidates for supporting *Noril'sk-Yakutsk* in the Great Extinction are

Moscow	3330 mi
Hudson	1200
Aleutian E	1340
Siberia 2	710

The order of impacts need not have started with *Noril'skYakutsk* but the distribution of sizes would be acceptable. The major offender could follow smaller fragments but not by much. Such a compound event would certainly devastate all life forms, especially in the northern hemisphere. This limitation would be evaluated by comparing the surviving species in the northern versus southern hemispheres. A great deal of uncertainty is included here but the question should be pursued. It will take a lot of research. It took two maps in Chapter 1 to indicate the most significant craters in Russia/Siberia without committing chaos on paper. See Fig. 1.21. Siberian Cluster and Fig. 1.22. Russian Mammoths. But that is not all. Many more can be detected from orbit but their boundaries can seldom be associated with specific features viewed from the ground. But most of the largest ones have been noted here for the record. Their assigned coordinates are not as accurate as implied by the number of digits.

North Island. This crater pertains to the Tasman Sea west of North Island, New Zealand. Coordinates are given for the present location although it has migrated very far in the last 33 mil yr. About half the rim includes the long peninsula past Auckland while the western portion remained in Australia. Matching the rims provides an exact reconstruction of the primeval continent and an estimate of when the missile struck. Most of it is now flooded. For details about the origin of this continental island refer to *Artesian* and compare it with the birth of another continental island, namely, *Madagascar N*.

North Sea. A small sea between islands of the United Kingdom from Norway and Denmark of interest primarily as a petroleum reserve. Its center lies on the rim of another crater, originating in central Ireland that opened the English Channel among other accomplishments. Probable remnants of the rim are the mountains north of Glasgow to the tip of Scotland and the island groups of Orkney and Scotland. The Hebrides group to the west indicates a double rim with ratio of 1.259 whose abnormally large value suggests failure of the crater wall on the east of the Denmark peninsula. This is one of several examples sharing a unique feature. Vast oil fields are found concentrated at the point where one crater is created on or near the rim of another that is much larger. Consider

North Sea	*North Sea* on *Ireland*
Middle East	*Persian Gulf* on *Iran*
Eurasia	**Caspian S** on **Caspian N**
India	*Aurangabad* on *Iran*
South America	*Maracaibo* on *Amazon*
Southeast Asia	Many rims in Indonesia and Burma

In addition, Southeast Asia is a critical source of oil from a battered landscape. Oil from Indonesia is on the inside of the rim where expected. A prospector armed with this information has an advantage over competitors.

North Slope. Buried under the Arctic Ocean north of Alaska. Rim is the Brooks Range with Rim Ratio of 1.166. The coastal region, known as the North Slope became the focus of public attention during construction of the Alaskan Pipeline. Grave issues were raised about oil spills, damage to tundra, migration of caribou, and spawning of graylings. A thorough and expensive monitoring program was successfully conducted by contractors and government teams. Prudoe Bay was the source of oil that was delivered to Valdez on Prince William Sound. A tragic accident of an oil tanker there required many years and millions of dollars to restore the ecology. East of Prudoe Bay up north is the U.S. Naval Reserve that is under attack for commercial development. Farther east, the U.S. Wildlife Preserve is in a flurry of debate. Again, a major oil field is found to lie inside the inner rim of a megacrater, in this instance, with a diameter of 790 mi. It would be an interesting project to determine what fraction of all major fields is linked to megacraters. Whatever the figure, it is high.

Northern Australia. Large crater centered 100 mi NW of Alice Springs. Nearly circular uplift including Mt. Ziel. Includes Ayers Rock 150 mi south whose vertical layers could indicate a crater rim or an enormous block that was spaded up and tipped over.

Panama. In the Gulf of Panama south of Panama City. The change in curvature in the middle of the country is typical for adjacent craters, in this case in the ***Pacific*** and ***Caribbean***. A convenient place to dig and construct the Panama Canal was provided at this flexion. A fuller explanation of the geography of northern Panama is provided in ***Antilles***. Uplifts within the crater are the Archipelago de las Perlas. Rim encompasses nearly all of northeastern Panama from Santa Catalina in the west to the eastern border with Columbia. It extends another 50 mi into Colombia at the west coast of the Gulf of Uraba. Primary rim features include Cordillera de San Blas and its extension by Serrania del Darien. Probably a three-rim crater with the outer two being the San Blas Islands and the 30-mi projection of islands beyond Point San Blas. No subduction is implied by the lack of earthquakes.

Panatal. In Mato Grosso Do Sul of Brazil 160 mi NW of Campo Grande. Lies inside of crater ***Brazil*** adjacent to the border with Bolivia. Inconsequential but suspected as plains lying above a lava-filled lake. Named for Panatal Matogrossense National Park.

Pasco. While hardly worth mention, a crater only 80 mi across is located at 46.59N 118.32W some 40 mi up the Snake River from the city of Pasco, Washington. The southern half of the rim is vivid from space with a smooth and level interior befitting a lava lake. That would explain the extended plains across the river at the government installation near

Richland. This crater is probably the usual reference for the traditional "Pasco Basin". Properly named, of course, but it really is much larger with an estimated area of 46,000 sq mi. It seems that many, closely-spaced craters everywhere are aligned generally toward the northwest. The implication is that fragment chains are common. They must have been created during a difference regime when earth's inclination was far from the present value. Otherwise, the chains would string out E-to-W along the same latitude.

Paton Plateau. A small, intrusive plateau in Siberia beyond the north end of Lake Baikal. Of interest primarily because it blocked inundation by the Siberian Traps as illustrated in Fig. 1.22. Russian Mammoths.

Peak Hill. In Western Autralia 115 mi NW of the town of Peak Hill. Carved west coast on Indian Ocean with no signs of overthrust. Volcanoes, earthquakes, and linear rift are absent. Did not fracture the continental basement. A multi-ring structure, nevertheless, caused by internal collapse. Radii of rims are 220, 260, and 290 mi. Much further research is required to catalog all known examples and tabulate their radii. Clues to the mechanics of formation would certainly evolve. Outer two rims are displayed by a peninsula in the Indian Ocean and their projected island chains. These include Quoin Head and Pelican separated by Denham Sound and the southern part of Shark Bay. North along the coast, further evidence of the outer rim is another peninsula of Cape Range National Park ending at Northwest Cape. Separated from mainland by Eonouth Gulf 50 mi long by 36 mi wide. No discernable trace inland.

Pelly River. Center in the wilds of the Canadian Yukon with few local landmarks. East of Dawson Range and 540 mi ENE of Anchorage. Immediately to the east of this crater are three more whose rims are touching,

A.	60.37N 126.07 W	450 mi
B.	60.32N 120.25W	220 mi
C.	59.17 N 115.22W	160 mi

Only ***Pelly River A*** is registered in the catalog but the group should be studied as a possible fragment chain.

Pelly River A. The largest of three craters in Canadian Yukon that may be a fragment chain. See ***Pelly River***.

Persian Gulf. One of three craters that created, then detailed, the Persian Gulf as illustrated in the sketch. The original fault is on the rim of a massive event in northern Iran. See details under **Iran**. This modest explosion later shaped the southern shore from A Qatif in Saudi Arabia past Bahrain and Qatar to the tip of the Oman Peninsula that pokes into Asia's belly. Comparison of this sketch with the distribution of oil fields throughout the region gives one pause. They are certainly related and suggest locations for further prospecting. Beyond the Strait of Hormoz, ***Jask*** finished the job. The specific shape of the peninsula was determined by the combined efforts of ***Persian Gulf*** and ***Jask***.

Philipine. To be distinguished from the vast ***Phillipine Sea***. Center is in Sulu Sea with primary achievement of raising the Philippine Islands south of Manila on Luzon and continuing south through thousands of islands to Davao, Mendanao. Confirmation of a crater is provided by the Philippine Trench running 700 mi on the Pacific side. Continuing subduction accounts for the many earthquakes including a 7.9 R that killed 8000 people on August 16, 1976. Another strike of that magnitude killed 1600 people on July 16, 1990. One of the nastiest volcanoes in the region is Mayon on Luzon. Since first recorded in 1616, it has erupted at least 40 times with total deaths of 60,000. Torrential rains from the plume caused devastating mudflows. During one week in April 1968, the ejected material amounted to 46 million cubic yards. Nature plays rough even in kindergarten.

Philipine Sea. In the far western Pacific near Hong Kong, this sea of ill-defined boundaries covers at least 5.5 million sq mi. The cause of the basin had to be powerful because the entire sea sits on top of its own tectonic plate. Asian shores in a circular pattern start with the southern Japanese island of Kyushu, famous for its Inland Sea SW of Kobe. Ranges bracketing the sea show a rim ratio of 1.059. The arc continues along the Ryuku Islands including Okinawa and the separate island nation of Taiwan. Original rim probably included the Philippine island of Mindanao that was distorted by other explosions. The eastern boundary is not so clear but reaches eastward to the Marianas although they were raised separately by a smaller event. The circle is completed by the Bonin Islands.

Polar Islands. Center is large Victoria Island in the Canadian artic. Defined primarily by waterways of Amundsen Gulf, Coronation Gulf separating Wollaston Peninsula from mainland, McClintock Channel between Stefansson Island and northern Victoria, and Prince of Wale Strait. The strait is a perfect arc with uniform width. Both shores are used to calculate a Rim Ratio. Victoria is the 10th largest island in the world, 92% the size of Great Britain at 81,930 sq mi. The entire island may be the central uplift of the crater.

Prague. A small crater covering most of Czechoslovakia with center near Prague. Nearly all the rim consists of Erzgebirge Mountains and Bohemian-Moravian Heights. A perfect example of a 190-mi baby. Some scientists have argued that it was, indeed, created by impact of an asteroid. Their current activity is to search for spherules, tektites, and shocked quartz to prove the point while other tasks might be more challenging.

Quebec. Located in the Laurentian Mountains of Canada's Quebec Province north of the capital, Quebec. Strong indication is Lac Mistessini to NW that is 840 mi long. It is a perfect arc associated with concentric ridges. One, in particular, is very narrow running the full length along the axis. Other lakes on the rim to the south are Lac St. Jean that drains into the Saguanaya River. Resistance of the ancient Canadian Shield prevented overthrust so no Rim Ratio exists. An asteroid struck squarely on the rim that had already been deeply plowed. Aerial views show an annular lake with a central landmass. With an original diameter of about 100 mi, it is one of the largest asteroid craters still preserved. This is another example of a small object striking the rim of a larger one. Located at 51.38N 68.70W and dated at 210 ±4 My, this asteroid crater, Manicouagan, has become famous. It displays an annular lake between rims that is especially striking when frozen and covered with snow.

Queen Maud. One of four megacraters that sculpted Antartica as depicted in Fig. 1.8. Seventh Continent. Center is 650 mi from South Pole in the direction of Madagascar (Sri Lanka). Established Atlantic coast of Queen Maud Land over 80° of longitude from Cape

Norgevia to Cape Darnly at the Amerg Ice Shelf. A continuous arc of mountain ranges follows the coast with varying setbacks up to 100 mi. None are volcanic. Glaciers creep to sea between the peaks. This geography is the unmistakable signature of a megacrater with the mountains as the inner rim and the coast as an inter-ring fault. Antarctica rides isolated on its own plate so the coastal fault created the Southwest Indian Ridge. Seafloor spreading on the ridge has moved the outer rim 2200 mi away on the far side. Maybe some day, this missing rim will be found.

Rain Forest. In the Amazon Basin of Brasil's Mato Grosso coincidental with the overlap of *Manaus* and *Brazil*. Closest city is Alta Floresta. Younger than the overlapping neighbor. Part of the Amazon jungle, primarily detectable by different vegetation. Rims are faintly seen but cannot be identified with recognizable landmarks. Candidate for a prominent chain of craters across South America that may have started in Africa as described in Chapter 7.

Resistencia. In central Paraguay near the city of Resistencia in Gran Chaco. Vast plains of 280,000 sq mi devoted primarily to cattle and cotton. Poorly drained region is largely defined by water as Bovien Swamps north and Parana River east. Western boundary, however, is Sierra de Cordoba, a secondary range in the Andes. Large but of limited interest.

Rio Grande. Not the river. This geological feature is a large circular intrusion in the South Atlantic about 880 mi off the Brazilian coast. Alternately known as Bromley Plateau. Covered briefly in Chapter 1 and shown in Fig. 1.14. Manaus And Andes. A major blemish on a smooth bottom covering 160,000 sq mi. Its cosmic source cut the bowed bay from Rio de Janeiro to Flarianopolis. Steep rise of coastal mountains at Sao Paulo correspond well with the western face of *Rio Grande*. Something strange took place there. No overthrust was achieved as in *Antilles* and *South Sandwich* because this formation was still part of South America. Continental drift continued in these latter examples, leaving pairs of latitudinal faults separating the undisturbed slabs from the moving, ocean floor on both sides. No such details are seen at *Rio Grande* although in all three cases the central portion of the craters became anchored as stable structures. Long-term lava flow built the plateau in contrast to spreading out along the ocean floor. Part of an explanation must be that the island arcs were caused by an impact at sea whereas *Rio Grande* struck a thick continent. Because *Rio Grande* has always been under water, it has never been exposed to erosion as in plateaus on land. So the pristine formation has been preserved. Studies are in order to compare chemical composition of the Sao Paula escarpment with samples for this plateau.

Ross. Named for the Ross Sea in the Southern Ocean adjacent to Antarctica. It was discussed in some detail in Chapter 1 with the map of Fig. 1.9. Birth of Australia. Little more is required here except for some details about New Zealand. The linear rift running northward has made the entire nation intensely volcanic. It runs along the Southern Alps of South Island with its peak of Mt. Cook over 12,000 ft. A short branch enters North Island that is only a truncated spur from a separate source. The main fault travels through the island's axis. This active strip is typical of linear fault systems with a width from 18 to 40 mi. It is known as Taupo Volcanic Zone for its volcanoes concentrated along a path of 160 mi. White Island off the northern coast is also included. The valley is bracketed to the east by Ikawhenna and Kaimanawa Ranges while on the west by Hanuhugarcoa Range. A host of hot springs steam in the valley including the Rotorua District only 120-mi bus ride for tourists visiting Auckland.

Ryukyu. Large crater in China midway between cities of Shaaxi and Qingdo. Created the southern Japanese island of Kyukyu. Also the Ryukyu Island arc from Shatsunan Shoto through Okinawa to Irimote off Taiwan. This 700-mi chain displays the ubiquitous double rim but only in the Okinawa group. Distance between rims of 60 mi corresponds to a rim ratio of 1.087. As in many similar cases, overthrust of the Pacific Ocean floor generated the Ryukyu Trench.

Sahara. Center in desolate region of northeastern Africa on the border between Libya and Egypt. Most evidence has been covered by sand with notable exceptions. The rim includes the volcanic Tibesti plateau in northern Chad. It contains many mineral deposits but is too remote in hostile territory for serious exploitation. Geographical features along the lower Nile are highly suspect. The stretch below Lake Nasser follows a gentle bend as becomes a rim-controlled course. The site of Aswan Dam was selected where the river, 1800 ft wide, cut through intrusions of igneous and metamorphic rock. The river snakes along a groove in a sandstone base in a rift 10 mi to 14 mi wide. Both sides are scarps rising to as high as 1500 ft. These features re-enforce the boundaries of this crater. The original course above Lake Nasser was shown in ***Arabia*** to be controlled by another crater.

Santa Cruz. Small crater in Gran Chaco of northern Parguay 160 mi SE of Santa Cruz, Bolivia. Occupies northwest quadrant of larger ***Chaco***. Of little consequence.

Sardinia. Epicenter near northern tip of this major island in the Mediterranean Sea west of Italy. One of six craters discussed here that completely determined the shape of the Italian Peninsula. They raised all the mountains, triggered all the volcanoes, established the pattern of drainage, and created its minor islands. For ease of reference, the whole group in no particular order in a geological sense consists ot ***Sardinia***, ***Volcano Belt***, ***Tyrrhenian***, ***Alps***, ***Venice*** and ***Adriatic Sea***. They are so intertwined that little appreciation could be obtained from text alone. So a sketch is required. A circular fault from Sardinia was responsible for the bulging "calf" of the Italian "leg." The corresponding inner rim is the entire arc of the Apennine Mountains that constitute, for the most part, the backbone of the country. A linear rift departed the site to the NW where the results are impressive. The fault struck along the border of what is now France and Spain then continued to produce the remarkably straight, northern coast of Spain. The Pyrenees Mountains developed along the fault with a half dozen principal peaks close to the Mediterranean. A reverse curvature is noted among the volcanoes of western Italy. Individuals are Lardereblo, Amiata, Veslini, Cimini, Sabatini, Alban Hills (Rome), Roccamonfina, Vesuvius (Naples), Vulture, and hills beyond to top of the "heel." The slight curvature of the volcanic belt implies a long radius. And, indeed, it is, with the center 445 mi away in the Hungarian basin near Budapest. The west coast was carved by a fault providing overthrust toward the west initiating subduction. The site at that time must have been a large sea whose thin crust was susceptible to rupture. Subduction provided the pathways for molten rock to build the volcanoes in accord with the many other examples around the world. Long dormant volcanoes in the north indicate that lava flow above the subducting slab has been shut off. However, southern volcanoes have remained alive into historical times. Vesuvius splashed a pyroclastic flow of hot gas and debris to bury the wealthy Roman suburb of Pompei. Hot springs are still present and Vesuvius threatens the population by rumbles and burps. The offset of Vulture is made clear in the context of ***Tyrrhenian*** off the coast of southern Italy north of the Aeolian Islands. This smaller crater continues the Apennines in a tighter arc to the "toe" of the boot and beyond. The north coast of Sicily was established along with the creation of Mt. Etna. An inner rim is indicated by the Aeolian Islands along with Isle di Ustica and the volcanic Stromboli. An extremely high rim ratio shows a collapsed-wall crater. Near Livorno in the north, another crater raised the French, Italian, and Swiss mountain arc that is so popular among skiers, photographers, and royalty. The gap between *Alps* and *Sardinia* carries the Po River. Still other craters designed the eastern coast. *Venice* struck across the water near Pulo, Yugoslavia. It shaped the entire north end of the Adriatic Sea including the curved coast at Ravenna. It is also likely to be responsible for a linear rift to the SE that opened the Adriatic Sea. Evidence is provided by the complex cluster of long, straight islands off the Yugoslavian coast. They lie closer to shore than many others nearby that are associated with the double rim of *Adriatic Sea*. The outer arc includes Cres, Losini, and Dugiotok among the largest. The bowed shoreline of eastern Italy from Ancona to Monte Saint Angela indicates its original position adjacent to Yugoslavia before severance by the outside arc of Adriatic Sea. A 120-mi crater near Prejedor, Bosnia-Herzegovina at 45.00N 16.48E is too small to bother with.

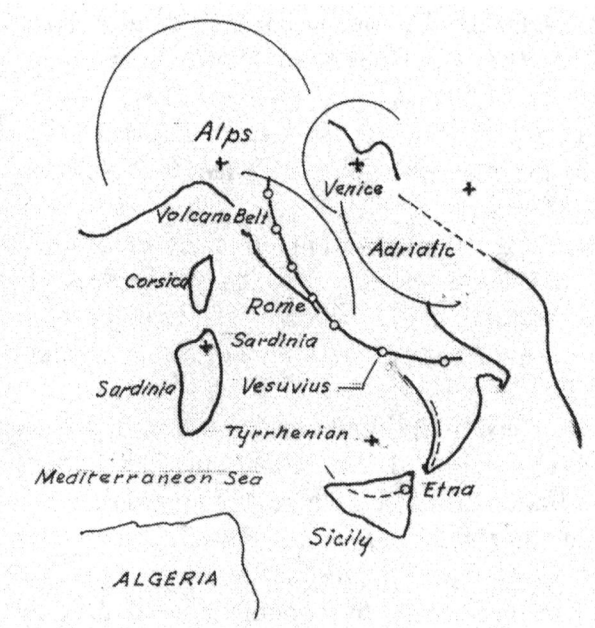

Saskatchewan. At Cree Lake in northern Sascatchewan, Canada. It is essentially marked by water between rims as follows: Lake Athabasca, Athabasca River, Churchill Lake, Lac La Ronge, Raindeer Lake. A linear rift is suggested by the wedge-shaped bay in Great Slave Lake and the narrow, linear Lake Winnipeg, equal to a rift angle of - 35.2°.

Saudi. The first clue for this central Arabian example was a photograph from an Apollo Spacecraft. To a practiced eye, it resembles a stain of tan gravy on a beige tie. The western rim is displayed by a slight bulge on the coast of the Red Sea. The linear rift from *Aegean* detoured around the crater following its weakened arc. The outboard record is the long indentation across the sea at Bur, Sudan. Widening of the Red Sea has caused Arabia to drift 190 mi NE relative to Africa. A curving range of mountains defines the inner rim between Mecca and Jidda. Clockwise from that point, the rim passes close to Medina although that detail is absent from most atlases. A semicircle to the east follows the Ad

Dahna past Riyadh that is inside the crater. An outboard valley runs 700 mi about midway between Riyadh and Dhahran on the Persian Gulf. A central, mountainous uplift is found near Halaban. Recent announcements by Saudi officials indicate discovery of a field of light oil there. While it could be burned directly in internal combustion engines, it will be refined. These fields are believed to be the largest ever found! Because they are associated with the crater, this find could have been predicted.

Scotia. A little known, tectonic plate in the South Atlantic east of South America's tip has this name. While minute compared to continents, it includes about 750,000 sq mi. The east end is marked by a typical island arc known as South Sandwich or Falkland Islands. These features draw immediate recognition of an impact crater. It is analogous in many ways to the island arc in the Caribbean. The recorded coordinates of the epicenter refers to the present location of the islands but the event actually occurred about 1500 mi farther west. The explosion was at the edge of the South American plate where little damage was done. Overthrusting at sea raised the islands. This phenomenon explains why many island arcs are semicircles. Their position remained fixed relative to earth's mantle while South America continued its journey. Ocean floor drifting as a thin ribbon was lost down the abyss. The primary distinction between *Caribbean* and *Scotia* is that the Caribbean Plate was paved with lava while the Scotia Plate was not. Their respective sizes may be the reason. Antilles at 560 mi created the Caribbean Plateau but no lava flowed into the Scotia bowl at 360 mi. Some threshold for opening the crust to magma is implied. Strike-slip faults were created to accommodate relative motion between the ocean floor moving past the island arc and the stationary floor in the crater. At both ends of the Scotia islands these faults, both known as Scotia Ridge, raised islands that are not members of the arc. The north fault built the South Georgia Islands while the south fault built South Orkney Islands and Shetland Islands off the Antarctic Peninsula. For the geological effect of drag along the ridges consult the write up accompanying Fig. 1.16. Continental Trimming And Antilles. Data from ages of the ocean floor provide an estimated age of 84 mil yr at the western edge of the plate.

Siberia 1. Near the city of Seymchan in the Magadan region of Siberia north of the Sea of Okhotsk. A multi-ring structure bearing little resemblance to most as it appears to have three, major rims. The inner is the Khrebet Mountains paralleling the coast. The middle is indicated by the Polostron Taygonos Pensinsula to the northeast. Southward as part of this rim is a weird looking peninsula shaped like an inverted **T** with a thick stem. The cross bar is consistent with the exposed sections of the ridge. Clincher to this argument is the island of Zavyalova and its short chain near the city of Magadon. An outer rim is probably the root of Kamchatka Peninsula beyond the Gulf of Penzhinskay. Respective radii of these rims are 140 mi, 190 mi, and 400 mi or 35%, 71%, and 100%. Such data from other craters should be assembled to study internal collapses. Perhaps, the numbers are related to the crater size and variations may depend upon local geology. A few others are noted herein but that pursuit is side tracked. These craters are found on continents that resist overthrusting. Rebound is then confined within the detectable, outer rim. They usually fail to produce earthquakes, lava flows, volcanoes, and trenches. When the rim invades ocean areas, however, they rupture along the outermost, circumferential fault producing island arcs. That is why island arcs are only crescents and why most are close to continents.

Simpson. Desert of 280,000 sq mi in central Australia close to Birdsville. Displays the typical pattern of continental drainage with moisture seeking Lake Eyr before flowing to the

Indian Ocean at Spencer Gulf. The rim is prominently displayed by parts of MacDonald Range and Musgrave Range, to the south along Gawles Range at Eyre Peninsula, and finally, Beal Range. This crater is nearly concentric with **Great Dividing** that is larger, reaching the Great Dividing Range in the south and the Pacific Ocean. The drainage pattern shows that **Simpson** is the younger.

Singkian. Modern spelling is Xinjang Uygur. A shallow basin in Western China containing the Takla Maka Desert that has been badly distorted by the Tien Shan Range. The southern boundary is a steeply rising Shan Range including Mt. Muztagata at 24,757 ft. It seems that all the world's highest peaks are on rims of the numerous craters. A radius of 850 mi swung around the center picks up three oddities of possible significance. Immediately inboard to the east is the Turfan Depression to a depth of 505 ft. The Altai Mountains in western Mongolia follow the rim for the entire length of 300 mi. Finally, Lake Balkhash inside the rims exhibits the correct crescent with its concave side directed toward **Singkian**.

Small Bight. In the southern corner of Australia forming the coast of Indian and Pacific Oceans from about Adelaide to Brisbane. Extends from southern portion of Great Dividing Range to Flinders Range in the west. Extensive drainage from the Great Dividing Range converges to feed the Darling River. Coal is abundant in the coastal mountains but petroleum is scarce. Name selected because this crater lies just east of another, together creating the south coast of the continent known as Great Australian Bight. The latter may be recalled as the rim of **Wilkes Land** in Antarctica. See Fig. 1.9. Birth Of Australia.

Snake. In southern Idaho with the Snake River and Columbia River flowing in the gap between rims of an internally collapsed crater. Center 130 mi NE of Boise, Idaho near Sawtooth National Recreation Area and Barak Peak. Eastern rim includes geothermal wonders of Yellowstone National Park and dramatic scenery of the Gran Teton National Park. Remarkable details, indeed. The outer rim continues south as the boundary of Columbia Plateau. The central part of the crater has been invaded by a massive, lava flow ending at the Salmon River Range. This plateau ranks among those recognized worldwide although being only 10% of the largest in area and volume. It has been studied exhaustively by geologists down to mapping of individual flows. An age of less than 20 My probably identifies it as the youngest megacrater on record. The Snake River Plain, about 370 mi long and typically 50 mi across, is too wide to be a conventional, double-rim. The very high Rim Ratio of 1.527 is the sign of a collapse with the Snake River Plain being an annulus full of silt. The river follows the front of the Columbia River Plateau from Idaho Falls to Caldwell. Then it crosses the plain at Twin Falls only to recross downstream. This crater, reaching north to Missoula, Montana, is substantially larger than a casual glance at Snake River Plain. Three other craters abut or overlap **Snake.** One of them pushed the western rim into an oval. Still another to the NW deserves individual comments under **Pasco.**

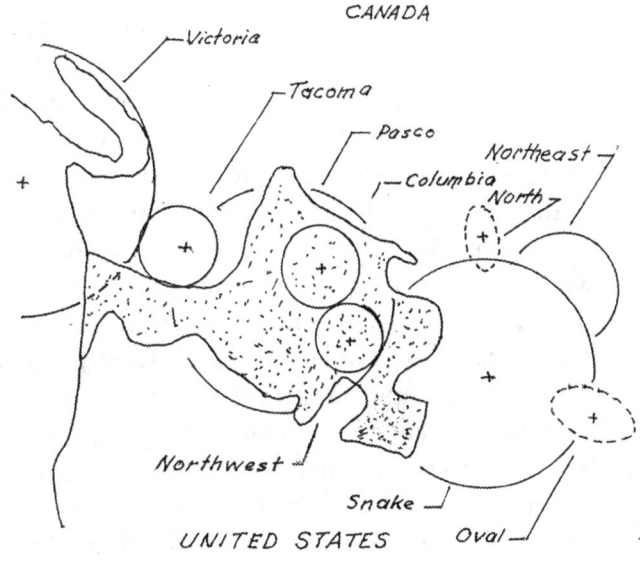

CANADA

—Victoria

—Tacoma

—Pasco

Northeast —

—Columbia

North —

Northwest —

Snake —

Oval —

UNITED STATES

⟨image⟩ –Columbia Plateau (lava) after Hooper (1997).

Somolia. Located in the horn of Africa 140 mi NW of Mogadishu. Deep in the desert on Audo Range close to the town of El Kere. Inner rim is eastern scarp of the Great Rift Valley, Ahmar Mountains, from the apex of the Afar Triangle in Djibouti southward to Lake Abaya between double rims. Outer rim is the high ground from Addis Abba to Arba Minch on the lake. Rim ratio of 1.185 may be exaggerated by subsequent widening of the Great Rift Valley.

South China. Established complex system of the Yantze River whose outlet implies a rift angle of 69.7°. With center just west of Wuhan, carved the buldging coast for 1350 mi from Lianyungang to Zhanjian. Western half of the crater has been virtually destroyed by a jumble of countless mountain ranges. The highest peak in the west, Gonga Shann (Kung-Ka Shan) at 24,790 ft, may be on the rim flanked by rivers Yalong Keng and Dau He. This suspicion comes from an observation that all or most of the world's highest peaks are so located. Microspherules, small tektites, have been found in South China dated to 365 My. Researches have associated them with a faunal extinction and possible impact of a bolide. The crater might actually be found from these spheres of molten rock that crystallize in air. But it would be too small to register among the giants and is most unlikely to be a big killer. On the other hand, the spherules could well have been produced in the mushroom cloud of a megaburst from any of several sites in that part of the world. If so, the date of 365 My becomes important.

Spitzbergen. A cluster of islands in the Arctic Ocean above Norway. Well known only to the 3000 residents concentrated along the shores of 24,000 sq mi. Sits on the Mid-Atlantic Ridge along with Iceland 1170 mi south. Iceland is extremely volcanic with continuous lava flowing at about the same rate that created the global, igneous plateaus. Natural steam is so abundant that it heats buildings and houses. Because cosmic explosions have been found as progenitors of other mid-ocean ridges, one can guess the same pattern holds for the North Atlantic. (South Atlantic was explained in **Timbuktu**.) North America was removed from

199

Eurasia by the forces that opened the original fault zone. As the new continent receded, the entire ridge released lava building parallel mountain ranges on the ocean floor. Results from the Deep Ocean Drilling Project show the oldest floor off the continents at 34 mil yr. Because drilling did not probe through the continental slopes, the underlying floor can only be estimated at about 50 My. That age would indicate the time of the explosion provided that subduction had already begun in the Pacific. Without some place to get rid of Pacific floor, the continental separation could not begin. Either it dove under the shores of North America or the western trenches had been formed. There appears to be no way of telling how long the continents waited at the starting gate. Another cluster of islands in Mid-Atlantic must have been involved. The Azores are not on the ridge but another rift system crosses nearby. Details of their shapes and orientation are strongly suggestive. Parallel row of long islands are spaced about 40 mi apart, appropriate to circular fault zones of megacraters. Their alignment is consistent with the idea of an explosion near Spitzbergen. Turning back the geological clock to reconstruct the original positions is also revealing. The traditional assembly of ancient Godwanaland appears to be in error. Prevailing concepts have been refined by computer analysis to obtain the best fit of continental margins. Minor overlaps are observed along with major gaps. Under the present paradigm, the east coast of the United States allegedly fits the northern part of Africa's western bulge. (Remember *Timbuktu*.) A totally different approach to this problem places the north coast of Labrador, Canada against the west coast of Spain as shown in the accompanying sketch. Backward running of time brings them together neatly. They originally separated along the Labradore - Agoros and Gibraltar transform faults. Pushing Canada back to Spain is best done by matching Battle Harbour and Cabo de Sao Vicente. Only the cape has remained stationary so it gives the most reliable distance from Spitzbergen as 2930 mi corresponding to the crater radius for a diameter of 5860 mi. The Gilbraltar Fault Zone goes through the Strait of Gibraltar and along the coast of Algeria and Libya. Due to relative motion of Africa, it then crosses the Mediterranean Sea, loops Sicily, then separates from Italy. It passes close to Mt. Vesuvius and fashions the sole of the boot. Much can be learned from the present positions of Battle Harbour and Cabo de Sao Vicente. Computer software can give accurate azimuths and distances from Spitzbergen. The annular separation between these landmarks amounts to 56.8°. The rotation of North America about its own axis is indeterminate. The continent experienced a northerly component given by the difference of latitudes as 15.16° or about 1050 mi. Recall the many instances of excessive damage at the intersection of a linear fault system and the circular fault of the same crater. Thus, the location of Azores out in the middle of the Atlantic is normal. All of these factors support the concept of a huge explosion with its epicenter not far from Spitzbergen although the rim remains hidden in Asia.

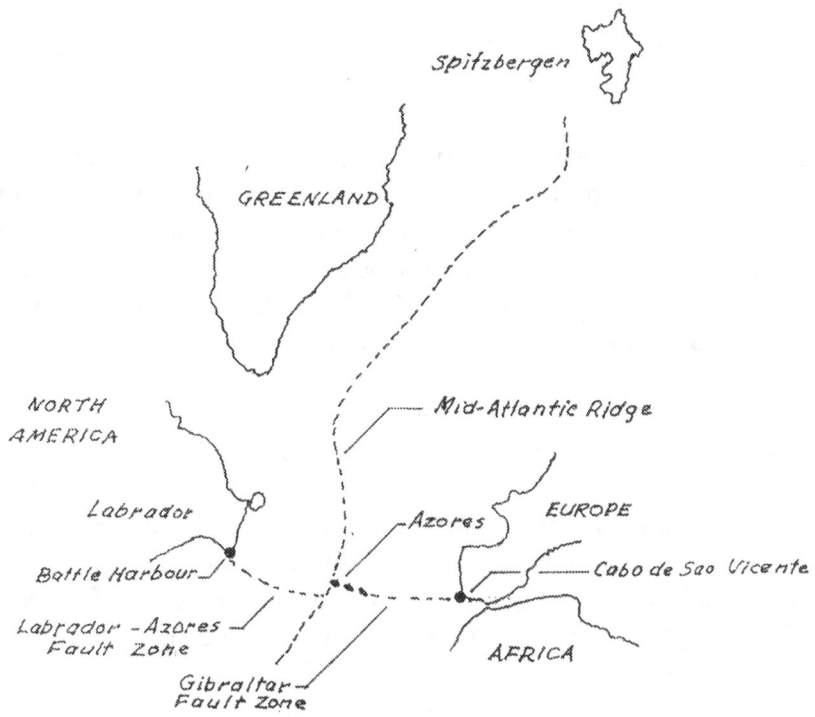

Spokane. Named for well known city although center is 160 mi south at Sacajawea Peak (9,838 ft). Tightly surrounding the peak as part of an uplift structure is a crushed rim 90 mi in diameter. It was mentioned as touching Snake on the NW but this crater is much larger. With a diameter of 380 mi, it reaches west to the Coastal Range and east to Glacier National Park. Detection is difficult because of intrusive mountains and massive lava flows of the Columbia Plateau. The latter crosses the States of Washington and Oregon EW with an average width of about 150 mi. The area is estimated at 60,000 sq mi. This crater must have contributed to complex ruptures releasing lava along with **Pasco** in Washington.

Springfield. An especially interesting 400-mi example is 70 mi southeast of Springfield, Missouri that can be identified full circle. A prominent section of the rim is the arc of Mississippi River from St. Louis to Memphis followed by Hot Springs, Arkansas and the arcing Gouache Mountains. The western portion passes close to Tulsa and Kansas City, controlling the local drainage. One or more serious ruptures has released lava forming the internal Arkansas Plateau of about 50,000 sq mi. A small run of the Mississippi River passes along the back side of the "heel" in the State of Missouri. That is the exact location whose earthquake hazards are unusually high. Epicenters of historical earthquakes at that location form a Christian cross whose transverse beam is slightly canted. This arrangement is compatible with the creation of the fault system by **Springfield** alone or it could have received some help. Somewhere nearby a vastly larger explosion left no local clues because of burial under miles of sedimentary rock. See **Joplin**.

St. Lawrence. East of New Brunswick, Canada with an uplift of Grindstone Island. Has earmarks of a collapsed bowl with radii of 65, 115, 270 mi or 24, 43, and 100%. Outer most rim is Prince Edward Island separated from coast of New Brunswick by semicircular Northumberland Strait. Outer two rims are suggested along the northern part of Nova Scotia

201

above Mulgrave to Cape Breton. Parallel highlands are separated by a complex waterway including Bras d' Or Lake. A linear fault along bearing 30.6° probably opened Strait of Belle Isle separating Long Range Mountains from Labrador.

Sudan. Structure of 900-mi is easily identified by mountains. Central uplift appears as Nuba Mountains. Rim shown by Jabal Marah (Chad), Ironstone Plateau (Central Africa Republic), Didingu Hills (Kenya). Northern arc is covered by sand. Many local craters of 80 mi to 200 mi have driven the Nile River around their rims. Too small to deal with. Not negligible, however, is one of 300 mi west of the Nuba Mountains to the border of Chad at 12.00N 27.16E to be called *Sudan Jr*. Had profound effect on the course of the Nile forcing a detour of 250 mi around a mountain range to the NE.

Sudan Jr. Contributed to shaping the Nile River. See *Sudan*.

Sulu Sea. Small as seas go but includes 210,000 sq mi of open water north of Borneo. Eastern boundary is Jolo Island, Basilan Island, and western peninsula of Mindanao at Zamboango. Northern limits are Negros and Cuyo Islands. Western rim appears to have been disturbed by a linear rift creating Palawan except for Balabac Islands that are still in place.

Sydney. Covers most of New South Wales in southern Australia west of Sydney. Primary evidence is the Great Dividing Range especially the Australian Alps east of Melbourne. Established the typical pattern of continental rivers, namely Billabong, Murrum, and Darling. A small crater, *Bass Strait*, created the island of Tasmania. It was not severed from the Australian Plate so the strait is only the flooded crater. Another is too small to treat formally but should be pointed out. Referred to by the author as *Spencer*, the basic data are 34.77S 137.22E, 150 mi. Nevertheless, it opened the combined Spencer Gulf and Gulf of St. Vincent at Adelaide. The rim is easily traced by the North Mount Lofty Ranges, Fleurieu Peninsula, Kangaroo Island, Eyre Peninsula, and the west coast of the gulf. Extraneous islands and the York Peninsula are not unusual.

Tanzania. The east African nation virtually covered by a lava plateau of an estimated 750,000 sq mi that is noted for the variety and great number of ungulates, predators, and birds. Rising high above the plains in a tight cluster are the three extinct volcanoes, Loolmalasin (11,967 ft), Meru (14,978 ft), and Kilimanjaro (19,340 ft). They are located about 230 mi south of Niarobi, Kenya. Kilimajaro, so often seen in movies and magazines, is of special interest in the present context because it is at the center of this megacrater! Prominent features are the crescent of Lakes Tanganyika, Kivu, Edward and Albert. This arc continues along the Victoria Nile to Murchison Falls. Tanganyika fills a fault zone 50 mi across that is typical for spacing between rims of large craters. At 400 mi long and nearly 1 mi deep, it is the largest lake in Africa by volume. Thorough research in the lake by geologists has shown a highly complex pattern of many faults with uplifted and fallen blocks. Similarity to another great lake on the rim of a megacrater in Asia is striking. See *Baikal*. Two radiating faults raise a possible ambiguity. The most prominent and familiar is the Great Rift Valley or East African Rift north to Addis Ababa, hosting Lakes Rudolph, Chev Bahir, Chaim, Shala, and Zhba. It is traditionally and correctly associated with an extension through the Red Sea, Gulf of Aquaba, and the Jordan River. However, the portion in northern Africa is not straight but follows the curvature of Lake Rudolf on the rim of

Somalia. That leaves a linear rift system to the south that originates from *Victoria*, a small crater inside of *Tanzania* holding Lake Victoria. It is worth following the southern fault through the 340-mi long Lake Hyasa that is the correct width. It proceeds on dry land to the coast where it interacted with Madagascar before it broke away from the continent. See *Madagascar*. The azimuth from Victoria to Blantyre of 190.3° corresponds with a rift angle of 10.3°. While the trio of volcanoes near the center of *Tanzaniya* have retired, one on at the north end of Lake Kivu is still in business. The offender in January 2002, Mount Nyiragongo, repeatedly threatened Goma, Congo by destroying 10,000 homes, food warehouses, and sending 400,000 people fleeing to safety.

Territories. In the far north of Canada's Northwest Territories 300 mi south of Victoria Island. Center is very close to Lac de Gras. By no means obvious, the rim is defined by Coronation Gulf, McTavish Arm of Great Bear Lake, Hottah Lake, chain of lakes to Faber Lake, North Arm, Simpson Islands, Christie Bay, and Artillery Lake. Other megacraters have held volcanoes near their centers or pipes when the mountains have been worn away. In the featureless lake country, four diamond-bearing pipes were discovered underwater in 1991. At enormous expense for building a dike and pumping out the water, the Divik Diamond Project was found to economically viable. This discovery opened up one of the most valuable sources of diamonds in the world. Production, beginning in 2002, is expected to continue for 20 yr with a reserve of 138 million carats. Also nearby is the existing BHP/Dia Met Ekati mine. Once again, a megacrater brought mineral wealth to the surface where volcanoes once stood.

Texas. With the center 100 mi east of Lubbock, a double rim is apparent. The San Luis Valley of northern New Mexico, flanked by Sangre de Christo Mountains and San Juan Mountains, is the headwater of the Rio Grande River. Its great arc passes through Albuquerque between rims down to Boquillas del Carmen. There it takes a right-angle turn around Big Ben National Park by switching to the inside fault. This offset of 118 mi also indicates a rim ration of 1.135. Rio Grande picks up the Pecos River along that fault at Amistall Reservoir then flows along the fault to the Gulf of Mexico. The hill country from San Antonio to Dallas reveals the rims that swing through Altus, Oklahoma. The so-called Raton Hot Spot in northern New Mexico lies on the inner rim. The sharp turn at Big Bend is most intriguing because it loops around an extinct volcano, Emory Peak (7835 ft). The inner rim at Big Bend is dotted with four more volcanoes higher than 7000 ft. Volcanoes are supposedly created only along continent margins associated with subduction. Here they rose on a crater rim. They happen to be common near the centers and rims of megacraters completely within continents. This crater is assuredly responsible for the legendary fields of petroleum and natural gas in West Texas.

Timbuktu. This story began in Chapter 1 when Pangea began to shatter from a mighty blow over Mali in Africa. A complex, rift system shaped the entire west coast of the continent from Gibraltar to Cape Town with minor assistance before and after. Oil, becoming more significant in world trade, is found on the coast of Nigeria which, as noted in many other locations, is the intersection of a crater rim with its own linear rift. This rupture did not launch South America immediately because the existing, tectonic plates were jammed. Nothing moved until a similar event created the west coast of South America from Panama to Tierra del Fuego along with the Peru-Chile Trench. Then South America, driven by forces other than escaping lava, could glide over the western ocean as it began a long, slow trip.

Relevant times are indicated by ages of the Atlantic floor closest to South America and Africa. Measurement of about 35 My were obtained to the edge of continental slopes. So the fracture itself can be estimated at about 50 My. *Amazon* created the northern bulge and the straight coast to the south in one stroke. The analogy with the west coast of Africa is shocking. As the Atlantic floor spread, an equal amount of the Pacific floor was lost down the Peru-Chile Trench. Common opinion holds that the loss was swallowed by the connected trenches in the West Pacific. But elsewhere the author shows such movement to have been negligible. The oldest Pacific floor close to South America is about 21 My as anything older has been lost. Subduction has been progressively larger for northern latitudes as South America rotated clockwise during its journey. For the latitude of 35°S at Santiago and Montevideo, the oldest floors are roughly 13 My and 50 My, respectively. So the loss of Pacific floor must equal the spread of the Atlantic floor over the same period. Since this portion of South America moved 42° of longitude at 57 mi/deg, the corresponding distance of 2400 mi must be the amount of Pacific floor lost under Santiago. While considerable drilling has been performed at the western trenches, published finding do not permit estimates of their ages. Such information is, of course, vital to reconstructing the tectonic behavior of the Pacific Plate. The trenches must have a short history because their volcanoes rise high above the ocean and many are still active. The islands are young and the impacts are geologically recent. Loss of Pacific floor under the islands simply can not account for the facts. Study of seamount chains is Chapter 1 destroyed the concept that the entire Pacific Plate has moved according to gospel geology.

Trinidad. Near the town of Trinidad, Bolivia. A circular flat with many rivers connected by hundreds of miles of canals. Constructed by an unknown, defunct civilization where intensive agriculture supported an estimated 30,000,000 people. Only recently discovered by advanced society from an airplane, exploration has barely started. Probably a lava-filled crater of 90,000 sq mi.

Tunguska. One of the very large craters in Siberia shown in Fig. 1.21. Siberian Cluster. It overlaps *Noril'sk-Yakutsk* and probably contributed to the Siberian Traps. Recent research indicates an igneous province extending much farther to the west than previously thought. Perhaps all the way to the Ural Mountains. This crater should not be confused with the destructive but negligible explosion of an asteroid in 1906. A paucity of landmarks is the region forced the ambiguity. *Tunguska* refers to a river and a recent explosion whereas *Tunguska* is the cometary blast that formed the crater millions of years ago. So many craters are scattered across Siberia that keeping book on them is not easy. Of course, that problem applies to all continents.

Turkey E. As frequently observed, the center is in a depression that has collected water. This one is at the north end of Lake Tuz Golu. Rims are revealed by coastal arcs in Black Sea and the Tores Dagear range facing the Mediterranean. See *Turkey W*.

Turkey W. The western shore of Turkey on the Aegean Sea along with many islands indicates a crater centered near Usak. The outer rim is the arc of islands north of Rhodes including Patmos. A rim ratio at Rhios amounts to 1.122. A clear track of the annular fault is the waterway of Dardanelles with the Gallipoli Peninsula outboard. This rim continues along the Balkan Peninsula to Istanbul and across the Bosporus into Asia. Subduction of the Aegean floor is lifting western Turkey with attendant earthquakes. An interesting link may exist among *Turkey E, Turkey W*, and *Aegean*. They have similar latitudes and overlap their neighbors about 40%. A possible chain of fragments should be examined having the following sequence

Turkey E + 10.5 min + *Turkey W* + 22.0 min + *Aegean*

They would, of course, have to be the same age. Their sizes also are suspiciously similar at 390 mi, 500 mi, and 350 mi, respectively.

Twin Bridges. Twin Bridges, Montana sits near the center of a basin surrounded by Beaverhead Mountains and Big Belt Mountains. On the south, the Red Rock River is concentric with the curve of the mountains and, as is rather common elsewhere, makes a right angle turn along a course in a radial fault as Beaverhead River. To the northwest, Big Hole River is similarly disposed. Linear mountains ranges to NE and SE, Swan Range and Wind River Range, run straight through the crater for a total of 1000 mi. This feature is unique because nearly all linear faults begin at the outer rim.

Tyrrhenian. One of six craters that shaped the Italian Peninsula. All are discussed and illustrated under ***Sardinia*** to clarify each contribution and put the puzzle pieces into the full picture.

Under Klondike. A very old crater nearly buried under ***Klondike***.

Ungava. A summary of proven or suspected impact craters publish in 1961 indicated diameters 0.95 mi to 273 mi. No megacraters were mentioned. The latter figure pertained to Belcher Bay and the next largest to Ungava Bay at the northern end of Canada's Quebec province. It lies about 275 mi east of Hudson Bay, reckoning between shorelines. The shoreline diameter of 150 mi is only a ragged approximation of a circle. The depth is 1190 ft and the rim rises 330 ft above sea level. Surrounding *Ungava* to the west and south for 350 mi is an expanse of lake-studded lowlands. *Ungava* is at the center of a great but shallow depression toward which five rivers flow like spokes of a wheel. The bay could well be an impact crater but important clues are missing. No inner rim can account for the flooding. Also, remote damage is not found. Resistance to damage by the primeval Canadian Shield in that area would have prevented a larger crater. Judging from the extent of erosion in the scalloped shore, *Ungava* must be very ancient. Furthermore, it was periodically protected from weathering over eons by the ice cap. But grinding glaciers probably honed down the high spots. While the flooded portion of the basin at sea level is only 150 mi across, the whole structure was probably much larger.

Uraguay. Located at Goya on the Parana River 390 mi NNW of Montevideo. Remnant of a rim is seen to SE as the short ridge of Sierra de Carvera. Similar detail seen near Cordova along short ridge of Sierra de Norte that is not connected to Sierra de Cardoba. A faint but definite, small crater.

Vancouver. Center in Pacific Ocean opposite Vancouver, British Columbia. Opened Strait of Jaun de Fuca and raised about 400 mi of coastal mountains including Mt. Raineir in Washington State. Part of the Cascade Range extends to Canada and, after a gap, proceeds along the Coastal Mountain Range to about Mt. Washington. Lower half of Vancouver

Island is inner rim and annular gap is represented by Georgia Strait and numerous islands including San Juan group and Texada Island. The topography from Kelsey Bay to mouth of Queen Charlotte Strait was formed by a small, unnamed crater near Rivers Inlet.

Vanuatu. In South Pacific between Fiji and New Hebrides of which Vanuatu is a prominent island. The New Hebrides are exposed peaks of the rim going south 350 mi to Tanna. Beyond that island, an unbroken, submerged ridge arcs to Fiji. Outboard is the south end of the New Hebrides Trench. An outer rim can be seen underwater that gives a rim ratio of 1.145. This whole region has been seriously modified by tectonic motion.

Venezuela. A 1060-mi crater covering all of Venezuela, most of Colombia, a large fraction of the Amazon Basin, and all of Guyana. The western rim is Cordillera de Merida, the eastern branch of the Andes around Lake Maricaibo. An offset of the Andes at San Cristobal gives a rim ratio of 1.071. The Caribbean coast is noted for "parallel" ranges called *coast* and *interior*. Their separation distance would refine the rim ratio but the information is not at hand. The rim continues east along the Caribbean shore for 1000 mi to Surinam. The center is near the southern-most point of Venezuela within a small basin of 260 mi that resembles a caldera. Adjacent on the north is the oval-shaped, Guiana Highland. This igneous intrusion is a world-class lava plateau around which flows the Orinoco River. Another detail is a 170-mi crater near the mouth of the Orinoco centered about 50 mi north of Cuidad Bolivar that supplies a name, that is, *Bolivar*. It is too small to treat separately but is mentioned here for a unique reason. It is second only to Lake Marcaibo as a producer of oil. Notice that *Maracaibo* struck the rim of *Amazon* whereas *Bolivar* struck the rim of *Venezuela*. Further examples of this phenomenon should not be required to attract attention of oil companies.

Venice. One of six craters that shaped the Italian Peninsula. Discussed under the single heading of *Sardinia* to clarify each contribution and consolidate the fragments of an interesting tale.

Victoria. This is 30,000-sq-mi Lake Victoria in the African country of Tanzania that is inside the crater, *Tanzania*. The epicenter is not at the center of the lake but along the eastern shore near Ushashi. The crater is much larger than the lake that only fills to lowest portion. Western rim is the escarpment of the East African Rift at Lake Kivu, and Edward And Albert. The eastern rim involves the Mau Escarpment and Highlands of Kenya. Curving lakes Eyasi and Natron fill the circumferential fault zones. A linear fault to the south may have created the 360-mi Lake Nyasa and continued along a dry valley to the Mozambique Channel. A competing cause of this lake and valley is *Madagascar N* that exploded while the island was still connected to the continent. Available evidence has not resolved this dilemma. Another small Lake Niavasha, also in the Eastern Rift Valley, is 45 mi NW of Niarobi. It partially fills circular basin 30 mi across that must be an impact crater. Other examples where smaller explosions attacked the rims of larger one have been the sites of major oil fields. See **Venezuela** and *Maracaibo*. This little target could be worth billions.

Volcano Belt. One of six craters that shaped the Italian Peninsula. Discussed under the single heading of **Sardinia** to clarify each contribution and consolidate the pieces into the whole picture.

Weddel. An Antarctic sea at the root of the Antarctic Peninsula east of South America. As everyone knows, the continent was originally attached to Africa but the location has been uncertain. Most reconstructions of Pangea show it somewhere along the *east* coast of southern Africa. The concept of megacraters, however, provides an invaluable tool for clarifying this geological problem. The Antarctic Peninsula originally nestled along the *west* coast of South Africa. Recall that Kalahari raised the Drakenburg Mountains and carved the southern coast of Africa from Cape Town to Maputo, Mozambique. But the original arc of the mountains continued out into what is now the Atlantic Ocean re-entering Africa some 800 mi north in Namibia. That missing piece of Africa can be identified as the Antarctic Peninsula. As South America pulled away, Antarctica went south and rotated counter clockwise. The shore of the Weddel Sea was originally alongside the Cape Town-Maputo coast. The corresponding coast of Antartica is Edith Ronne Land from Coats Land to Ellsworth Land containing mountain range of Theron, Heritage, and Sentinal. They are the ones that once belonged to the Drakenbergs. The beginning arrangement had Vinson Massif in Antarctica adjacent to Cape of Good Hope. So precise values of the continental drift become available. From their respective coordinates,

Vinson Massif	78.58S	86.07W
Cape Town	33.65S	18.37E
Difference	44.93°	104.44°

In other words, Antarctica went south relative to Africa by 44.93° while it rotated counterclockwise 104.44°. This finding is somewhat puzzling because continents tend to approach the equator from centrifugal force of earth's rotation. That would have required Antarctica to seek the tropics in ancient times. Perhaps it succeeded. Evidence of fossilized leaves in Antarctica corresponds with fossilized palm trees and coral reefs in the Arctic. This information requires an earlier inclination of earth's axis to approach 90° as was found regarding Rift Angles. For *Weddel* and *Kalahari* to be parts of the same crater their diameters would have to be identical. They are. *Weddel* (1140 mi) versus *Kalahari* (1120 mi). In addition, Ellsworth Mountains are part of a double rim with the inner being Sentinal Range. The sequence of two great African explosions becomes clear. *Kalahari* preceded *Timbuktu*. The southern coast of Africa had been weakened before *Timbuktu* liberated South America and Antarctica at the same time. Lava in the Kara Traps must be older than Timbuktu and some of it might be found under Weddel Sea. Since *Weddel* and **Kalahari** are parts of a single crater, one of them should be dropped. But retention of both for stronger reasons should not significantly disrupt the statistics.

West Pacific. A lava plateau of 25,000 sq mi spreads across the Pacific floor half way between Wake Island, Hawaii and Tokyo, Japan. A westward projection of the Hawaiian chain crosses this Shatskiy Rise at the designated coordinates for this crater. The geography is standard. A huge basin with linear rift creating islands. This portion of the western Pacific is identified as a megacrater of 1800 mi whose linear fault set the stage for step-wise development of volcanoes along the Hawaiian chain. So a viable alternative comes to the rescue as the hotspot theory expires. Volcanoes coming down the Emperor chain found an easy trail for making the bend near Wake Island. Westward projection of the rift to Fuji San is probably meaningful.

Wilksland. A megacrater of 2200 mi that sliced Australia away from Antarctica about 50 My. Situated in Wilksland with the rim fault along the continental edge from Knox Coast to Victoria Land. The corresponding arc in southern Australia, known as the Great Bight, extends across that entire continent from Perth to Hobart, Tasmania. The center is 650 mi from the South Pole along East Longitude 115°. Fig. 1.8. Seventh Continent shows how the shape of Antarctica was completely determined by four megacraters. The net result is a continent ringed by mountain ranges. As the basin collected snow that compressed into ice, the increasing depth brought weight into play. An ice layer covering 5 mil sq mi would influence the isostatic balance of the continent. Pressure from ice at the thickest point of 13,000 ft, would amount to 740,000 lb per sq ft, certainly enough to bend basement rock. Loss of moisture by evaporation is less than the gain from new snow, so the continent can expel water only by glaciers. They flow into the sea between peaks of the encircling mountain ranges. With landforms other than the sum of today's megacraters, Antarctica would be quite different.

Williston. In North America plus Colombia and Venezuela, 88 basins in bedrock and uplifts were recognized in 1989. No satisfactory explanation of their origins has been forth coming but their causes are now clear. A prominent basin, Williston, is near the town of Williston in the northwestern corner of North Dakota. It, along with many others, has been scrutinized by geologists so details of its cross section are available. The depression is 2.98 mi deep from the present surface, 2.48 mi below sea level, and 680 mi across. It was created more than 500 My then filled with 14 layers of sediment of different composition that were transformed into rock. The present, depth-to-diameter ratio of 0.00365 applies only to the terminal phase of rebound from greater depth. Other basins have other ratios, such as, Hudson (0.00287), Michigan (0.00862), and Illinois (0.00583). What residual depth might be expected for a megacrater 3000 mi across? Using the smallest ratio above predicts the final depth of 8.61 mi. But calculations from overthrusts suggest the maximum depth of a 3000-mi bowl, when explosion pressure and heat are removed, to be as much as 600 mi and possibly more! Larger craters are known to have much greater ratios like *Aegean* at 1.051. This complex question is saturated with uncertainties and begs for thorough analysis.

Yellow Sea. Found in the Yellow Sea close to the coastal city of Qingdao, China. Shaped the east coast of Korean Peninsula and, immediately south, the island of Nagasaki Ken. Includes mountain range west of Beijing where part of the Great Wall was built. The SW quadrant has been smothered by alluvium from the Huan River.

Yellowstone. The famous National Park in the NW corner of Wyoming. Many geothermal attractions. Mammoth Hot Springs Terraces, hot mud caldrons, and Geysers. Among the 200 geysers, Old Faithful is the most popular. This location ranks among the three most active in the world. Others are Gesyers in Iceland and Waimangu that have been associated with megacraters. A small crater only 93 mi across encircles the entire park. The center at Lake Yellowstone is surrounded by Abscaroka Range, Madison Plateau, and Grand Teton National Park. That spectacular range, including Gran Teton (13,771 ft), lie on the Yellowstone rim. Most of the features in the park form an inner arc on the west that includes Tower Falls, Steamboat Geyser, Gibbon Falls, Firehole Falls, Imperial Geyser, Great Fountain Geyser, Mystic Falls, Old Faithful Geyser, Lone Star Geyser, Lewis Lake, Lewis Falls, and Heart Lake. This crater appears to have collapsed internally. The region has been severely battered by missiles whose own craters are too small to be registered in the catalog, namely, a) *Yellowstone E*, 43.98N 109.07W, D = 100 mi with eastern rim is Bighorn Mountains, and b) *Yellowstone W*, 44.00N 111.82W, D = 160 mi at east end of Snake River Plain. Recall that the rim of the much larger *Snake* also passes through the park. No wonder that Yellowstone National Park is rich in geothermal activity. The whole thing has been permanently bound together in contrast to the continent drifting over a fictitious hotspot in the mantle.

Yukon. In far north of Alaska on Porcupine River at international border with Yukon Territory, Canada. Porcupine goes through center of crater and changes name to Yukon River. Meanders along a generally straight path to Tapana on bearing 160.8°. Other factors control the river downstream. Outer rim is weakly defined by British Mountains to north. Probably collapsed bowl exhibiting a central uplift and a single, inner rim at 57 mi radius. Relative rim spacing is 47.4% and 100%. While not thoroughly investigated in this work, a general observation should be reported. Very small craters are able to retain the shapes of their original depressions after debris has fallen to the bottom. Somewhere in a range of intermediate sizes, perhaps 50 mi to 500 mi, original depressions suffer failure of inner walls that may be singular or several. This process generates internal rims that are widely spaced. Larger craters somehow avoid this behavior because their basins are relatively shallow.

Zacatecas. In central Mexico near the city of Zacatecas flanked by Sierra Madre and Sierra Madre Oriental. Lake Chapala near Guadalahara lies between double rims that also define Bahia de Banderas at Puerto Vallarta. The inner rim continues to sea being exposed by Islas Tres Marias.

Zambia. Bowls of mega-explosions seldom recover completely but leave shallow depressions that collect rain. Depending upon the weather, the amount takes violent swings over time. The center may consist of extensive lakes, swamps, or dry flats. This one in northern Zambia is at or near Lake Bangusulu that is partially surrounded by Banweulu Swamp. Together they cover roughly 600 sq mi. Diameter of 1350 mi nearly crosses the continent in southern Africa. It reaches west into Angola near Luanda, crosses Lake Tanganyika at Burun, Burundi, and reaches south to cross Zimbabwe at Bulawayn 250 mi SW of Harare. The eastern rim is of such interest that it must be examined carefully. The continental coast was formed from Dar es Salaam, Tanzania to Beira. A coastal arc of short radius extends down to Maputo where a second indentation is noted. Farther south, the shore follows the circumferential fault of *Kalahari*. Fig. 1.2. South-Central Africa makes this clear enough although this crater had to be excluded to avoid a complete muddle. The strip of coast under inspection is precisely where the major island of Madagascar originated. Further data in this context is shown by three small craters still recorded on the island. They are described and sketched under *Madagascar N*. Now pieces of a large puzzle come together in understanding how Madagascar was set adrift. Consider the present island joined to Africa and the following sequence.

1) The linear rift from *Indian Ocean* sliced off a piece of the African coast forming the eastern coast of the Madagascar before its release,
2) The entire region was peppered by comets of intermediate energy leaving rims that have been seriously weathered,

3) ***Zambia*** with all its might sent its own circumferential fault through the region being deflected by taking the path of least resistance around all the pre-existing caters. That event was the driving force for severance of Madagascar, and

4) The continental fragment drifted to its present position away from a spreading fault for which no evidence remains.

Zimbabwe. A southern country in Africa essentially encompassed by this 680-mi crater whose absolute give away is the Zambezi Gorge on the northern border. One of the world's great rivers flows along its arc for 750 mi from Livingston near Victoria Falls. A dam was completed on the river in 1959 in the narrow Kariba Gorge after much controversy of resettling people, rescuing of animals, and determining the beneficiaries from Project Noah. The resulting Lake Kariba of 2000 sq mi is the second largest, manmade lake in the world by volume. It provides for extensive irrigation, a bountiful fishery, and cheap electricity for the copper industry. Downstream one finds the Caborru Basa Reservoir and, crossing the Great Rift Valley, a final run in the gorge to sea. Similar gorges are common among moderate-size craters on continents where they are thoroughly confirmed by other evidence. It is tempting to say that all significant gorges on earth are produced by mega-explosions. The physics of opening them is a mystery. This crater rim has unexpected implications regarding the island of Madagascar. When it still belonged to Africa, it was also damaged by the crater rim. That clue is indispensable in finding the exact position of Madagascar vis-à-vis the continent. That point is expanded and sketched under ***Madagascar N*** and a sequence of explosions is described under ***Zambia***.

Appendix B. Previous Research.

"North America's sedimentary saucers...are deceptively simple, and yet...there is a dearth of adequate explanations for the origin..."
- Albert W. Bally

Many years were devoted by the author to researching and writing a previous book that had an unsuspected link to this one. Prehistoric civilization and ancient history were thoroughly explored along certain clues of archaeoastronomy. Computer analyses tracked the rising and setting sun and planets back 10,000 yr. Solstices and equinoxes were naturally most important because they were especially meaningful to the ancients. Before the civilization of Babylon and Egypt arose, somebody was taking precise measurements of celestial affairs and recording them. To be indelible forever, the astronomical alignments were used to specify holy mountains and to select templesites, all disposed along a Great Circle for thousands of miles. These sites evolved into worship of Apollo with the primary shrine on Delos Island that has now been revealed as the epicenter of the megacrater *Aegean.* At the time of the earlier studies, a crater that large was unthinkable but its discovery must be attributed to a daughter, Nancy. Standing back far enough for details to fade out, she commented on the perfect arc formed by the southern islands of Kithra, Crete, Karpathos, and Rodhos. That did it! Geological studies confirmed the idea of a crater and lead to the long string of discoveries reported herein. Each one was exhilarating as its prodigious magnitude became apparent.

Toward completion of the present work, several authors were found to have anticipated some of the megacraters. Their work simply identified them as candidates and, on the whole, no further investigations were reported. These ideas were apparently ignored by the scientific community. That previous work is acknowledged here although it had little influence on the author's discoveries.

Gallant

A thorough and mathematical treatise on super-sized craters was published in 1964. Energy requirements were calculated for *Hudson Bay* at 210,000,000 MT. Experiments by the Canadian government with explosives up to 1.7 KT showed how that energy would be distributed among cratering, fracturing rock, and heating. As those experiment were underground, adjustments were made for explosions on the surface. A few microseconds after impact, conditions would instantly change to 1,000,000° C and 7,000,000 atmospheres. Global effects of large explosions were examined in calculations for changes in length of days, changes in length of years, and changes in inclination of the axis.

Island arcs were recognized as possible debris from large explosions. Areas suspected as impact craters, in addition to Hudson Bay, were

Location	Diameter (mi)
Gulf of Campeche	500
Lesser Antilles	600
Argentina - Tierra del Fuego	800
Sea of Japan	780
Antarctic Weddell Sea	125

Other candidates discussed were Becher Islands, Ungava Bey, Gulf of St. Lawrence (all in Canada), Gulf of St. George (Argentina), Mosquito Gulf (Belise), and Gulf of Samborombon (Brazil).

High temperatures at meteorite impacts could initiate fusion and fission reactions.

A meteorite the size of asteroid Juno would theoretically dig a crater 2050 mi across. This author wondered if such a large crater would ever be found.

This important reference provides a wealth of information although some of the calculations might be debatable.

Gilvarry

Some ocean basins may be very large craters.

Dauville

Possible craters up to 3100 mi in diameter were considered.

Norman

A key article in *New Scientist* in 1977 was summarized in a much more accessible source by Corliss in 1980. Questions were raised regarding northwest Africa's arcuate coast, small circular structures detected from aircraft, and very large ones found in satellite photos. Review of giant craters on the moon was the basis for suggesting similar structures on earth up to 1860 mi across. Suggested configurations that might be craters were identified as

West Coast of U.S.	Indonesia	Western Iran*
Gulf of Mexico	Northern Australia	Tanzania
Northern Andes	East Coast of Australia	Western Iran*
Aleutian Islands	Great Bight of Australia	Northern Europe
Amazon Basin	Kazakustan	Himalayas
Canary Islands*	Kalahari	

* All the others correspond with the present megacraters.

Selection of these craters was strengthened on evidence of intense impacts including tiny diamonds, compressed quartz, shatter cones, impactite, and tektites. It was further suggested that lava escaping through crustal fractures would generate ore deposits.

Saul

Corlis (1980) also summarized a paper from the British scientific weekly, *Nature,* by Saul (1978). An area under study lay northeast of Phoenix, Arizona with sides of 2° of both latitude and longitude. An innovative and simple approach to the issue of craters was based upon plastic maps in which topography was exaggerated by a factor of three. Such maps on a table were illuminated by a strong beam just grazing the surface. This arrangement simulated the conditions one would have just before sunset. Higher elevations stood out while lower elevations were naturally in shadow. Coming clearly into view was a pattern of circles resembling the surface of the moon! It covered the entire area with circles sometimes abutting, operlapping, or contained within larger examples. It was concluded that these features were remnants of asteroid impacts.

Next, this map was compared to one showing the locations of productive mines in the region that was published by the U.S. Geological Survey. Twenty-one mines were found that produced a wide range of metal ores and minerals including copper, gold, lead, molybdenum, silver, tungsten, vanadium, zinc, anhydrite garnet, chrysocolla, serpentine limestone, and turquois. All of them lay on the circles and particularly at their intersections. Even at this small scale, the location of productive mines was determined by damage to rock below the surface from celestial objects.

Further studies included northern Mexico, Appalachians, Alaska, Madagascar, and Corsica. The research identified 1,170 circles of which 55% could be traced completely around. Diameters ranged from 4 to 430 mi.

Also noted was the arc of South Africa with a diameter of 1370 mi that was traced through central Angola, Limpopo Valley, coastal Mozambique, Agulhas Plateau, seamounts, and Walis Ridge.

Buthman

In 1998, Buthman published his recognition of several, large arcs and depressions. His Sumatra-Bali arc is the same as *Indonesia*. Philippines was noted as arcuate. South China Sea was seen as a central depression of 1900 mi.

Hilldebran

An asteroid was suspected as the cause of the Colombia Basin, 186 mi in diameter. Timing was judged to be close to the Cretateous-Teriary boundary that has been accurately established as 65 My. This information is compatible with the idea that *Chicxulub* might have been one of a chain of fragments that hit the earth within a span of a few hours. It appears that an abnormal number of asteroids bombarded earth at that time.

Hartnady

In 1986, this author speculated that the 190-mi Amirandi Basin in the Arabian Sea was created by an asteroid that caused a major extinction. It supposedly reorganized tectonics and triggered lava flow that became the Deccan Traps. This instance is the only one known to link impacts with lava flows.

Comments

Many earlier authors clearly ascribed very large, circular structures to impacting asteroids. Perhaps two dozen of them anticipated those discovered by the present author with research dating back to 1951. Their findings received very little, serious attention by the majority of geologists. One explanation is that geologists were saturated with the concept of gradualism. A principle of geology was established when that discipline attempted to elevate itself into a true science. It stated that every feature of topography was produced at the slow, steady rate that has always prevailed. That concept later clashed with scientists who believed that speeding objects from space disrupted the gradual changes in terrain. After decades of debate, the religion of gradualism is terminally ill if not already dead. Its burial does not, however, alter the slow changes in topography and life forms between the cataclysms.

It is now time for geology to wake up to a new paradigm. Quite the opposite of earlier orthodoxy, overwhelming evidence has been presented to illustrate the significance of cosmic explosions. Every aspect of geomorphology and evolution has been controlled by them. The battle of opposing camps will surely continue but the impactists now have an overwhelming arsenal.

This state of affairs has developed over a period of 50 yr as authors conducted investigations and reported new data and concepts. Proper recognition should be extended to the pioneers: Gallant, Givarry, Dauville, Norman, Saul, Buthman, Hilldebran, Hartnady, and others unknown.

Appendix C. Megacraters On Venus.

"Curiosity is one of the permanent and certain characteristics of a vigorous mind."
- Samuel Johnson

An excellent way to confuse the public is to invent a new word when an old one is perfectly adequate. Highlands should be called *highlands.* Plains should be called *plains.* But unusual names have been assigned to surface features on Venus that are meaningless without translation. A brief dictionary of Venus language is given below.

Regio	region
Terrae	highlands
Regiones	moderate-sized highlands
Planitae	low plains
Plana	high plains
Montes	mountain ranges
Chasma	canyons
Rupes	steep cliffs

Ordinary English will be used here unless the professional designation is necessary.

Until the space age, essentially nothing was known about the surface of Venus because it was, and still is, completely obscured by its dense atmosphere. Then experiments with a huge radio telescope at Arecibo, Cuba broke through. A few regions of continental proportions were found to be very rough and about 2 mi above the general surface. Subsequent research from two spacecraft orbiting the planet revealed further details. But a real map of the topography was finally provided by radar echoes from Magellan. They are so sharp that photo-quality images of the surface can be produced.

Asteroid Craters

Fewer asteroid craters were found on Venus than had been expected. A total of about 900 did have a size distribution comparable to other bodies but with a lower threshold. None were found smaller than about 2 mi in diameter because small asteroids that would have created them burned up in the thick atmosphere. An exact analogy is the flaming tail of a terrestrial meteor that comes to an end when the grain-sized particle has been consumed. The largest crater was 168 mi in diameter, perhaps, about the upper limit for asteroid craters corresponding to the largest ones that have been nudged out of the asteroid belt.

Highlands

The most prominent of the previously discovered highlands, Ishtar Terra, was found to be about the size of Australia. It was distinguished by a great plain in the western half and rugged mountains in the eastern half. Planetary scientists and geologists have explained its origin as being uplifted by vertical movement of underlying magma. Whether the flow is upward toward the surface or downward toward the center is still debated because the physics is ill defined. As it turns out, neither explanation is likely to be correct.

The best image of this highland is available in a perspective drawing based upon the radar data with exaggerated elevations. The plain has all the characteristics that have become so familiar. The boundary is a sharply defined arc around two thirds of a complete circle. Its edge falls steeply down to the planet's geoid, that is, sea level if it had any water. A circumferencial fault has allowed a vertical displacement with the disk of the highlands slipping upward past the undisturbed crust. A cluster of mountains rises at the center of the arc as common among large craters. Lava filled the entire basin then froze to become the central plains. The center is located at about 46.8 N 315 ° longitude with a diameter of about 1300 mi. This configuration almost certainly marks a cosmic-scale explosion resembling a thin plug of toothpaste emerging from a tube.

The very thick crust of the planet severely delimited cratering so the maximum depth was much less than a crater of the same size on earth. It also prevented overthrusting because the vertical displacement was only a small fraction of the crustal thickness. The internal disk simply slid upward a relatively short distance along the circumferencial fault before coming to rest. In a thin crust, expansion of the disk would have created a subduction trench. Subduction trenches are, therefore, very unlikely on Venus in contrast to the terrestrial examples. The latter are seen on earth as to be predominantly associated with ocean areas where the crust is relative thin. One structure on Venus is an arc of volcanoes that resembles an island arc. They would be islands if an ocean were present and imply an example of overthrusting. Overthrusting exists on Venus but it is very rare.

The eastern half of this highland is a linear, mountain range culminating in a huge peak of volcanic origin. It is higher than Mt. Everest and bulkier than Hawaii reckoning from the Pacific Ocean floor! This pattern appears to mark a linear, rift system radiating from the center of the crater.

Volcanoes are powerful critters but extremely dumb. They can not decide to distribute themselves randomly about the surface. They have no freedom to choose their locations. They form only where major faults open avenues for magma to escape to the surface. Only the explosions of cosmic scale are powerful enough to fracture the crust to sufficient depth. Consequently, all volcanoes appear at or close to a) centers of megacraters, b) their rims, and c) their linear faults. It should be no surprise to find volcanic mountains at the center of the western plains and along a straight line in the eastern highlands. Ishtar Terra appears to be a megacrater of intermediate size.

Circular Canyons

Another striking feature on the surface is a perfect circle 1600 mi in diameter known as Artemis Chasma. This canyon of unusual shape stands out brightly against a dark background. Also associated with this circle are at least four others that are nearly contiguous, generally with smaller diameters. Taken together, they resemble a chain of craters produced by fragments of a single comet. Curiosity forces a closer look at the data in Table C.1. Chain Of Fragments. Fragments are identified in order of their impacts with letters of the alphabet as in Shoemakerthat hit Jupiter in 1994. Levy 9.

Table C.1. Chain Of Fragments.

Identity	Diameter (mi)	Latitude (deg)	Longitude (deg)	Delay (hr)
A	1600	31.3 S	128.2	
				1.60
B	1500	21.0 S	134.5	
				1.48
C	1350	23.6 S	151.0	
				0.906
D	950	20.0 S	156.6	
				1.032
E	2160	20.0 S	168.5	
			Total	5.02

A progression toward smaller diameters is obvious with the glaring exception of fragment E that is notably large. That sequence of fragments is not uncommon and may mean that E simply resisted breakup in the gravitational field. Taking into account the size of Venus and the length of its day, one easily determines the relative times of arrival as shown in the last column. They landed at intervals close to 1 hr between fragments for a total period of 5.02 hr. It is likely that additional craters of smaller size will be found along the axis of the group. A general drift toward the north in observed in the table that is comparable to a gentle, southerly drift of the S/L9 fragments on Jupiter.

A chain of craters on earth would stretch out to the west because the surface rotates to the east. Quite the opposite is seen on Venus that is exactly appropriate. Unique among the planets, the rotation of Venus is retrograde, that is, the surface moves to the west so a chain of craters would line up toward the east.

Also, the relative sizes of all the circles is about what would be expected.

Population

Because Venus appears to be deficient in asteroid craters, the number of megacraters might also be limited. As for asteroids, the reasons are twofold. The strong field from the nearby sun probably draws many of them to itself that otherwise would have hit Venus. It acts as a giant catcher's mitt collecting wayward objects and protecting the planet. The second reason is that the source of asteroids is farther away than for Earth. The asteroid belt lies outside the orbit of Mars while Venus is closer to the sun than the earth.

The same may be true regarding megacraters. Some found by the author are assembled in Table C.2. A few candidates include five associated with Artemis Chasma.

Table C.2. A Few Candidates.

Identity	Number	Diameter (mi)	Latitude (deg)	Longitude (deg)
-	1	3400	40.0 S	210.0
-	1	2400	17.4 N	2.4
E	1	2160	20.0 S	168.5
-	1	2000	40.0 N	180.0
-	1	1600	27.9 N	70.9
A	1	1600	31.3 S	128.2
-	1	1500	1.0 N	181.0
-	1	1500	29.5 S	330.0
B	1	1500	21.0 S	134.5
C	1	1360	23.6 S	151.0
-	1	1300	49.0 N	47.1
-	1	1200	5.0 S	56.6
-	1	1000	24.6 S	308.5
D	1	950	20.0 S	156.5
-	1	800	45.5 N	93.6
-	1	700	27.6 S	44.5
-	3	600	48.3 N	128.5
-	1	500	29.7 S	27.2
	2	350	60.0 N	270.0

The range of diameters among these 22 examples is in close agreement with that for earth up to 3400 mi. Also, the locations on the planet are roughly random with equal numbers in the northern and southern hemispheres. Many others probably exist but are hard to find in an available map. Also, craters smaller than 350 mi are invisible in a map covering the entire surface. A massive effort will be required for detailed studies of enlargements of limited regions on the surface to establish a complete list of candidates. The number of megacraters tabulated above is only a small fraction of the 208 found on earth. Even the latter number is known to be deficient. Also, the projected total of 1000 craters above 300 mi is even larger. It appears at this point that Venus is very deficient in megacraters compared to earth and the reason should be clear.

The low mass of Venus, 0.815 times that of earth, weakens its ability to capture comets passing in the neighbor. On the other hand, the force from solar gravity is 1.89 times that at the surface of earth. This data alone could predict a deficiency of craters on Venus. The size of the planetary orbit would contribute to this disparity. Passing comets could circle the sun at distances greater than the orbit of Venus with no possibility of impact.

Spacing Of Craters

The distance between centers of craters in a chain can provide some clues to the history of the parent comet. At least five fragments, it is recalled, created a chain of craters on Venus within the period of 5.02 hr. Breakup in a gravitational field imparted an initial velocity of separation that would not change in the absence of additional forces. The time interval between impacts and typical speed of comets lead to their separation in space. Many such details of the original string of fragments can be calculated from such data.

Appendix D. Lessons From Jupiter.

"...happened long ago, when the solar system was young; similar calamities cannot happen again now that the planets have evolved."
- Clark R. Chapman

To a first approximation, a stream of comets is considered to travel along the same trajectory. So impacting a rotating globe would be similar to shooting it with a machine gun as the bullets spaced themselves out along the same line of latitude. The globe is also assumed to be stationary. These concepts would naturally direct any search on earth for lines of craters recording damage from comet streams. Scars on Jupiter left by the Shoemaker-Levy 9 conglomerate reveals a weakness of the model. The following analyses relies principally upon data from West (1994).

Shoemaker-Levy 9

Table D.1. Data For S/L9 Impacts shows the individual fragments, their impact dates, times, and latitudes expressed as decimal fractions to the second place. The event started with fragment **A** on July 16th at 20.18 hr Universal Time (UT same as Greenwich Mean Time) and lasted until fragment **W** struck on July 22[d] at 08.10 hr UT. The simple model would predict all fragments landing along the same latitude as **A** but substantial differences are noted. Latitudes between **A** and **W** were a whole degree. Taking the radius of the planet at that latitude as 85,750 mi, the surface displacement would be 747.65 mi, say rounded to 750 mi. There was a pronounced trend for each fragment to hit farther south than the previous one, as illustrated in Fig. D.1. Trends On Jupiter.

Table D.1. Data For S/L-9 Impacts.

Fragment Number	Day July 1994	Latitude S Minus 43.00°	Cumulative Time (hr)
A	16	.15	0
B	17	.17	6.7
C	17	.38	10.84
D	17	.46	15.72
E	17	.48	1900
F	18	.55	28.37
G	18	.60	35.37
H	18	.74	47.35
K	19	.80	62.06
L	19	.92	74.10
N	20	1.39	86.30
P2	20	1.64	91.20
Q2	20	1.26	95.55
Q1	20	1.05	96.04
R	21	1.07	105.39

S	21	1.16	115.09
T	21	1.99	117.99
U	21	1.43	121.74
V	22	1.44	128.20
W	22	1.15	131.92

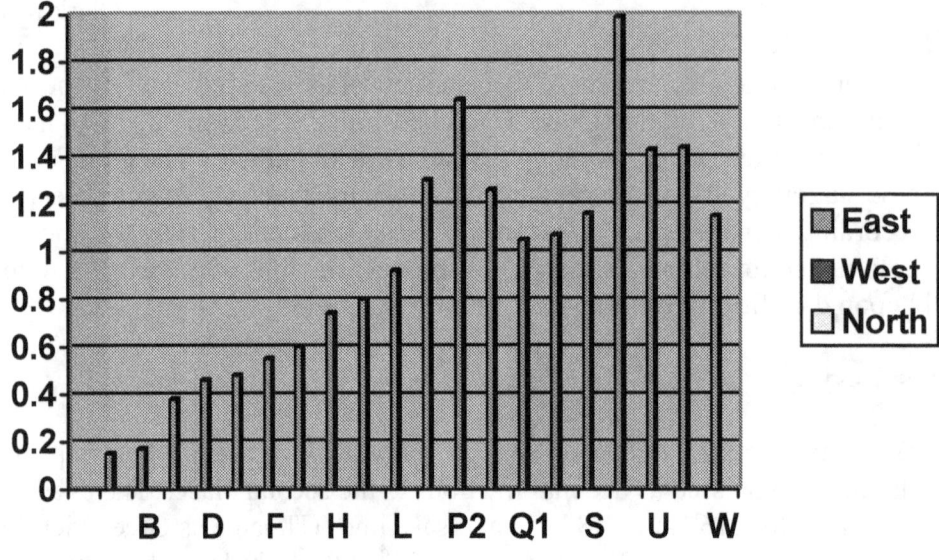

Fig. D.1. Trends On Jupiter.

The gradual increase in south latitude becomes more clear by plotting the decimal fractions starting at 43° S instead of the values directly. Numbers above 44° S appear as *one plus a decimal fraction.* The bar graph shows the strong and nearly uniform trend toward more southerly impacts from **A** to **U**, amounting to 1.43° over a period of 121.74 hr. If all fragments had landed at 43.15° as established by **A**, the vertical bars would have all been the same height.

Several rogue fragments in the string of flying comets were substantially offset from the main axis. Among them were **N, P2, T,** and **P1**. Fragment **T** was traveling with the greatest offset of 1.99 - 1.35 = 0.64° of latitude that translates into 480 mi off the track. On the surface of Jupiter its displacement from the simple prediction would be 1.99° x 750 mi/° = 1490 mi.

As to longitudes, the scars could have been created anywhere. The huge planet completes a rotation on its own axis at the unreasonably short period of 9 hr 50 min 30 sec. Several Jupiter days passed during the entire event. Specifically, the number of rotations was 131.92 hr / 9.842 hr/rot = 13.404 rotations, that is nearly thirteen and a half Jupiter days. If the scars had been permanent craters, it would have been impossible by studying them to establish the order of their positions in the original chain.

Why, one may ask, was the pattern created? Answers appear to lie in two areas.

Jupiter was not a stationary target but moved along in solar orbit at the good clip of 29,520 mi/hr. During the period of bombardment, it moved a total of 29,520 mi/hr x 131.92 hr = 3,894,000 mi! So the landing point on the surface of any delayed fragment would not be likely to hit on exactly the same latitude as the first one.

Secondly, longer exposure of more distant fragments to the gravitational field probably contributed to the observed pattern. But complications in this area are too great to address.

A short string of asteroid fragments landing on earth within an hour would leave a chain of craters less than 730 mi long. Within reasonable limits, they would be on approximately the same latitude depending upon their initial positions relative to the axis of the aerial string. Two such fragments flying only 0.5 hr apart, however, would land close to the same latitude as *Aleutian E* and *Aleutian W.* Similar results should not be expected of a chain of comets stretched out far enough to rain down for a few days.

Mechanism Of Impacts

Since Jupiter has no surface, fragments of S/L-9 never struck anything solid. Explosions took place in the atmosphere. Nuclear explosions must have been triggered when compressed gases ahead of each fragment reached several million degrees required for spontaneous fusion. Chemical composition of comets is not completely known and they may vary considerably. Nevertheless, they contain large amounts of hydrogen that can undergo fusion according to the steps described in another appendix.

Fragments were so large and traveled so fast that they left vacuum trails through the atmosphere. Local gases and plasma did not have enough time to close off the evacuated tube. Much of the explosive force was directed backward and upward whose plumes had been observed over the Jupiter horizon. Debris from the explosions spread out at high altitude forming dark circles up to the size of earth. In all likelihood, material in those clouds was supplied exclusively by the fragments.

Similar mechanics would be expected for comets exploding in earth's atmosphere. Shock waves, heat, enormous pressures, and neutrons would interact with the surface. The fireball, however, may not have reached the ground.

For asteroids penetrating the atmosphere and the surface to some depth, gravity would dominate the dispersal of rocks of all sizes. Not only would they be influenced by earth's gravity but also the attractive forces between each pair whose size could be extremely large. Canup (2001) at Southwest Research Institute is working through this process of staggering difficulty with the most sophisticated, computer simulations.

Explosions high in earth's atmosphere would be very different. As in S/L-9, an evacuated track would come down to the point of explosion. Most debris from the comet would project upward except for the hydrogen that was consumed. The plume of plasma would cool and condense into solid particles of every size. Larger one would fall immediately. Small glass spherules would rain down over vast areas while those of microscopic size, along with dust particles, would linger and drift afar. Solar energy would be entirely blocked from a few days to many years then slowly return to normal.

Particles of cometary explosions must have been deposited on earth in boundary layers between all geological periods. Rare and stable isotopes in those boundaries that were created by absorption of neutrons would confirm the nature of the explosions. An organized search for these clues is certainly justified.

Appendix E. Analysis Of Rim Ratios.

"I grow old learning something new everyday."
- Solon

Because earth's crust is stretched when it lines a crater, one suspects a relationship between its size and the extent of overthrust at the periphery. That relationship was converted to numerical values, called Rim Ratios, being the outer radius divided by the inner radius. This subject can be examined in a *scatter plot*. They are easily created by entering data in a computerized spreadsheet, selecting Scatter Plot, the displaying the graph as in Fig. E.1. Rim Ratios. One advantage of this format on a computer screen is the ability to point the cursor at any data point and recover the corresponding values. They imply a particular event so subtle questions about the database can be studied. About a quarter of all craters (26%) displayed double rims whose separations are provided in the 7th column of the tables in Appendix A. They have a mean value of 1.114 of the radius that is equivalent to an overthrust of 11.4% of the radius. All values lie between 1.03008 and 1.211. A standard deviation of 0.043669 means that half are confined to the band from 7.0% to 15.8%.

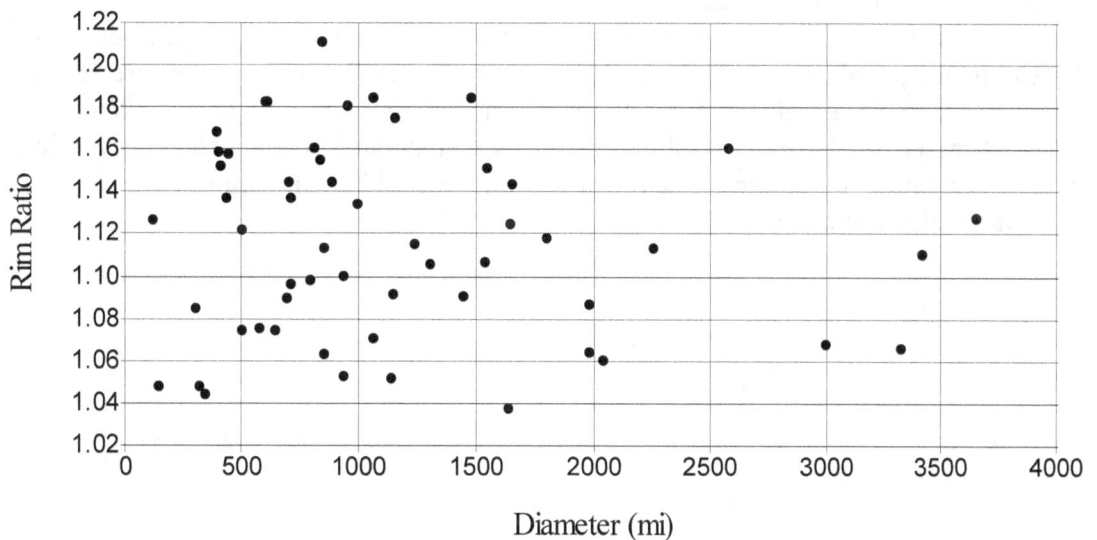

Fig. E.1. Rim Ratios.

Results are confusing. Nevertheless, some meaning may be extracted from the diagram. The largest crater with a diameter of 3640 mi produced an intermediate overthrust of 1.128. Whereas the opposite end of this spectrum, a small crater with a diameter of only 320 mi produced one of the largest ratios of 1.183. Very large overthrusts appear to be created by craters *under* 1000 mi in diameter. That may, however, be illusory because relatively few are larger than that arbitrary cutoff. Variation among the larger group, the range of values, is greater than that of the smaller group.

Of special interest are examples of large overthrusts represented by 1.211 and five others having values above 1.180. They lie in a rather restricted zone of diameters, about 600 mi to 1500 mi. Do they represent cases when the craters were formed in thin, ocean floor? That is a tough question because their ages are generally unknown and the pattern of continents versus oceans has changed over time. At any rate, the craters under consideration are *Phillipine**, *Baikal*, *Montana**, *Somalia**, *Aleutian E**, and *Iran.* Asterisks indicate the likelihood of oceanic creation. Two small craters on the North American continent have almost the lowest Rim Rations. They are *Belcher* (140 mi, 1.048) and *Navajo* (320 mi, 1.048).

Something is obviously wrong with *Chile* at 3.2% in the company of other values averaging 11.4 %. The rounded shoreline at the southern tip of South America is easily interpreted as part of a megacrater but an alternative explanation was suggested in Section 1.15 of Chapter 1 and illustrated in the accompanying map. Creation of that curve, attributed to drag along the slip faults from the South Sandwich Islands, is supported by the bastard value of its rim ratio. Chile, therefore, should be stricken from the list. It has not been because the data above is essential to the argument.

Some megacraters were omitted from the master catalog because the pattern of their concentric rims was far from typical. Their 2 to 4 rims were widely spaced and overthrusting was not apparent. Several examples in this category are compared in the Catalog Of Megacraters of Appendix A. Alternative mechanisms to produce theses anomalies must be explored. Firstly, these craters may not have been generated by aerial explosions. They might have been dredged up by huge asteroids that survived the atmospheric plunge and penetrated deep within the earth before exploding. Perhaps, sloughing on the internal faces of the bowl created the inner rims.

Some numbers are interesting here. Asteroids approaching at a typical speed of 40,000 mi/hr would be equivalent to 11 mi/sec. So a delay of 2 sec after striking the surface would have placed the explosion at a depth of 22 mi. Extraordinary masses of rock to be moved would react rather slowly to the resulting forces. It seems reasonable that a crater might be excavated to a depth of a hundred miles. Such events would be compatible with the idea of failing inner walls of the mature crater.

Appendix F. Kinetic Versus Nuclear.

*"It is well known from the history of science
that many great revolutionary works...would
would never have been published if the writer's
peers had been asked."*

- Immanuel Velikovsky

Fragments of Shoemaker-Levy 9 were estimated to be in the range of 3 to 5 km across but, for the present purpose, the average size will be treated as 4 km or 2.49 mi. That object would be small compared with the 10-km comet, or more likely asteroid. that excavated **Chicxulub**. There appears to be a disparity between the sizes of objects striking Jupiter and Mexico compared to the extent of the damage they caused.

Kinetic

Energy of a moving object depends upon its mass and speed or, in the shorthand and precision of mathematics, $E = \frac{1}{2} mv^2$. Calculation of that value is most convenient using the cgs (centimeter-gram-second) system of units that can be converted at the end into more familiar terms. Because the numbers are very large, the scientific convention will be used where a number close to 1.0 is multiplied by some power of 10 as in $2.34 \times 10^3 = 2340$. The general reader may experience rough sledding here but is encouraged to proceed anyway to understand the concepts if not the numbers. The primary audience for this appendix is the scientist or engineer for whom the equations and calculations would be second nature.

Taking the diameter of a spherical fragment of Shoemaker/Levy 9 as 2.49 mi converts into centimeters as follows

$$2.49 \text{ mi} \times 1.609 \times 10^5 \text{ cm/mi} = 4.01 \times 10^5 \text{ cm}$$

whose volume would be

$$V = (4/3) \pi r^3 = 3.40 \times 10^{16} \text{ cm}^3$$

For a typical density 40% greater than water, the weight would be

$$M = V \rho = 3.40 \times 10^{16} \text{ cm}^3 \ 1.40 \text{ gm/cm}^3 = 4.76 \times 10^{16} \text{ gm}$$

The designated speed converts to

$$134,000 \text{ mi/hr} \times 44.7 \text{ (mi/hr) / (cm/sec)} = 5.99 \times 10^6 \text{ cm/sec}$$

Now the input data are available to calculate the energy as

$$E = \frac{1}{2} \ 4.76 \times 10^{16} \text{ gm} \times [5.99 \times 10^6]^2 = 1.71 \times 10^{28} \text{ gm cm}^2/\text{sec}^2$$

which, by definition, is the same as 1.71×10^{28} erg, the fundamental unit of energy in the cgs (centimeter gram second) system.

In this form, the result is hardly useful because the present interest is in relating the energy to the explosive power of nuclear bombs that have units of *millions of tons of trinitrotoluene (TNT)* usually express as x MT for any value of x. The Air Force in Williams (1963) defined a megaton as approximately 4.2×10^{15} Joule where the Joule is the unit of energy in The International System of Units equal to 10^7 erg so that

$$1 \text{ MT} = 4.2 \times 10^{22} \text{ erg}$$

That conversion factor provides the energy of the fragment as

$$1.71 \times 10^{28} / 4.2 \times 10^{22} = .407 \times 10^6 \text{ MT} = .407 \text{ MT}$$

So the kinetic energy of the hypothetical comet would be .407 MT, a most impressive figure when compared to a measly 15 MT of a hydrogen bomb. The number of such bombs required to produce the same energy is simply

$$.407 / 15 = .027 \text{ bombs.}$$

So the kenetic energy in this example is equivalent to only 2.7% of single, Nominal HBomb.

Nuclear

Many factors support the idea that the Jupiter explosions and megacraters on earth were caused by nuclear fusion of hydrogen as in an H-bomb but on an incomparable scale. The essential requirements are the presence of ordinary hydrogen and the heavier isotope deuterium in an environment of extreme temperatures and pressures. All of these factors exist when a comet crashes through earth's atmosphere. In addition, the nuclear reaction can be triggered by a large spark between the comet and earth in parallel with the application of an atomic bomb to set off a hydrogen bomb.

The basic data for Jupiter include the impact velocity of about 134,000 mi/hr, energy release between 6 and 200 million MT, and gas temperature about 200,000 °C.

It is learned from the authoritative Gladstone (1957) that the fusion of 1 lb of deuterium is somewhat more powerful than a standard atomic bomb with explosive energy of 26,000 tons of trinitrotoluene (TNT). That value should be viewed in comparison the energy of the most powerful fragment of S/L9 at 200,000,000 million MT. That staggering number is so large that it must be examined in the light of personal experience. The author worked as a scientist for 5 yr on the effects of nuclear weapons. It took 2 yr to fully comprehend the destructive power of an atomic blast of 20,000 tons of TNT that is usually abbreviated to 20 KT with K signifying kilo = 1000. Photographs of post-attack Hiroshima and Nagasaki are stunning illustrations. Another 2 yr were required to comprehend the power of a weapon 1,000 times stronger, an H-bomb of 15 MT witnessed from close range on shipboard at Eniwetok. But the estimated energy in the Jupiter explosion was equal to the combined energy of 13,000,000 million hydrogen bombs! The mind boggles at such numbers and the full meaning may exceed the mental capacity of humans. Nevertheless, some idea of the violence can be roughly gauged by studying the megacraters that have been described.

The next task at hand is to examine the composition of comets at the atomic and nuclear levels to see how much energy could be released. A starting point is the figure from Gladstone (1957) where one sees immediately that an explosion on Jupiter could be accomplished by fusion of 200,000,000 million MT / 0.26 MT per lb = 769.2 x 10^{12} lb of hydrogen.

Can there be that much hydrogen in a comet? Can it be ignited? Equations in Chapter 6 showed the fusion reaction starting with the heavy isotope of hydrogen, that is, H^2 (deuterium). So ignition would very likely require a resident supply of H^2 that happens to be known in measured amounts. Meier (1998) and a team of other astronomers used highly sophisticated techniques with the Clerk Maxwell Telescope on Mauna Kea, Hawaii, for the experiments. They found the ratio of deuterium to normal hydrogen in water on Hale-Bopp to be D/H = (3.3 ± 0.8) x 10^{-4} where the uncertainty is the standard deviation. This result closely matches previous observations of Halley's Comet and Hyakutake. So H^2 is, indeed, present in Hale-Bopp but it only weakly contaminates the normal hydrogen in water and methane gases and, doubtless, in their ices. The amount is a small fraction of the total with only 33 lb of H^2 per 10,000 lb of hydrogen. A higher concentration was later found in hydrogen cyanide (HNC) that occurs in smaller quantities and will be neglected here.

Recall that the idealized model of a comet contained 25% water, 25% methane, and 50% other solids. Hydrogen, therefore, is available in H_2O and CH_4. Each molecule would contain two atoms of hydrogen in water and four in methane. The fractions, 2/18 and 4/16, account for the ratio of Hyd in Water & Methane the sources of hydrogen would be:

$$\begin{aligned}
\text{Water:} \quad & 0.25 \, (2/18) = 2.78\% \\
\text{Methane:} \quad & 0.24 \, (4/16) = \underline{6.25\%} \\
& \text{Total} = 9.03\% \qquad \text{or about 9\%.}
\end{aligned}$$

In that much hydrogen, the amount of deuterium becomes.

$$H^2 = 0.0903 \text{ W } 3.3\text{x}10^{-4} = 2.98 \text{ x } 10^{-3} \text{ W, where W is the weight of the comet.}$$

In other words there would be 2.98 lb of deuterium in every 1000 lb of material in the hypothetical comet. It will be recalled from chap—that FUSION of 1 lb of hydrogen corresponds to the energy release of 26KT.

Earlier equations for fusion also showed the release of 3 neutrons in the consumption of two atoms of deuterium. Clearly the potential exists for the creation of more deuterium than the original amount. Escape of a significant number of neutrons from a large body is very unlikely although some will be taken out of play through capture by other elements. Capture of one of them by a hydrogen nucleus converts it into deuterium. So a chain reaction can be set off by fusion of resident deuterium, sustained by consumption of available deuterium, and augmented *ad infinitum* by birth and consumption of new deuterium atoms. The chain reaction is made easier because the fusion of deuterium is spontaneous under the right conditions and does not need to capture one of the free neutrons.

Next consider how the minimum size of the fragment can be estimated. The required weight of hydrogen in the fragment was shown to be 7.69 x 10^{12} lb. The corresponding weight of the fragment follows from the knowledge that hydrogen accounted for 9.03% of the total. The corresponding weight of the fragment in the adopted model would follow from an earlier equation of H = 0.0903 W so

$$W = H / 0.0903 = 7.6 \times 10^{12} / 0.903 \times 10^{-1} = 8.52 \times 10^{13} \text{ lb.}$$

And how large would a comet of that weight be?

For a typical density, ρ, 40% greater than water and a spherical comet with an unknown radius, r, the relationships are

$$W = (4/3) \pi r^3 \text{ from which } r = [(3\rho / 4\pi) W]^{-3}$$

Substituting known values yields

$$r^3 = (8.52 \times 10^3)(3)(1.4)/4 \ (3.141)$$

so that

$$r = 53,340 \text{ ft} = 10.1 \text{ mi.}$$

and the diameter becomes **20.2 mi**. This remarkable result should be compared with the typical size of S/L9 fragments, **2.49 mi**. The implications are profound.

If the analysis is approximately correct, comets must, in fact, consume all their hydrogen in nuclear explosions. The speed and mass of a comet do not contribute significantly to its explosive power. Inert constituents are vaporized. Nothing hits the ground except air under extreme temperature and pressure. Escaping neutrons would be captured in rock leaving evidence of the event. Radioactive decay of the activated material with very long half lives might pinpoint the time of the explosion.

Comparison

Asteroids plow into earth vaporizing rock and blasting out solid and molten material in vast quantities. A large bowl is left in the surface. Comets, on the other hand explode high in the atmosphere, also vaporizing and melting rock, but exert great pressure on the underlying bedrock over extended areas. The signature is a huge, multi-ring structure.

There is clearly a fundamental difference between craters caused by asteroids and comets. The first factor is a disparity in size with the maximum for asteroids represented by *Chiczulub* or maybe the larger *Aegean.* Crater *structures* are also characteristic. Simple bowls are left by asteroids where internal surfaces may suffer landslides. Multi-ring structures leave much larger circles defined by faults, uplifts, and drainage patterns. Larger sizes also initiate subduction of ocean floors under continents and fracture the crust along the existing terminator. Asteroids strike at any time whereas the largest comets strike at local dawn.

Finally, a relatively small comet consuming its own supply of hydrogen in fusion reactions can pack quite a wallop and raise hell on earth. It will be recalled that the explosive power derived from kinetic energy of the model comet was equal to *2,710 H-bombs* whereas fusion of hydrogen in the same model equaled *13,000,000 H-bombs*. Kinetic energy, therefore, is inconsequential and can be neglected except for one consideration. It is expended in compressing air to produce the environment required for igniting the nuclear reactions.

Appendix G. Size Of Tunguska.

Something flattened a Siberian forest by exploding at high altitude in 1908. The debate over its nature, comet vs asteroid, has been resolved for most scientists in favor of a comet. The present work has suggested that energy from aerial explosions of comets derives from nuclear reactions, in particular, the fusion of hydrogen. A method of analyzing the explosive power of an idealized model of a comet was presented earlier. Naturally, the composition of a comet can be adjusted according to scientific taste. That same approach can be applied to estimating the size of the Tunguska comet.

Fusion of 1 lb of hydrogen would yield the equivalent of 26 KT in the vernacular of atomic weaponry, 26,000 tons of TNT (Glasstone 1957). The amount of energy of Tunguska has been compared with that of a hydrogen bomb about 1000 times more powerful. The Tunguska explosion at about 26 MT would have required 1000 lb of hydrogen. Using the idealized model of a comet in which hydrogen accounted for 9.03% of the total weight and the density was 40% greater than water, one finds the weight of Tunguska as 10/0.0903 = 1110 lb. If the comet were spherical, not a bad assumption, its radius would be

$$r = [3 \, W \, / \, 4 \, \pi \, \rho]^{1/3}$$

where ρ is the density. Substituting the values gives

$$r = [3 \, (1110) \, / \, 4 \, \pi \, (1.4) \, 62.4]^{1/3} = [3.03]^{1/3} = 1.45 \text{ ft}$$

or a diameter of only 2.89 ft, say roughly 3 ft across! Very small but not out of line with the size of a hydrogen bomb.

Whatever reasonable composition of the comet is assumed, the calculations show that it must have been very small. For example, assume that hydrogen amounted to only 1/10 % of the total weight and the comet was very tenuous with density of only one hundred[th] that of water. (Astronomers have seen stars behind comets.) The adjusted result would be

$$r = [3 \, (1110) \, / \, 4 \, \pi \, (0.001 \, (62.4)]^{1/3} = [4247]^{1/3} = 16.2 \text{ ft}$$

with a diameter of 32.4 ft, about the size of a nice living room.

Extensive damage to the Tunguska forest was probably caused by a comet less than 100 ft in diameter, or on the order of 10 ft, or possibly smaller.

And Megacomets

This analysis for Tunguska can be extended to estimate the size of megacomets responsible for the largest megacraters. It is a simple matter of prorating the amount of hydrogen required to release the energy estimated for the largest fragment of S/L9, namely,

200,000,000 H-bombs. Discussion will be suppressed to emphasize the calculations. They then become easier to check along with the logic. Readers averse to all mathematics can simply look at the answer and conclusion.

26,000 tons of TNT	~	1 lb hydrogen
1 atomic bomb @ 15 KT	~	0.577 lb hydrogen
1 H-bomb is 1000 time larger	~	577 lb hydrogen
200,000,000 H-bombs	~	1.15×10^{11} lb hydrogen
Comet weight	~	12.4×10^{11} lb
Comet radius = $[3\,W\,/\,4\,\pi\,\rho]^{1/3}$	~	5950 ft = 1.13 mi
Comet diameter	=	**2. 25 mi**

While these calculations include many uncertainties, they should provide a reasonable clue to the required sizes. And there is no *a priori* reason to believe that such comets do not live undetected in the Oort Cloud under the influence of solar gravity or haunt the cosmos.

Appendix H. Candidate Craters.

"What I have discovered is really very simple, looking back. But it took me a long time to get it because I I wasn't thinking that way."

- Leslie J. Freeman

The practiced eye detects clues to megacraters even in areas where they had formerly been overlooked. Once suspected, further evidence usually becomes apparent along with a growing conviction that a true example has been found. On the other hand, a crater may not be backed up with sufficient evidence to warrant recognition as such. These judgments are not always easy. Faint records of very ancient explosions may never reveal further evidence and, of course, many of the oldest will remain hidden forever. At least until new methods of geological research can see below the surface with improved detail. Maybe the necessary details are already available in the archives of the oil industry. And oil prospectors may recognize them when they learn what to look for.

It seems improper to ignore cases with only weak evidence so a few are recorded here for further examination by interested parties.

Iceland   65 N 19 W, D = 2800 mi

A large, island nation in the North Atlantic. Famous for its intense volcanism and geothermal resources that are believed to be caused by the Mid-Atlantic Ridge on which it squarely sits. It is unique as no other place on that ridge has anywhere near the mass of lava that has flowed there and continues today. The rate is comparable with the flows that created the major Lava Plateaus throughout the world. Land area of 20,000 sq mi still grows. It may be the location of a great explosion related to the western rim carved along the northeast coast of Canada. That shoreline extends from St. John's, Newfoundland to Ellsmere Island at the northwest corner of Greenland. While quite large, such a crater would fit nicely in the range of sizes that have been found. Supporting evidence remains elusive and may never come to light. Even so, ***Iceland*** may eventually be confirmed as a megacrater.

Laramie   41 N 105 W, D = 1660 mi

A possible crater large enough to be of interest may join several others in the western states of the U.S. The epicenter would be near Laramie, Wyoming. Mountain ranges along the Pacific coast may indicate the rim whereas any such evidence to the east could have been hidden by deep, alluvial deposits of the Great Plains. Primary clues are the proximity of the Rocky Mountain National Park and alignment with the massive, lengthy, and straight fault system in Canada. This Canadian Trench with a width of several miles is quite similar to many others that have been found. It shows up prominently on most maps passing west of Calgary.

Petrified Forest   35 N 109 W, D = 280 mi

Lies due west of Albuqerque, New Mexico and is possibly related to the Petrified Forest. The inner rim to the east would be the first mountain range west of the Rio Grande River. Indications of this crater are so strong that it could have been entered into the Catalog.

Vancouver   46 N 134 W, D = 1230 mi

This famous, elongated island in southwestern Canada may be part of the rim that shaped the Pacific coast down to Cape Mendocino, California. While concave that far south, the coast switches at that point to convex that has been identified as the rim of other prominent craters in North America.

Shanghai   31 N 121 E, D = 1130 mi

Near the Chinese city. Rim may be mountain ranges from Changshe to Beijing plus Korean Peninsula. Also a submerged ridge and island groups from southern tip of Kyushu, Japan to Taiwan including Satsunan-Shoto, Nansei-Shoto, Okinawa, Hirara, and Omotot-dake. Probable rim ratio of 1.079.

Appendix I. Methods.

"I was reprimanded and told to accept the traditional interpretations."
- Zechariah Sitchin

Previous investigations of megacraters have relied almost exclusively upon noting circular patterns on the landscape. It has been moderately effective but additional approaches to finding them are needed. The present study has uncovered some novel methods for conducting the search.

A large fraction of the megacraters whose evidence can still be seen has been identified in Appendix A. Others will undoubtedly be found as the prospect of their existence has been introduced. However, a large number in the smaller sizes remain undiscovered. Even smaller craters may eventually be found to be essentially without limit. The present work rejected some very substantial craters because the catalog would have become unwieldy and because they were too small for global significance. Nevertheless, they would have raised havoc locally with very serious consequences hundreds of miles in all directions. Dust in the atmosphere from these 'pygmies' probably changed the climate to the detriment of all creatures on land. Those swimming in the oceans and seas close by were destroyed by heat and shock waves. Tidal waves of truly enormous proportions inundated islands and devastated coasts of surrounding continents. Creatures at these remote locations did not fare well.

The complete history of earth can only be understood when craters of all sizes have been found and described. Fortunately, a fresh variety of detective tools can now be applied to this monumental task.

Arcs And Circles
Special attention should be directed toward
- Shorelines - small bays are strong candidates
- Mountains - secondary and smaller mountains and hills
- Sea Mounts - underwater volcanoes
- Faults - concentric with small elevation changes
- Earthquakes - perfect record of major faults
- Volcanoes - created by lava usually flowing under overthrusts
- Trenches - overlapping bedrock suppressed the lower portion
- Mineral Resources - with mountains that follow crater damage
- Vegetation - changes in color and shade are sometimes revealing

Linear Faults
- Rift Systems - project the alignments to crater centers
- Volcanic Islands - built by lava from major faults
- Coral Reefs - surface remnants of extinct volcanoes
- Sea Mounts - sunken remnants or young volcanoes
- Submerged Ridges - lava flow from linear cracks
- Gorges - associated with major events

Curious details became apparent in many countries. Some of the most scenic and dramatic landscape had been set aside as national parks. Centers of other craters were frequently found close by, suggesting a multiple hit. So another avenue to be explored is the geology of all the national parks in the world.

The Future

Hopefully, further interest will develop in examining the megacraters of Appendix A. Geology in the regions should be thoroughly reviewed. Each one should be verified with additional evidence or rejected for lack of evidence. Experts may also be motivated to search for additional examples.

Every country might keep track of research projects conducted by their citizens and coordinate missions of government agencies, universities, and independent laboratories. A few of the participating organizations should take lead roles in maintaining copies of research reports, technical articles, conference speeches, television footage, and books on the subject. These depositories should become the primary links to a world-wide program with an identifiable location. The United Nations is an appropriate place for this function. The staff could expect a steady flow of data from the U.S. Geological Survey and many American universities. Parallel flows could be expected from other developed nations with the required strength in scholarship and finance. Worthy projects should be supported by grants.

Appendix J. Speed Of Crustal Cracks.

"In regions such as the Indian Ocean, where geophysical data are largely lacking, recon- struction must be a compromise."

- D. H. Tarling

At great distances from explosions, the first indication is telegraphed by seismic waves under ground and by shock waves in the atmosphere. These messengers are quite slow compared to the speed at which major rift systems develop. Rifts are generated by the transverse forces from the expanding crater that pry the crust open. In splitting firewood, the crack progresses much faster than the maul. Fortunately, the geological record provides sufficient data to calculate the approximate speed of fracturing. The following diagram pertains to creation of the Ninety East Ridge that was created by **Bengal**. A computer and supplemental reference materials may be required to verify the analysis as sketched below.

Global Positions
A	5.43N 89.56E
B	30.08S 89.56E
C	30.08S 87.13E

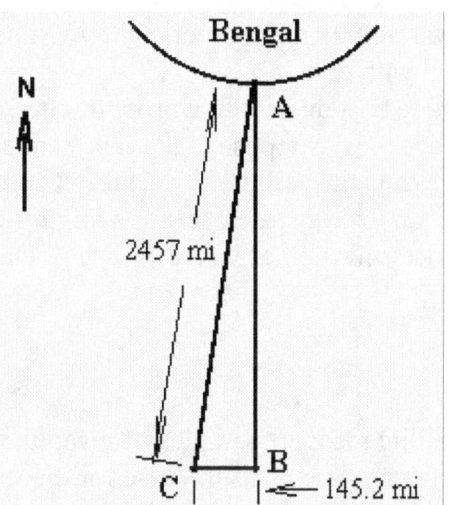

The rift began on the rim of **Bengal** at the point designated **A**. The longitude line of 90 ° is represented by **A-B.** The rift proceeded along the gently curved path **A-C**. During the time for the rupture to reach the end of the road at **C**, rotation of the earth toward the east had been displaced by 2.43 ° according to the longitudinal values above. At earth's rotation rate of 15 deg/hr, the time interval was 2.43 deg / 15 deg/hr = 0.162 hr. This surprisingly short period was only 9 min 43 sec! In order to traverse the actual distance of 2457 mi, the propagation rate had to be 2457 mi / 0.162 hr = 15,167 or about 15,000 mi/hr.

Sudden Stop

It has become axiomatic that advancing rifts can not extend beyond existing rifts that they encounter. Forces opening the growing fracture are instantly relieved by lateral movement of rock on the near side of the target fault. Fracturing along the initial direction comes to a dead stop.

Ninety East Ridge turned sharply toward Australia along the East Indian Ridge. Following the clue that rifts radiate from megacraters, it is simple to pick the one of many explosions in Australia that most likely ripped open the East Indian Ridge. It was clearly the huge *Artesian* on the eastern half of the continent with a diameter of 1130 mi. Recall that the Emperor-Hawaiian chain also veered east along a fault created by *Mid-Pacific* with a diameter of 2580 mi.

Determining the angles of these changes in direction can improve understanding of the process. The most accurate approach is to deal with applicable coordinates. Points are selected along the northern and eastern paths and the *knee* of the course change. Then the angles are calculated with results shown below.

Rift System	North Leg	Knee	East Leg	East Veer
90 E to E India	14.97S 88.00E	30.23S 87.92E	32.10S 98.88E	76.17 °
Emperor-Hawaii	41.05N 170.55 E	31.73N 172.35E	28.83N 178.63E	54.00 °

The Ninety East Ridge executed an easterly turn of 76.16 ° whereas the comparable figure for Emperor-Hawaii was 54.00 °. Encounters with existing rifts at rather small angles allow the transition to proceed with ease.

Some further details of these encounters are worthy of attention because the configuration at the knee is revealing. Suppose that a junction is shaped something like a capital **T**. The cross bar represents the earlier rift because it stopped growth of the stem. In the Pacific case, *Mid-Pacific* must have struck before *Aleutian W*. The configuration in the Indian Ocean, however, is indeterminate because the knee is shaped like a reversed, capital **L**.

Crater Ages

Whenever deep-ocean drilling recovers samples of lava, they can be dated in the laboratory with excellent accuracy. Such was the case for the Emperor Seamounts between the rim of *Aleutian W* and the bend near Midway Island. From the experimental data, it was possible to determine the rate at which the sequence of volcanoes proceeded along the fissure. It amounted to 50 miles per million years (mi/mil yr). Similar data for Ninety East Ridge would yield the appropriate rate that the sequence of volcanoes marched southward through the Indian Ocean. Absent that data, it can be reasonably assumed that the rate would have been approximately equal to the Emperor Chain. As Ninety East Ridge traveled a distance of 2457 mi before veering east, the time required would have been about 2457 mi / 50 mi/mil yr = 49.1 mil yr. That amount of time is an estimate of the latest date that *Bengal* struck. It would also give the latest date for other megacraters whose linear rift crossed the primary as for *Artesian* in Australia. As new information becomes available, this method can be useful in dating some of the megacraters. Confirmation may come from studies of the local geology.

Appendix K. Cordillera Neovolcanica.

"Growth in science is a series of understandings with each one based upon data of the former and brought about by new data or a broader perspective."

\- Author unknown

Some doubt might be raised about the Trans Mexico Volcanoes being on a straight line as required for a linear rift. To inquire into that question, the latitudes and longitudes of critical points were measured as follows

Name	Latitude	Longitude
*Mexico *	18.90 N	94.90 W
Arizaba	19.10 N	97.27 W
Malinche	19.23 N	98.03 W
Popocatepetl	19.00 N	98.62 W
Iztaccihuati	19.17 N	98.62 W
Paricutin	19.45 N	102.25 W
Colima	19.57 N	103.60 W

* Coordinates here are more accurate than those in the Catalog.

The above data were used to calculate the azimuth or bearing from the crater center to each of the volcanoes as shown in Table K.1. Test For Linearity.

Table K.1. Test For Linearity.

Name	Angle To (deg)	Distance To (mi)	From Average	Absent Popocatepetl
Epicenter	—	—	—	—
Drizaba	275.48	155.3	+ 0.25	- 0.28
Malinche	276.87	205.2	+ 1.64	+ 1.04
Popocatepetl	272.23	243.0	- 3.00	—
Iztaccihuati	274.73	243.5	- 0.24	- 0.84
Paricutin	275.73	480.8	+ 0.50	0.00
Colma	276.09	569.0	+ 0.86	+ 0.26
Average	275.09		1.08	0.70

Results are listed in the second column with the average value of 275.09 ° at the bottom. The spread in numbers is seen to be small but further examination is required to appreciate the full meaning. The figure of interest is the deviation of each bearing from the mean as shown in column four. Positive and negative signs indicated displacements to the north or south, respectively. Again, the average is shown at the bottom amounting to 1.08 °. The chain of volcanoes can certainly be considered to lie on a straight line with such a small spread of bearings. Popocatepetl is seen to be a major contributor to the error and might not be part of the family. So the deviations are calculated again disregarding that volcano. To do so requires finding a new average at the bottom of column two. New results are listed in the last column where no entry is shown for Popocatepetl. The new average is found to be less than one degree, namely, 0.70 °. With this small variation, the volcanoes are clearly on a straight line that originates at the center of the crater, **Mexico**. Such a straight line of volcanoes emanating from the center of a crater is clear evidence of a linear, rift system. Close examination of the geography shows that Drizaba lies just outside the crater rim that is the common origin of linear rifts.

The computed bearings provide a means for accurately determining the rift angle. That distance from true north is simply 275.09 ° - 360.00 ° = - 84.9 °. Although very large, this angle is comparable with the theory of unstable earth. It also compares favorably with the values of **Yukon** (80.8°), **Mexico Gulf** (78.6°), and **India** (78.6°). None of these craters could have been formed during the present regime of earth's inclination.

Appendix L. Surprise Alignment.

*Normal science…often suppresses fundamental
novelties because they are necessarily subversive
of its basic commitments."*

- Thomas S. Kuhn

Three, high-end megacraters dominate all of upper South America: ***Amazon***, ***Brazil***, and ***Brazilia***. On a map of that continent, one notices that the centers of all three lie close to a straight line. That relationship seems odd and raises the question if they could have been fragments of the same comet that struck in sequence. The alignment, however, is severely skewed from a parallel of latitude. If they are a true chain of craters, they had to be created during some other regime of earth's orientation.

Continental Drift

The Atlantic coast of South America hides the key for access to the facts. The bulging coast at Rio de Janeiro was originally attached to Africa at a shallow indentation on its coast. This snug fit ties down the positions along the coast of Angola at Luanda. The two continents were joined there during the integrity of Pangea. While racing south to split Pangea, the rift system from ***Timbuktu*** found a soft spot. It was the rim of a younger crater that offered easy passage. Upon completing a slight detour along that previous rift, it continued southward along its original course. ***Brazilia,*** as it is now identified, drifted away on the back of South America. Reconstruction of Pangea presents some startling details. Two great craters across Africa are in unique positions. The centers of ***Congo*** and ***Tanzania*** lie close to the straight line projected southeast from the South American chain. A coincidence, perhaps. But shocking, nevertheless.

All of them are large.

Amazon	1800 mi
Brazil	2180
Brazilia	1230
Congo	1890
Tanzania	1100

The African portion of a map displays the true lines of latitude and longitude. Those in South America today, of course, are very far off. A graphical solution shows South America has rotated clockwise by about 38.0 ° since leaving home and settling at its present address. If the entire line of craters were caused by a string of fragments, their alignment would have to lie along a parallel of latitude. The fact that they don't can be explained by two possibilities. Firstly, the spin axis of earth might have been different. An important distinction must be emphasized here. The cited difference has nothing to do with the inclination, that is, the angle by which the axis differs from the perpendicular to the ecliptic plane. The possible difference implies an entirely new spin axis without regard to the inclination. Perhaps it was caused by some cosmic collision far greater than any that produced the observable megacraters. But it is hard to imagine any event that could effect such a change without completely destroying earth. The enormity of earth's angular momentum, however, weakens that interpretation.

The alternative explanation may be correct. Pangea as a whole must have been rotated by 12.0 ° counter clockwise from the present orientation of Africa. If the full chain of five craters was a coordinated bombardment, then it was completed in 3 hr and 27 min. This line of investigation is too far afield for deeper investigation here. It is, however, a concept that deserves much further study.

Where Else?

The relationship between Africa and South America has been studied for a long time and is well known. The concepts are verified by data from magnetism in rocks, junctions of geological formations, continuity of animal species, continuity of plants, and matching of original coasts by computer analysis. Because most attention has been directed toward the Western World, corresponding data for the East is not as precise. Gradual displacements of China and Southeast Asia are more uncertain.

No rule of nature would restrict the impacts to Africa and South America. Some craters in the Pacific Ocean could be involved. Mintz (1972) presents the most accurate positions of primordial landmasses available to the author. During the Late Triassic about 200 My, South America and Africa were joined. Pangea had already been split east-to-west with North America and Eurasia departing to the north while South America-Africa remained in the south. Separation was along a waterway parallel to the east coast of North America, passing through the Strait of Gibraltar and the Mediterranean Sea. Farther east, Asia was drifting away from Africa along the spreading Red Sea. At any rate, this global landscape provides a timely, hunting ground for craters that might be aligned with those in South America and Africa. India would be a candidate except that it was never in the indicated position. A clear but uncertain projection points to the heart of Australia beyond which is open ocean. Two candidate craters come to mind. *Artesian* (1130 mi) and *Alice* (930 mi).

The problematic nature of these ideas has not escaped the author. The whole package could have been mere chance. Final answers will only be known when ages of the craters have been established with some accuracy. One or more may be the same age. If so, they could belong to the family. Any that are far off in age must be eliminated.

It will take a lot of research over several decades before the questions raised here can be settled. Other possible chains in Pangea should also be studied, such as, *Mozanbique - Brazil - Rain Forest - Manaus - Venezuela - Maracaibo.*

Appendix M. Geological Myths.

"With the collision of the Indian subcontinent against the southern shelf of the Asiatic land mass, the uplift of the Himalalya mountains began."

- J. Tuzo Wilson

Certain seeds have been planted in geology that took root and grew into giant, fictitious oaks. The causes are not entirely clear but seem to be related to introduction of new concepts, especially by scientists of some standing. Others elaborated upon them building layer upon layer of insulation against any challenge. The new ideas entered the curricula of universities and took on the air of holy writ. Learned professors taught this gospel to students who, upon earning their doctorates and taking academic positions, passed the material on to the next generation of students. As the chain lengthened, the original concepts that may have had some merit crystallized into inviolable doctrine. Upon achieving a loftly status, any errors were simply transmitted without question or careful review of the available data. A few such myths are examined here.

Emperor-Hawaiian Chain

As this subject was treated earlier, only a few more comments are required. The worthy concept of tectonic plates was not immediately endorsed by the establishment. With accumulation of more evidence, however, it was adopted as a key factor in earth's history and rightly so. Plate movement over hot spots easily explained long chains of volcanic islands, coral reefs, and sea mounts. A difficulty arose when application of these ideas required movement of tectonic plates over unconscionable distances along with wild gyrations. Some geologists may have felt uncomfortable about these problems but any challenges apparently went unheeded. A new paradigm was needed but none came forth. So the myth of long-range migration of the Pacific Plate survived. An alternative hypothesis was presented and defended in Chapter 1.

Island Arcs

Similar problems arose regarding formation of island arcs. Some explanation was essential. As tectonic movement had proven to be so successful, it was naturally applied again. The official position has been that island arcs were created by the subduction of ocean floor into trenches that are universally found on their convex sides. Beyond that, nobody tried to work out the physics. What mechanism could force the islands into such perfect arcs? Pursuit along this line would have been futile anyway because subduction was not the driving force. It should now be clear enough that island arcs represent overthrusts of earth's crust on the periphery of megacraters.

Multi-ring Structures

At this writing, craters on the moon have been thoroughly studied by experts in organized science. Very detailed analyses of their size distributions, however, have

243

neglected the multi-ring structures. Perhaps there are not enough of them or they presented an unwieldy break in the smooth distribution of all the others. The largest scar on the moon, South-Pole Aiken with a diameter of 1300 mi, was thought to be the largest in the entire solar system. It was pronounced with authority that nothing like the lunar mares has ever existed on earth and never will. *Au contraire*, they are the subject of this book.

Multi-ring structures have scored high marks in the error department. Following a misconception about the history of impacts on the moon, the giant scars were judged to be very ancient. A prevailing axiom was that all the objects floating in space had been captured about 4 billion years ago. So the multi-ring structures, mares, had to be very ancient. Of course, those ideas are balderdash.

Early termination of bombardment by asteroids and comets throughout the solar system has been a reassuring concept. People need not worry about large-scale disasters on earth. Observations in the last decade, however, have been alarming. Too many asteroids have been whizzing past the earth at troublesome distances. Perhaps the earth is not exempt from further attacks in the immediate future.

The comet, Shoemaker-Levy 9 put this issue to rest forever. The world would have really been plowed up and fried if the twenty-some-odd fragments had swung around Jupiter into a collision path with earth. There is no assurance that the next comet to invade the solar system will not crash on earth. A frightening aspect of this lesson is the short lead time of less than 3 yr for defensive action. But that really doesn't matter because nothing could be done about it anyway. Just hope for the best, pray, and carry on.

Wayward India

India broke off from Africa and Antarctica some time in the distant past. It then migrated 5000 mi to the north at considerable speed and slammed into the southern flank of Asia 55 My. Resulting forces in earth's crust subsequently raised the Himalayas. This argument is supported by linking a chain of hot spots and volcanic islands that must be the trail of long-range, tectonic drift.

Not a word of this myth is valid. As in other instances, the geological cart pulls the horse. The double arc of the Himalayan Mountains was established on the circular rim of a massive explosion in Asia then gained altitude from tectonic forces. Very little movement of India is implied. All of the mountain ranges probably continue to reach for the sky at about one inch per year.

The concept of continental drift was a brilliant contribution to geology. Many complex issues could be explained by it. So its popularity lead to excesses when invoked beyond the bounds of reason. In the case of India, special circumstances assisted adoption of the myth.

Geological knowledge of the Himalayas is far from complete. In fact, it has just begun in earnest. Uncooperative governments have stifled investigations for a hundred years. High altitude and low temperatures discourage field work. Every year or so a hand full of adventurers 'conquer' the peaks but see more snow and ice than geology. To be sure, untold numbers of local residents continue to observe the terrain but few have any interest or training in geology. If some have published scientific papers, none are available in English. In short, the amount of information on the Himalayas is woefully limited. Scholars have objected to unwarranted extrapolations from local studies in the region to the mountain ranges as a whole.

Crumbling Bedrock

A startling bit of scientific chicanery has been uncovered by Talent (1998) whose research was summarized by Lewin (1998). An Indian geologist at the University of Punjab, Dr. Viswa Jut Gupta, is the focal point of this bizarre story. For two decades, it appears that Gupta enlisted the aid of paleontologists around the world to study certain fossils he provided and to write joint articles for technical journals. Gupta claimed on numerous occasions that he had found the fossils in the Himalayas but, as it turned out, they had come from many other sources. Several dozen experts from the U.S., Europe, India, and Australia were drawn into the scheme resulting in the publication of more than 300 articles! The geology of the Himalayas became a house of cards based upon the analysis of fossil species that never lived there. An American expert alleges that the bulk of the material published by Gupta is totally false. Representatives of Science Journal eventually were able to contact Gupta who lightly passed off the problem as minor disagreement among experts. Nevertheless, this grand deception has thoroughly clouded any understanding of Himalayan geology. It also raises suspicions on the long-range drift of the so-called subcontinent of India.

History Of India

A huge explosion near Mashad, Iran ripped open the Persian Gulf and sent a linear fault system down the west coast of India. It actually created that coast by cutting off a thin slice that drifted away to the southwest. That rift became the mid-Indian Ocean ridge from which both sides slowly drifted apart. For India to have started as part of Antarctica and traveled north involves unacceptable concepts. It had to swim upstream against the ocean floor that was spreading southward from the mid-ocean ridge. Furthermore, the idea of drifting requires another cute trick. The sub-continent would have to jump over the ridge itself while firmly attached to the southern side. All of this is so ridiculous that wide-spread acceptance of the Indian myth over many decades is astounding.

To appreciate India's true, geological history requires a reconstruction of the original conditions. That is done by simply turning back the clock of the spreading floor of the Indian Ocean. Much has been learned by such analysis regarding the Mid-Atlantic ridge but this instance seems to be have been overlooked by geologists.

The key to unraveling this mythological mess is an obscure cluster of islands in the Indian Ocean known as the Seychelles. They cover 170 sq mi composed of 40 main islands in the Mahe group plus 52 atolls. A very small population swells with seasonal tourists seeking entertainment by fishing, snorkeling, yachting, and sunbathing on the beaches. A little-known, underwater ridge of 700 mi supports the islands that are anomalous. It must be a continental fragment because its principal component is granite. While insignificant in the culture and economics of the region, this ridge is fundamental to understanding the geology.

Sketch A on the left in Fig. M.1. Indian Ocean Birth And Growth shows the present arrangement of the continents, islands, and submerged features relevant to this inquiry. The Mid-Indian Ocean Ridge shown by a dotted line in Sketch A runs generally northwest to southeast but many transverse fractures have produced numerous, short offsets. Wide arrows show how opposite sides are spreading away from the ridge, a process that has been going on for a very long time. Here, again, one sees that ocean floor close to the ridge must be very young while portions grow older at greater distances. The oldest floor must be at the west coast of India and the east coast of the Seychelles Ridge. The distance between

India and the Mid-Ocean ridge is 870 mi compared to the 1030 mi distance between Seychelles and the Mid-Ocean Ridge. Normally, these figures would be the same but the difference of 160 mi can be attributed to the offset shown in sketch A along the Owen Fracture Zone. Clearly, the Seychelles Ridge originally snuggled up to and was contiguous with the Indian coast. In other words, it was part of Asia at the beginning of ocean spreading.

What happened? A linear fault system separated the two entities that began to drift apart to their present positions. And some powerful forces were certainly required. The source can be found near Mashad in the country of Iran at the center of a megacrater named *Iran.* This crater with a diameter of 1470 mi was previously discussed in the context of creating the Tigris-Euphrates Rivers, opening the Persian Gulf, and putting into place the greatest petroleum reserves in the world. Details included in Sketch B highlight the initial relationship between the crater rim and the linear fault. The original path of the Mid-Indian Ocean Ridge was along the Persian Gulf, through the Strait of Hormuz, and to the coast of Asia where it turned 90° to follow the linear rift from *Iran.* This configuration has changed over time to a fault along the 'foot' of Arabia whose origin was previously found to be *Ethiopia*.

The place where the original ridge changed direction is critically important regarding instability of the crust. That is where the crater rim intersects the linear fault from the same explosion. More damage to the crust should be suspected there because of the simultaneous disturbance from forces creating the rim and those creating the linear fault. Geology far below the surface at that spot must be in disarray on an unprecedented scale.

It is. Powerful earthquakes repeatedly devastate that area. On January 26, 2001, the capital of the Indian State of Gujarat, Ahmedabad, was flatted with the death of 2,300 people and 14,000 sent to hospitals.

Because that point is exactly the same as originally produced by the explosion near Meshad, Iran more than 50 My, nothing has moved. Instead of drifting from Antarctica to its present position, India has remained attached to Asia as it always has been. It never moved anywhere, is not a separate continent, does not subduct under Asia, and never did.

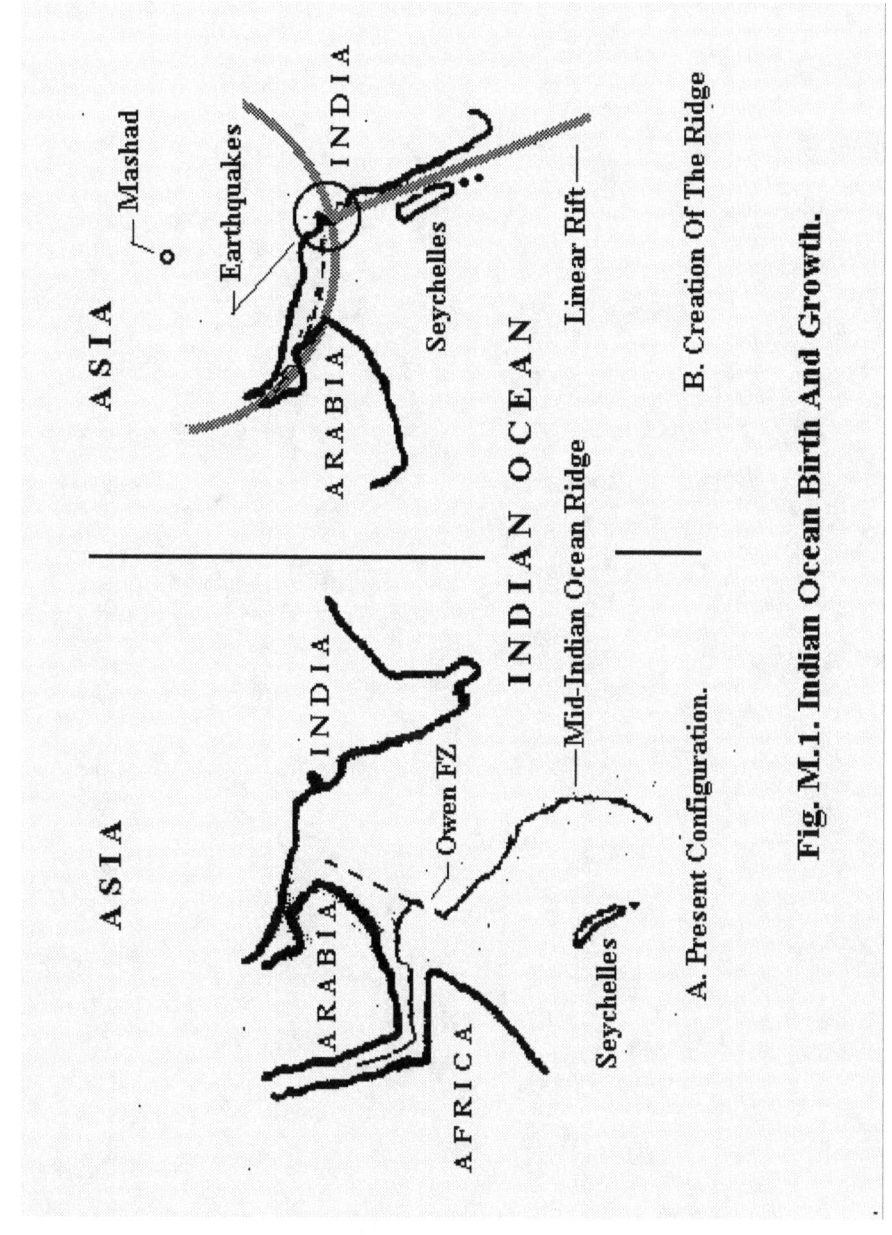

Fig. M.1. Indian Ocean Birth And Growth.

Bibliography.

*"All that mankind has ever done,
or thought, gained or been: it is lying
as in magic preservation in the pages
of books."*

- Thomas Carlyle

Abantt, Straher and Arthur N. Strahler, *Modern Physical Geography*, Fourth Edition, John Wiley & Sone, 1992.

Abelson, Philip H., *Isotopes in Earth Science*, Editorial, Science Vol 242 No 4884, 9 December 1988.

Acton, Gary D. and Richard G. Gordon, *Paleomagnetic Tests of Pacific Plate reconstructions and implications for motion between Hotspots*, Science Vol 263, 4 March 1994.

Adams, G.I., Charles Butts, L.W. Stephenson, and C.W. Cooke, *Geologic Map of Alabama*, Alabama Geological Survey Map 7, 1:500,000, 1926.

Ahlbrandt, Thomas S., *The Sirte Basin Province of Libya - Sirte - Zelten Total Petroleum System*, Version 1.0, U.S. Geological Survey Bulletin 2202-F, Junly 20, 2002. http://greenwood.cr.usgs.gov/pub/bulletins/b2202-f/

Albright, Joseph, *Saudis discover Huge Deposits of rare Super-Light weight oil*, Cox News Service, San Francisco Chronicle, September 8, 1990.

Alexander, C.M. O'D., A.P. Ross, and R.W. Carlson, *The Early Evolution of the inner Solar System: A meteorite perspective*, Science Vol 293 No 5527, 6 July 2001.

Algermissen, S.T., Robert M. Hamilton, Arch C. Johnson, Grian J. Mitchell, and Steven G. Wesnousky, *Earthquake Hazard and Risk Assessment to the Central United States*, in Proceedings of Annual Meeting of American Association for the Advancement of Science, New Orleans, 15-20 February, 1990.

Allen, Jane E., *New clue to demise of Dinosaurs*, Associated Press, San Francisco Chronicle, December 29, 1994.

Alter, Dinsmore, *Pictorial Guide To the Moon*, Third Revised Edtioon, Thomas F. Crowell Company, 1973.

Alvarez, Luis W., Walter Alvarez, Frank Asaro, and Helen V. Michel, *Extraterrestrial Cause for the Cretateous-Tertiary Extinction*, Science, Vol 2008 No 4448, 5 June 1980. Reprinted in Trower, W. Peter, Discovering Alvarez, *Selected Works of Luis W. Alvarez with commentary by his students and colleagues*, University of Chicago Press, 1987.

Alvarez, Luis, *Experimental Evidence that an Asteroid led to the Extinction of Many Species 65 million years ago*, Proceedings of the National Academy of Sciences, Vol 80, January 1983.

Alvarez, Walter, Frank Asaro, and Alessandro Montanari, *Iridium Profile For 10 Million Years Across the Cretaceous-Tertiary Boundary at Gubbio (Italy)*, Science, Vol 250, 21 December 1990.

Anderson, Dan L., *Top-Down Tectonics?*, Science Vol 293 14 September 2001.

Anderson, Robert S., *Evolution of the Northern Santa Cruz Mountains by Advection of Crust Past a San Andreas Fault Bend*, Science, Vol 249, 27 July, 1990.

Anderson, Roger S., *Marine Geology, A Planet Earth Perspective*, John Wiley and Sons, 1986.

Angler, Natalie, *What Underlies Himalayas' Quick Rise to the Heavens*, New York Times, reprinted in San Francisco Chronicle, fall of 1990, exact date unknown.

Antarctica, Index to United States Topographic and Other Map Coverage of Antarctica, U. S. Geological Survey, July 1987

Apocalypse Then (Possible Asteroid cause for Great Dying 250 My), NASA, Marshal Space Flight Center, (suggests fullerenes linked to Big Five Extinctions), February 23, 2001.

Associated Press, *Arkansas Schools Take Quake Forecast Seriously*, San Francisco Chronicle, July 10, 1990.

Asteroid and Comet Impact Hazard Bibliography (1992-2000) - NASA Ames Space Science Division, 8/19/01.
http://impact.arc.nasa.gov/related/biblio/

Attanasi, E.D. and J.L. Haynes, *Future Supply of oil and gas from the Gulf of Mexico*, U.S. Geological Survey Professional Paper 1294, U.S. Government Printing Office, 1983.

Audley-Charles M.G. and A. Hallam, Editors, *Godwana and Tethys*, Geological Society Special Publications, Vol 37, from a meeting May 1986, University of Oxford Press, 1988.

Auger, Emmanuel, et al, Seismic *Evidence of an extended Magmatic Sill under Mt. Vesuvius*, Science, Vol 294, 16 November 2001.

Austin, Steven A., *Catastrophes in Earth History: A Source Book of Geological Evidence, Speculation, and Theory*, ICR Technical Monograph 13, Institute for Creation Research, El Cajon, California, 1984.

Bahat, D. and A. Rabinowitcer, *On the Initiation of the East African Rift System: A Quantitative Analysis*, in Illies, 1970.

Baikal, Lake, Encyclopedia Britannica, 24 July 2001.
file://c:\Program%20Files\Britannica\2001\cache\info_10_.html

Baley, Mark E., *Where have all the Comets gone?*, Science Vol 296, 21 June 2002.

Bally, Albert W. and Allison R. Palmer, *The Geology of North America; An Overview*, Geological Society of America, 1989.

Bangs, Nathan, et al, Structure, Tectonics, and Sediment Flow into the Lesser Antilles Subduction Zone, National Science Foundation.

Barnes, Virgil E., *Tektites,* Scientific American, Vol 25, No 5, November 1961.

Basu, Asish R. and Paul R. Renne, *Early and late alkali igneous pulses and a high-3He plume origin for the Deccan flood basalts*, Science Vol 261 Issue 5123, 13 September 1993.

Basu, Asish R. and Robert J. Poreda, *High 3He plume origin and temporal-spatial evolution of the Siberian flood basalts*, Science Vol 269 Issue 5225, 8 November 1995.

Batson, R.M., P.M. Bridges, and J.L. Inge, *Atlas of Mars*, Scientific and Technical Information Branch, National Aeronautics and Space Administration, 1979.

Battke, William, Jr. and Stanley Love, *Do crater chains exist on Earth?*, Astronomy Vol 25 Issue 4, April 1997. (Abstract at EBSCOhost.)

Baum, Rudy, *Fullerenes detected in crater on spacecraft*, Chemical & Engineering News Vol 72 Issue 19, May 9, 1994.

Baxter, John and Thomas Atkins, *The Fire Came By - the Riddle of the Great Siberian Explosion*, Doubleday & Co., 1977.

Baxter, Sarah, *The Making of the Red Sea*, Discover, October 1985.

Beales, C.S., M.J.S. Innes, and J.A. Rottenberg, *Fossil Meteorite Craters*, in Barbara M. Middlehurst and Gerard P. Kuiper, Editors, *The Moon, Meteorites, and Comets*, Chapter 9, University of Chicago Press, 1963.

Beatty, J. Kelley, *Killer Crater in the Yucatan?*, Sky & Telescope, July 1991.

Beatty, J. Kelley, *Space surveillance: Hazard due to comets & asteroids*, Sky & Telescope, Vol 87 Issue 2, February 1994.

Beckel, Lothar, Editor, *The Atlas of Global Change*, MacMillan Library Referenc USA, 1998.

Becker, L, R.J. Pareda, et al, *Extraterrestrial helium trapped in fullerences in the Sudbury impact structure*, Science Vol 272 Issue 5259, 12 April 1996.

Becker, L. and T.E. Bunch, *Fullerenes: fullerenes and polycyclic aromatic hydrocarbons in the Allende meteorite*, Meteoritics & Planetary Science, Vol 32 No 4, July 1997.

Becker, L., Editor, *Fullerenes: an extraterrestrial carbon carrier phase for noble gases,* Proceedings of the National Academy of Sciences, 97(7), 2000.

Becker, Luann, et al, *Impact event at the Permian-Triassic boundary: Evidence from extraterrestrial Noble Gases in Fullerenes*, Science Vol 291 No 5508, 11 February 2001.

Behrendt, John C. and Donald D. Blankenship, *CASERTZ aeromagnetic data reveal late Cenozoic flood basalts (?) in the West Antarctic. (Volcanic and tectonic activity in subglacial basin of South Ross),* Geology Vol 22 Issue 6, June 1994.

Behrendt, John C., Editor, *Petroleum and Mineral Resources of Antarctica*, Geological Survey Circular 909, 1983.

Belt, Don, photographs by Sarah Leen, *The World's Great Lakes*, National Geographic, June 1992.

Bilham, Roger and Paul Bodin, *Fault zone connectivity: Slip rates on Faults in the San Francisco Bay Area, California*, Science Vol 258, 9 October 1992.

Bills, Bruce G., *An oblique view of climate (Earth obliquity calculated decrease from 54° to 23.5° in period 750 - 500 My).* Nature Vol 396, 3 December 1998.

Bjerre, Jens, *Kalahari*, translated from Danish by Estrid Bannister, Hill and Wang, 1960.

Blecker, Wouter and Richard Stern, *Window on the early Earth*, Science Vol 288 Issue 5471, 6 February 2000.

Bollinger, G.A. and S.P. Nishenko, *Forecasting Damaging Earthquakes in the Central and Eastern United States*, Science, Vol 249, 21 September 1990.

Bonatti, Enrico, Cesare Emiliani, Gote Ostlund, and Harold Rydell, *Final Desiccation of the Afar Rift, Ethiopia*, Science, Vol 172, 30 April 1971.

Books & Series in Print, An Annotated Catalog, 1989-1990, Bishop Museum Press, Honolulu.

Brauchli, Marcus W., Staff Reporter, *Japan taps unlikely oil source- itself*, The Wall Street Journal, August 31, 1990.

Brimhall, George, *The Genesis of Ores*, Scientific American, May 1991.

Broad, William J., *Scientists urge defense against killer Asteroids*, New York Times, April 1, 1992.

Broad, William J., *Spy satellite photos come in from the cold - Gore helps lift veil from secret '60-'70 images*, New York Times News Service, February 27, 1995. http://edcwww.cr.usgs.gov/dclass/dclass.html

Brown, Barry, *Snowbound Region holds one of world's largest Gem Caches*, ABC News. http://abcnews.go.com/sections/business/Daily News/diamonds 990831.html

Brown, Peter L., *Comets, Meteorites, and Men*, Taplinger Publishing Company, 1974.

Buffett, Bruce A., Edward J. Garnero, and Raymond Jeanloz, *Sediment at the top of Earth's core*, Science Vol 290, 17 November 2000.

Buffler, Richard T., *Late Cenozoic rift sedimentation, volcanism, and tectonism in the North Afar: Geological setting for potential hominid sites in the Danakill Region, Eritea*, June 22, 2001.
http://www.ig.utexas.edu/research/projects/afar/afar.html

Burgess, C.F., et al, *The Structural and Stratigraphic Evolution of Lake Tanganyika: A Case Study of Continental Rifting*, Chapter 35 in Manspeiger.

Burgess, Eric, Venus: An Errant Twin, Columbia University Press, 1985.

Buthman, D., *Crater Tectonics' Role in exploration; How asteroid and comet strikes have shaped Earth*, Offshore, Vol 58 Issue 7, July 1998.

Campbell, I.H., et al, *Synchronism of the Siberian Traps and the Permian-Triassic Boundary*, Science Vol 258, 11 December 1992.

Canup, R.M. and K. Righter, Eds., *Origin of the Earth and Moon,* (book including 29 papers from conference in Monterey, California, December 1998), Science Vol 292, 15 June 2001.

Canup, Robin M. and Erik Asphaug, *Origin of the Moon in a giant impact near the end of the Earth's formation*, Nature Vol 412, 16 August 2001.

Canup, Robin M., William R. Ward, and A.G.W. Cameron, *A Scaling Relationship for satellite-forming impacts*, Accepted by Icarus, December 4, 2000.

Canup, Robin, *Impact Origin of the Earth-Moon system*, Technology Today, Southwest Research Institute, Spring 1999.

Canup, Robin, *Moon-Forming crash is likely in New Model*, (summary at EBSCOhost), Science Vol 283 Issue 5398, 1 January 1999.

Canup, Robin, Public Information Officer, *Research indicates Earth's Moon may have formed in year or less*, University of Colorado at Boulder
http://www.colorado.edu/

Carey, S. Warren, *Theories of the Earth and Universe, A History of Dogma in the Earth Sciences*, Stanford University Press, date unknown.

Carlson, Richard W., *Do continents part passively or do they need a shove?*, Science Vol 278 Issue 5336, 10 October 1997.

Carr, Michael H., *The Surface of Mars*, Yale University Press, 1981.

Carver, John B., Chief Cartographer, *Earth's Dynamic Crust (map)*, National Geographic Magazine, August 1985.

Cattermole, Peter, *Venus, A Geological Story: A New Geology*, John Hopkins, 1994.

Chain of Craters Provides Clues, USA Today Magazine, Vol 126 Issue 2637, June 1998.

Chamberlin, Wellman, *Top of the World (map),* National Geographic Society, Atlas Plate 3, 1965.

Chapman, Clark and David Morrison, *Cosmic Catastrophes*, Plenum Press, 1989.

Chope, *Horizontal Motions of Global VLBI Sites*, NASA, 6 June 2002.
http://bowie.gsfc.nasa.gov/926/Vlkihorz.html

Choubert, G., and A. Faure-Muret, General Coordinators, *Atlas geologique du monde* (Geological World Atlas), United Nations Educational, Scientific, and Cultural Organization (UNESCO), 1976.

Christensen, Ulrich, *Hot Spots*, Nature 391, 739-740, 19 February 1998.

Christeson, G.L., *Deep Three-Dimensional Structure of Chicxulub impact crater from wide-angle seismic data*, American Geophysical Union, Eos Trans, AGU, 81 (48), Fall Meet. Supple., 2000.

Chronicle News Services, *Quake Ravages India - death toll tops 2,300*, also, Peter Popham, London Independent, *Random destruction devastates city*, San Francisco Chronicle, January 27, 2001

Chui, Glennda, Mercury News Science Writer, *Find could uproot energy theories, (gas in meteorite crater)*, San Jose Mercury News, July 28, 1992.

Chung, Sun-Lin and Borming Jahn, *Plume-lithosphere interaction in generation of Emeishan flood basalts at the Permian-Triassic*, (In China, not recognized internationally), Geology, Vol 23 Issue 10, October 1995.

Chyba, C.F., P.J. Thomas, and K.J. Zahnle, *The 1908 Tunguska explosion: atmospheric disruption of a stony asteroid*, Nature 361:40-44(1993).

Clark, Thomas H. and Colin W. Stearn, *Geological Evolution of North America*, 2d Edition, The Ronald Press, 1968.

Clube, S.V.M. and W.M. Napier, *Close Encounters with a million Comets*, New Scientist, Vol 95, 1982.

Clube, Victor and Bill Napier, *The Cosmic Serpent: a catastrophist view of Earth's History*, Universe Books, 1982.

Coffin, Mike and Fred Frey, *Ocean Drilling Program Leg 183: future ODP drilling on the Kerguelen Plateau and Brown Ridge: Determining the origin, growth, and evolution of a very large igneous Province in the Southern Indian Ocean*, Ocean Drilling Program, 4 June 1999.
http://www.ig.utexas.edu/research/projects/kerg/leg183/leg183.htm

Coffin, Mike, *USA-Japan geophysical investigation of the Ontong Java Plateau*, National Science Foundation.

Comins, Neil F., *What if the Moon didn't exist?: voyages to Earth that might have been*, Harper Collins, 1993.

Condie, Kent C., *Plate Tectonics and crustal Evolution*, Pergamon Press, Inc. 1976.

Conference on *Large Body Impacts and Terrestrial Evolution: Geological, Climatological, and Biological Implications*, Snowbird, Utah, 1981.

Cook, T.D. and A.W. Bally, *Stratigraphic Atlas of North and South America*, Exploration Department of Shell Oil Company, Princeton University Press, 1975.

Coombs, Cassandra, *Portal to the ancient moon: Mare Crisium*, Astronomy Vol 23 Issue 10, October 1995. (Summary at EBSCOhost).

Cope, Michael J., *Tanzania's Mafia Deepwater Basin indicates potential on new Seismic Data*, Oil & Gas Journal, 10 October 2000.

Corliss, William R., *Unknown Earth: a handbook of geological Enigmas*, The Source Project, 1980.

Couper, Alastan, Editor, *The Times Atlas of the Oceans*, Van Nostrand Reinhold Co., 1983.

Courtillot, Vincent, translated by Joe McClinton, *Evolutionary Catastrophies: The science of mass extinctions*, Cambridge University Press, 1999.

Cowan, G.A., *A Natural Fission Reactor*, Scientific American, Vol 235, 1976.

Cox, Allen and Robert Brian Hart, *Plate Textonics, How It Works*, Blackwell Scientific, 1986.

Cox, Donald W. and James H. Chester, *Doomsday Asteroid: can we survive?*, Prometheus Books, 1996.

Cozaya, Eberhard, *Rivers of the World*, Van Nostrand Reinhold, 1981.

Croneis, Carey, Editor, *Structural Geology of North America*, Second Edition, Harper & Row Publishers, 1951.

Crovisier, Jacques and Therese Encrenaz, *Comet Science: the study of Remnants from the Birth of the Solar System*, Cambridge University Press, 2000. Reviewed by Dale P. Cruikshank, *Icy Displays from before Time*, Nature, Vol 407, 12 October 2000.

Curran, H. Allen, et al, Editors, *Atlas of Land Forms*, 2d Edition, John Wiley & Sons, 1974.

Curtis, Michael L., *Tectonic history of the Ellsworth Mountains, West Antarctica: reconciling a Gondwana enigma*, Geological Society of American Bulletin, July 2000.

Darwin, Charles, *The Structure and Distribution of Coral Reefs*. University of California Press, 1962.

Davies, John K., *Cosmic Impact*, St. Martin's Press, 1986.

Davies, Merton E., Chief Editor, *Atlas of Mercury*, National Aeronautics and Space Administration, 1976.

Daview, Geoff, *A channeled plume under Africa (young flood basalts on Ethiopian plateau)*, Nature Vol 395 Issue 6704, October 22, 1998.

De Grazia, Alfred, Ralph E. Juergens, and Livio C. Stecchini, Editors, *The Velikovsky Affair, Scientism vs. Science*, University Books, 1967.

De Sitter, L.V., *Structural Geology*, Second Edition, McGraw-Hill, 1964.

Deccan Traps, India (17.224N 43.47E), elevation 4,000 ft. December 3, 2001.
http://volcano.und.edu/vwdocs/volc_images/europe_west_asia/india/decann.html

Delaney, Paul, Richard S. Fisk, Asta Miklius, Arnold T. Okamura, and Maurice K. Sako, *Deep Magma Body Benerth the Summit and Rift Zones of Kilauea Volcano, Hawaii*, Science, Vol 247, 10 March 1990.

DeLorme, *Eartha Global Explorer: The entire Globe on one Disc* (software), 1998.

DeMets, Dennis, Tim Dixon, Paul Mann, and Eric Calais, *A GPS Study of Caribbean Plate kinetics and distributed deformation between the Caribbean and North American Plates.*
http://www.ig.utexas.edu/research/projects/car_gps.htm

Dence, M.R., G.R.A.F. Grieve, and P.B. Robertson, *Energy Relations, Terrestrial Impact Structures, Principle Characteristics and Energy Considerations* in D.J. Roddy, et al, Editors, *Impact and Explosion Cratering*, Pergamon Press, 1977.

Dence, M.R., *Meteorite Impact Craters and the Structure of the Sudbury Basin*, Geological Association of Canada Special Paper 10, 1972. Reprinted in McCall, G.J.H., Editor, *Astroblemes - Cryptoexplosion Structures*, Dowden, Hutchinison & Ross, 1979.

Dent, Brian Edward, *Studies of Large Impact Craters*, unpublished doctoral thesis No. 3781, Stanford University, 1974.

Derry, Duncan R., Committee Chairman, *Tectonic Map of Canada* (1 in = 60 mi), Geological Association of Canada.

Desert glass: is it baked Australia, Science News, Vol 160, November 24, 2001.

DeSitter, L.U., *Structural Geology*, Second Edition, McGraw-Hill, 1964.

Diavik Diamonds Project, March 22, 2002.
http://www.aber.ca/project/mine_plan_details.htm

Dietz, Robert S., *Are we mining an Asteroid?*, Astronomy, Special Issue, Vol 1 No 1, January 1991.

Dietz, Robert S., *Demise of the Dinosaurs: A Mystery solved?*, Astronomy, July 1991.

Donofrio, Richard R., *Impact Craters: Implications for basement hydrocarbon production*, Journal of Petroleum Geology, 3,3, 1981.

Ediger, Vernita L., *Scientist reports finding First Mass Extinction: many Species died in Cambrian period*, San Francisco Chronicle, August 6, 1992.

Editors of Astronomy, *Volcanoes: windows into the Earth*, Special Issue.

Ehrlich, Paul R., et al, *Long-Term Biological Consequences of Nuclear War*, Science, Vol 222, 23 December 1983.

Eicher, Don L., A. Lee McAlester, and Maria L. Rottman, *History of the Earth*, Prentice-Hall, 1984.

Eicher, Don L., *Geologic Time*, Prentice-Hall, Inc., 1968.

Encarta, Interactive World Atlas (software), Microsoft, 2000.

Encyclopedia of World Rivers, Rand McNally, 1980.

Ergun, M., R.W.C. Westaway, and G.R. Foulger, *A GPS Survey of an Area of Continental Extension: Western Turkey*, Paper G31A-18, Annual Meeting of the Geophysical Union, San Francisco, 1989.

Espenshade, Edward B., Jr., Editor, *Goode's World Atlas*, Rand McNalley, 1983.

Fairbridge, Rhodes W., *The Encyclopedia of Geomorphology*, Reinhold Book Corporation, 1968.

Fakhry, Ahmed, *The Oases of Egypt, Vol 1, Siwa Oasis*, The American University in Cairo Press, 1973.

Fenneman, Nevin M., *Physical Divisions (map of USA)*, U.S. Geological Survey, 1928.

Fiero, Bill, *Great Basin - Geology of the Great Basin*, University of Nevada Press, 1985.

Fisher, Robert L. and Roger Revelle, *The Trenches of the Pacific*, Scientific American, Vol 193 No 5, November 1955. Reprinted in Scientific American Resource Library, Readings in the Earth Sciences, Vol 1, W.H. Greeman, 1969.

Fisk, Harold N., *Mississippi Valley Geology - Relation to River Regime*, Paper No. 2511 in Stanley A. Schumm, Editor, River Morphology, American Society of Civil Engineers, 1952.

Florentin, J., M.R. Maurrasse, and Gautam Sen, *Impacts, Tsunamis, and the Haitian Cretaceous-Tertiary boundary layer*, Science Vol 252, 21 June 1991.

Ford, J.F., R.K. Dokka, R.E. Cripper, and R.G. Blom, *Faults in the Mojave Desert, California as Revealed on enhanced Landsat images*, Science, Vol 248, 25 May 1990.

Fremlin, Gerald, Editor in Chief, *The National Atlas of Canada*, Fourth Edition, Macmillan Company of Canada Limited, 1974.

French, B.M., *Shock Metamorphic Features in the Sudbury Structure, Ontario: A Review*, Geological Association of Canada Special Paper 10, 1972.

Frey, F.A. and D. Weiss, *The Ninety East Ridge in the Indian Ocean: A 5000 km Hotspot Trace With a DUPAL Isotopic Signature*, Paper V12C-09, American Geophysical Union, San Francisco, December 4-8, 1989.

Galtsoff, Paul, Coordinator, *Gulf of Mexico, It's Origin, Waters, and Marine Life*, Fishery Bulletin 89, Fish and Wildlife Service Vol 55, U.S. Government Printing Office, 1954.

Ganapathy, R., *Evidence for a Major Meteorite Impact on the Earth 34 Million Years Ago: Implication on the Origin of North American Tektites and Eocene Extinction*, Special Paper 190, Geological Society of America, 1982.

Garrett, Wilbur E., Editor, *Earth's Dynamic Crust (map)*, National Geographic Magazine, August, 1985.

Gee, Henry, *At full tilt (changing earth's obliquity)*, Nature News Service, October 24, 2001. http://www.nature.com/nsu/981210/981210-1.html

Gerlach, Arch C. (Ed.), *The National Atlas of the United States*, U.S. Department of Interior, 1970.

Girardeau, Jacques, et al, *Evidence for a Heterogeneous Upper Mantle in the Cabo Ortegal Complex, Spain*, Science Vol 245, 15 September 1989.

Girdler, R.W., *Processes of Planetary Rifting as seen in the Rifting and Breakup of Africa,* in P. Morgan and B.H. Baker, Editors, Processes of Continental Rifting Special Issue, Tectonophysics, International Journal of Geotectonics and the Geology and Physics of the Interior of the Earth, Vol 94 No 1-4, May 1, 1983.

Glasstone, Samuel, Editor, *The Effects of Nuclear Weapons,* United States Atomic Energy Commission, 1957.

Glen, William, *What Killed the Dinosaurs?,* American Scientist, Vol 78, July-August, 1990.

Global Catastrophes in Early History - An Interdisciplinary Conference of Impacts, Volcanism, and Mass Mortality, (Snowbird II), Snowbird, Utah, October 20-23, 1988.

Global Maps of historical earthquakes, National Earthquake Information Center, U.S. Geological Survey, September 20, 2000.

Goldsmith, Donald, *Nemesis - The Death-Star and other Theories of Mass Extinction,* Walker and Company, 1985.

Goldsmith, Donald, *Scientists Confront Velikovsky,* Cornell University Press, 1977.

Gorshov, Sergei G., Editor, *World Ocean Atlas, Vol 1, Pacific Ocean,* Pergamon Press, 1976.

Grady, James, *Photo-Atlas of the United States, First Compete Photographic Atlas of the U.S. Using Satellite Photography,* Ward Ritchie Press, 1975.

Greene, Owen, Ian Percival, and Irene Ridge, *Nuclear Winter - The Evidence and Risks,* Polity Press, 1985.

Grieve, R.A.F. and P.B. Robinson, *The Terrestrial Cratering Record - 1. Current Status of Observations,* Icarus, Vol 38, 1979.

Grieve, R.A.F., *Impact Cratering on the Earth,* Scientific American, April 1990.

Grieve, R.A.F., *The Record of Impact on Earth: Implications for a Major Cretaceous/Tertiary Impact Event,* Special Paper 190, Geological Society of America, 1981. (See entry under Silver.)

Grigg, Richard W. and David Epp, *Critical Depth for the Survival of Coral Islands: Effects on the Hawaiian Archipelago,* Science, Vol 243, 3 February 1989.

Grinspoon, David Harry, *Venus Revealed: a new look below the clouds of our mysterious twin planet,* Addison Wesley Publishing Company, 1997.

Grosvenor, M.B., Editor, *Eastern Soviet Union,* Atlas Plate 46, National Geographic, March 1967.

Grosvenor, M.B., Editor, *Atlantic Ocean (map),* National Geographic Society, Atlas Plate 62, June 1968.

Grosvenor, M.B., Editor, *Pacific Ocean Floor (map),* National Geographic Society, October 1969.

Grosvenor, M.B., Editor, *The World (map),* National Geographic Society, December 1970.

Grosvenor, M.B., Editor, *West Indies and Central America (map),* National Geographic Society, January, 1970.

Grunow. A.M., *Creation and destruction of Weddell Sea floor in the Jurassic,* Geology, vol 21 July 1993.

Guennoc, Pol, Georges Point, and Zohaia Nawab, *The Red Sea: History and Associated Mineralization,* Chapter 39, Part B in Manspeizer.

Gutschick, Raymond C., *Geology of the Kentland Structural Anomaly, Northwestern Indiana,* Field Guide for Earth Science Students, University of Notre Dame, presented at the 10th Annual Meeting, North Central Section, Geological Society of America, Western Michigan University, April 1976.

Guy-Bray, J.V. and Geological Staff, International Co., *Copper Cliff: Shatter Cones at Sudbury*, Journal of Geology 74: 1966.

Hack, J.T., *Interpretation of the Cumberland Escarpment and Highland Rim, South-central Tennessee and Northeast Alabama*, Plate 1, 1:24,000, U.S. Geological Survey Professional Paper 524-C.1966.

Hamil, Ralph, *Exploring Phobos*, The Futurist, June 1985.

Hance, William A., *The Geography of Modern Africa*, Second Edition, Columbia University Press, 1975.

Hancock, Lee, *Mid-America quakes over prediction*, Dallas Morning News reprinted in San Francisco Examiner, 5 August 1990.

Hardaker, Terry, Cartographic Editor, *Peters Atlas of the World, A Revolutionary new view of Earth - showing it in true proportions for the First Time*, Harper & Row, 1990.

Harrison, Richard Edes, *Shaded Relief* (map of United States), 1:7,00,000, U.S. Geological Survey, 1969.

Hassan, L.Y., Mineral occurrences and exploration potential of the East Kimberley, Report 74.
http://www.dme.wa.gov.au/geology/hot_data/report74.html>

Hawkesworth, C.J. et al, *U-Th Isotopes in arc mamas: Implications for element transfer from the subducted crust*, Science Vol 276, 25 April 1997.

Hayes, D.E., *The Tectonic and Geologic Evolution of Southeast Asian Seas, Part 1*, Geophysical Monograph Series, Vol 25, American Geophysical Union, 1980.

Head, J.W., *Origin of the Rings in Lunar Multi-Ringed Basins, Symposium On Planetary Cratering Mechanics*, Flagstaff, Arizona, 13-17 September, Lunar Science Institute, 1976.

Herbert, Sandra, *Darwin As a Geologist*, Scientific American, Vol 254 No 5, May 1986.

Hess, Wilmot N., Donald H. Menzel, John A. O'Keefe, Editors, *The Nature of the Lunar Surface*, Proceedings of the 1965 IAU-NASA Symposium, April 15-16, 1965, Johns Hopkins Press, 1966.

Hildebrand, Alan R. and William V. Boynton, *Proximal Cretaceous-Tertiary boundary impact deposits in the Caribbean*, Science, 18 May 1990.

Hinds, Norman E.A., Editor, *Evolution of the California Landscape*, Bulletin 158, Division of Mines, State of California, 1952.

Hodel, Donald Paul, Secretary of Interior, and Dallas L. Peck, Director U.S. Geological Survey, *National Energy Resource Issue, Geologic Perspective and the Role of Geologic Information*, U.S.G.S. Bulletin 1850, 1988.

Hodges, Kip, *Reconnaissance in Tibet, Review of the Geological Evolution of Tibet*, Science Vol 244, 9 June 1989.

Hoffman, P.F., *Speculations on Laurentia's First Gigayear, (2.0 to 1.0 Ga)*, Geology 17, 135, 1989.

Hooper, P.R., *The Timing of crustal extension and the eruption of Continental Flood Basalts* (Flow-by-flow mapping of Columbia River and Deccan basalt groups), Nature Vol 345 Issue 6272, May 7, 1990.

Hooper, Peter R., *The Columbia River flood basalt province: Current status*, in Mahoney 1997.

Hope Alive in India rubble (Bhuj), Chronicle News Service, San Francisco Chronicle, January 30, 2001.

Hoy, Steve, *Chicxulub Crater*, U.S. National Report to IUGG, Rev. Geophys, Vol 33 Supplement, American Geophysical Union, 1995.

http://www.man.acuk/Geology/special/

Huene, Friederich von, *A new Phytosaur from the Palisades near Near York*, Article XV, American Museum of Natural History, Novitates, 1913.
http://www.njpalisades.org/cnmar99.htm

Hurley, Patrick M., *The Confirmation of continental Drift*, Scientific American Vol 218 No 4, April 1968.

Huxley, Anthony, Editor, *Standard Encyclopedia of Mountains*, G.P. Putnam's Sons, 1962.

Ihinger, Pphillip D., *Mantle flow beneath the Pacific plate: Evidence from seamount segments in the Hawaiian-Emperor...*, American Journal of Science Vol 296 Issue 9, November 1995.

Illies, J.H. and St. Mueller, Editors, *Graben Problems*, Proceedings of an International Rift Symposium, Karlsruhe, October 10-12, 1968, E. Schweizerbar'sche Verlags-bachhandlung, 1970.

Internet Geology, (106 news letters on petroleum in countries of former U.S.S.R.)
http://geocities.com/internetgeology/

Isbell, Douglas and Mary Hardin, *Chain of Impact Craters Suggested by spaceborne radar images*, March 20, 1996.
http://www.jpl.nasa.gov/s19/news80.html

Jablonski, David, *Extinctions: A paleontological perspective*, Science Vol 253, 16 August 1991.

Jansen, Ramond E., *The History of a River*, Scientific American Vol 186 No 6, 1952 reprinted in Scientific American Resource Library, Earth Sciences Vol 1, W.H. Freeman, 1969.

Jha, V.C., *Himalayan Geomorphology: Study of Himalayan Ramganga Basin*, Department of Geography, \Vishva Bharti University, India.

Judge, Joseph, *Child of Gondwana*, National Geographic, February 1988.

Kallergia, G., Editor, *Proceedings of the VI Colloquium on the Geology of the Aegean Region*, Institute of Geological and Mining Research, Athens, 1977.

Kaula, William M., Venus: *A Contrast in Evolution to Earth*, Science, 9 March 1990.

Keller, G.R., E.G. Lidiak, W.J. Hinze, and L.W. Braile, *The Role of Rifting in the Tectonic Development of the Midcontinent U.S.A.,* Tectonophysics, Vol 94, 1983.

Kelley, Kevin W., *The Home Planet*, Addison-Wesley Publishing Company, 1988.

Kerr, R.A. *Good News for Volcano Watchers*, Science vol 245, 21 July 1989.

Kerr, R.A., *Another Movement in the Dance of the Plates*, Science, Vol 245, 5 May 1989.

Kerr, R.A., *Commotion Over Caribbean Impacts*, Science, Vol 250, 23 November 1990.

Kerr, R.A., *Huge Impact is Favored K-T Boundary Killer*, Science, 11 November 1988. Review of Global Catastrophes in Earth History: An Interdisciplinary Conference on Impacts, Volcanism, and Mass Mortality, Snowbird, Utah, 20-23 October 1988.

Kerr, R.A., *Impact-Geomagnetic Reversal Link Rejected*, Science, Vol 247, 23 February 1990.

Kerr, R.A., *Puzzling Out the Tectonic Plates*, Science, Vol 249, 16 February 1990.

Kerr, R.A., *Snowbird II: Clues to Earth's Impact History*, Science, Vol 242, 9 December 1981.

Kerr, Richard A., *Did Volcanoes drive ancient Extinctions?*, Science, 16 December 2001.
http://www.sciencemag.org/cgi/content/full/5482/1130

Kerr, Richard A., *Dinosaurs and friends snuffed Out?,* Science, Vol 251, 11 January 1991.

Kerr, Richard A., *Dinosaurs' death blow in the Caribbean Sea?*, Science, 18 May 1996.

Kerr, Richard A., *From Earth's Core to African oil*, Science Vol 294, 12 October 2001.

Kerr, Richard A., *Huge Impact tied to Mass Extinction*, Research News, Science Vol 257, 14 August 1992.

Kerr, Richard A., *New way to hit the hot spot hints at a complex Pacific,* Research News, Science Vol 276 Issue 5316, 23 May 1997.

Kerr, Richard A., *Paring down the Big Five mass extinctions*, Science Vol 294, 7 December 2001.

Kerr, Richard A., *When a radical experiment goes bust* (drilling an impact crater for oil), Research News, Science Vol 247, 9 March 1990.

King, B.C., *The Baikal Rift*, Geological Society London Journal, Vol 132:348-349, 1976.

King, Philip B., *Tectonic Features* (map), U.S. Geological Survey, 1967.

King, Scott D. and Jeroen Ritsema, *African Hot Spot Volcanism: Small-scale convection in the upper mantle beneath cratons*, Science, Vol 290, 10 November 2000.

Knoles, D.B., H.L. Reade, Jr., and J.C. Scott, *Geology and Ground Water Resource of Montgomery County, Alabama*, U.S. Geological Survey Water Supply Paper 1606, Plaate 1, 1:125,000, 1963.

Koeberl, Christian and Virgil L. Sharpton, *Terrestrial Impact Craters*. http://www.lpi.usra.edu/pub/publications/slidesets/impacts.html

Kopal, Zdenck, *New Photographic Atlas of the Moon*, Taplinger Publishing Company, 1971.

Kraft, Maurice and Katia Kraft, *Volcano*, translated from Icelandic by John Sheply, Harry N. Abrams, Inc. Publisher, 1975.

Kresak L., *Comet discoveries, statistics and observational selection*, in Wilkeing 1982.

Krinov, E.L., *Giant Meteorites*, translated from Russian by J.S. Romankiewiez, Pergamon Press, 1966.

Krinov, E.L., *Meteorite Craters on the Earth's Surface*, Chapter 7 in B.M. Middlehurst and G.P. Kuiper, The Moon, Meteorites, and Comets, University of Chicago Press, 1963.

Kuhn, Thomas S., *The Structure of Scientific Revolutions*, Second Edition, Volumes I and II, Foundations of the Unit of Science, University of Chicago, 1970.

Kulk, J. Lawrence, *Geological Time Scale,* McGraw-Hill Encyclopedia of Science and Technology, Vol 6, p 147, 1977.

Kumar, Senthill, et al, *Earthquake Recurrence and Rupture Dynamics of Himalayan Frontal Thrust, India*, Science Vol 294, 14 December 2001.

Kunk, M.J., G.A. Izett, R.A. Haugerud, and J.F. Sutter, *40 AR / 39 AR Dating of the Manson Impact Structure: A Cretaceous-Tertiary Boundary Crater Candidate*, Science, vol 244, 30 June 1989.

Kunzig, Robert, *All Is Flux, Nothing, including Hot Spots in the Earth's Mantle Stay Still*, review of an article by Peter Molnar and Joann Stock, Scientific American, September 1987.

Kunzig, Robert, *The Seafloor from Space*, Discover Vol 17 Issue 3, March 1996. (Available at EBSCOhost).

Kyte, Frank T., Lei Zhou, and John T. Watson, *New Evidence on the size and possible effects of a Late Pliocene Oceanic Asteroid Impact*, Science Vol 241, 1 July 1988.

Lacey, Marc, *Eruptions put Congo in chaos - Hundreds of thousands of refugees scrounging for food*, New York Times, January 20, 2002. (Reprint at www.sfgate.com).

Landes, Kenneth K., *Petroleum Geology of the United States*, Wiley-Interscience, 1970.

Larmer, Brook (in Lima), *The Rain Forest at Risk: Searching for oil in 'The womb of the world' (Andean Amazon)*, Newsweek, August 12, 1991.

Larson, Roger L., *The Mid-Cretaceous superplume episode,* Scientific American Vol 272 Issue 2, February 1995. (Available at EBSCOhost).

Laskar, J. and P. Robutel, *The chaotic Obliquity of the Planets. (Mars in chaotic region, 0° to 60°; Mercury and Venus stabilized by tides; earth may have been stabilized by capture of Moon),* Nature Vol 361, 18 February 1993.

Le Maire, T.R., *Stones from the Stars; the Unsolved Mysteries of Meteorites,* Prentice-Hall, Inc., 1980.

Le Masurier, W.E. and J.W. Thomson, *Volcanoes of the Antarctic Plate and Southern Oceans,* American Geophysicl Union, date unknown.

Lee, Der-Chuen, Alex N. Halliday, Gregory A. Snyder, and Lawrence A. Taylor, *Age and Origin of the Moon,* (Twenty one lunar samples were found to be 4.51 billion years old by radioactive decay of hafnium. Time placed in orbit not addressed.) Science Vol 278 No 5340, 7 November 1997.

Lee, Douglas B., photographs by Frans Lanting, *Okavango Delta: Africa's Last Refuge,* National Geographic, December 1990.

Leonardi, Piere, *Volcanoes and Impact Craters on the Moon and Mars,* Elsevier Scientific Publications Company, 1976.

Leonick, Michael D., *Whew! That was Close: Earth's Narrowest Escape from an Asteroid in 52 Years,* Time, 1 May 1989.

Levorsen, A.I., *Geology of Petroleum,* W.H. Freeman and Company, 1967.

Lewin, Roger, *The Case of the Misplaced Fossils,* Science, vol 244, 21 April 1989.

Lewis, John S., *Rain of Iron and Ice: The very real threat of Comet and Asteroid Bombardment,* Addison-Wesley, 1996.

Lightfoot, Peter C. and Chris J. Hawkesworth, *Flood Basalts and Magmatic Ni, Cu, and PGE Sulphide Mineralization: Comparative Geochemistry of the Norl'sk (Siberian Traps) and West Greenland Sequences* in Mahony, 1997.

Lin, Auning, et al, *Co-Seismic Strike-Slip and rupture length produced by the 2001 MS 8.1 Central Kunlun Earthquake,* Science Vol 296, 14 June 2002.

Lipman, Peter W., David A. Claque, James G. Moore, and Robin T. Holcombe. *South Arch volcanic field: newly identified young lava flows on the sea floor south of the Hawaiian Ridge,* Geology Vol 17 Issue 7, 1989.

Lisovsky, Nickolie, G.N. Gogonenkov, and Yuri A. Petzoukha, *Soviet Union's Tengiz field: a Pre-Caspian depression giant oil, gas accumulation,* Oil & Gas Journal, September 17, 1990.

Lissaner, Jack J., It's *not easy to make the Moon,* Nature 389, 1997.

Longwell, Chester R. and Richard F. Flint, *Introduction to Physical Geology,* 2d Edition, John Wiley & Sons, 1965.

MacDonaald, Gordon A., *Volcanoes,* Prentice-Hall, 1972.

MacDonald, Gordon A. and Agatin T. Abbott, *Volcanoes in the Sea, the Geology of Hawaii,* University of Hawaii Press, date unknown.

Macdonald, Gordon A. and Agatin T. Abbott, *Volcanoes in the Sea, the Geology of Hawaii,* University of Hawaii Press, date unknown.

MacDonald, Gordon A. and Will Kyselka, *Anatomy of an Island - A Geological History of Oahu,* Bernice P. Bishop Museum Special Publication 55, Bishop Museum Press, Honolulu, Hawaii, 1967.

Macneil, John S., *Mass Extinction Struck Suddenly,* Science Now, July 20, 2000.

Madsen, J., *Tectonic History of the Ocean Basins,* December 30, 2001. http://www.geology.udel.edu/~jmadson/tectonic-history.html

Magellan Reveals Venus, Astronomy, Vol 23 Issue 2, February 1995.

Magris, Claudio, *Danube,* Farrar Straus Giroux, 1989.

Mahony, John J. and Millard F. Coffin, Editors, *Large Igneous Provinces, Continental, Oceanic, and Planetary Flood Volcanism*, Geophysical Monograph 100, American Geophysical Union, 1997.

Major Earthquakes around the world,
(2001) http://www.infoplease.com/ipa/A0872591.html
(2000) <...A087157...>
(1999) <...A0778116...>

Mammerick, *Bathymetry of the North Pacific Ocean*, Plate 1A, Bathymetry, the Eastern Pacific Ocean and Hawaii, Vol n of the Geology of North America, Geological Society of America, 1989.

Mann, Paul and Ingo Pecher, et al, *Crust mantle interactions during Continental Growth and high pressure rock exhumation at an oblique arc continent collision zone: SE Caribbean Margin*, Institute For Physics, University of Texas at Austin, 22 June 2001. http://www.ig.utexas.edu/research/projects/venez_margin.htm

Maps (software, 8 discs), National Geographic Interactive and Broderbund, 1999.

Mark, Kathleen, *Meteorite Craters*, The University of Arizona Press, 1987.

Martin, Glen, Chronicle Staff Writer, *Oil to flow from island wilderness, (Papua New Guinea,* San Francisco Chronicle, May 9, 1992.

Marzoli, Andrea, et al, *Extensive 200-million-year-old Continental Flood Basalts of the Central Atlantic Magmatic Province* (CAMP), Science Vol 284.

Masaaytis, V.L., *Astroblemes in the USSR*, International Geology Review 18:1249-1258, 1976. Reprinted in McCall.

Mason, Brian and William G. Melson, *The Lunar Rocks*, Wiley-Interscience, 1970.

Masters, C.D., D.H. Root, and E.D. Attanasi, *Resource Constraints in Petroleum Production Potential*, Science Vol 253, 12 July 1991.

Matthes, Gerard H., *Paradoxes of the Mississippi*, Scientific American, Vol 184 No 4, April 1951. Reprinted in Scientific American Resource Library Readings on Earth Sciences, Vol 1, W.H. Freeman, 1969.

Matthews, J.L., et al, *Cretaceous Drowning of reefs on Mid-Pacific and Japanese Guyots*, Science Vol 184, 26 April 1974.

Maugh, Thomas N. II, *New Evidence that huge comet hit Earth*, Los Angeles Times, date unknown.

McCammon, Catherine, *Deep Diamond Mysteries*, Science, Vol 293, 3 August 2001.

McConahay, Mary Jo, Chronicle Foreign Service, *Dreams of a Guaatemala oil boom*, San Francisco Chronicle, October 12, 1990.

McConnel, R.B., *A Precambrian Origin for the Proto-Rift Dislocation Belt of Eastern Africa*, in Illies, 19780.

McConnel, R.B., *The Evolution of the Rift Systems in Africa in Light of Wegmann's Concept of Tectonic Levels*, in Illies, 1970.

McHone, John F., Ronald A. Nieman, Charles F. Lewis, and Ann M. Yates, *Stishovite at the Cretaceous-Tertiary Boundary*, Raton, New Mexico, Science Vol 243, 3 March 1989.

McLean, Dewey, *The Asteroid Impact vs Volcano Greenhouse Dinosaur Extinction Debate*, July 26, 2001. http://filebox.vt.edu/artsci/geology/mclean/Dinosaur_Volcano_Extinction/

McNutt, Marcie K. and Anne V. Judge, *The Superswell and Mantle Dynamics beneath the South Pacific*, Science, 25 May 1990.

Meier, Roland, et al, *A Determination of the HDO/H_2O Ratio in Comet C/1995 01(Hale-Bopp),* Science Vol 279 No 842, 1998.

Meier, Roland, et al, Deuterium in Comet c/1995 01 (Hale-Bopp): Detection of DCN, Science Vol 279 No 5357.

Melosh, H.J., *Impact Cratering, A Geological Process*, Oxford University Press, 1989.

Melosh, Jay, *Planetary science: A new model Moon*, Nature 412, 2001.

Menard, H.W., *Geology of the Pacific Sea Floor*, Experienta, 15:205-213, 1959.

Menard, H.W., *Islands*, Scientific American Library, 1986.

Menard, H.W., *Marine Geology of the Pacific*, McGraw-Hill, 1964.

Miller, Peter, paintings by William H. Bond, *The March toward Extinctions*, Supplement, National Geographic Society, June 1989.

Mintz, Leigh W., *Historical Geology - the science of a dynamic Earth*, Charles E. Merril, 1972.

Mitton, S., *High-altitude explosions caused odd craters on Venus*, New Scientist, Vol 134 Issue 1815, 4 March 1992.

Modeling the Moon's origin, Science News Vol 152 Issue 7, October 16, 1997 from annual meeting of Division for Planetary Science, American Astronomical Society, Cambridge, Massachusetts. (Abstract at EBSCOhost).

Molnar, P., P. Tapponier, and W.P. Chen, *Extensional Tectonics in Central and Eastern Asia, A Brief Summary*, Royal Society London Philos. Trans. A300:403-406, 1981.

Monastersky, R., *Eruptions Cleared Path for dinosaurs*, Science News Online, Vol 155, No 17, April 24, 1999.
http://www.sciencenews.org/sn_arc99/4_24_99/fob1.htm

Monastersky, Richard, Lessons and Question from Armenian quake, Science News, Vol 135, January 21, 1989.

Moore, Patrick and Peter J. Cattermole, *The Craters of the Moon, An Observational Approach*, W.W. Norton & Company, 1967.

Morell, Virginia, *How Lethal was the K-T Impact?*, Science Vol 261, 17 September 1993.

Morgan, P. and B.H. Baker, Editors, *Processes of Continental Rifting*, Special Issue, Tectonophysics, International Journal of Geotectonics and the Geology and Physics of the Interior of the Earth, Vol 94 No 1-4, May 1, 1983.

Morgan, W. Jasen, *Hotspot Tracks and the Early Rifting of the Atlantic*, in Morgan 1983.

Morrison, David and Clark R. Chapman, *Target Earth: It will Happen*, Sky & Telescope, March 1990.

Mullac, R. Dietmar, Walter R. Roest, and Jean-Yves Royer, *Asymmetric sea-floor spreading caused by ridge-plume interactions*, Nature, Vol 396, 3 December 1998.

Muller, Richard, *Nemesis—The Death Star-The Story of a Scientific Revolution*, Weidenfeld & Nicolson, 1988.

Murphy, Kim, *Major quake hits Egypt - Hundreds Dead*, Los Angeles Times, October 13, 1992.

Myth versus Mathematics; Velikovsky's Theory (Ancient records show a rapid roll of earth 1500 yr BC. Strong argument but controversial).
http://www.dreamscape.com/morgana/lysith2.htm>

Nalivikin, Vladimir Dmitrievich, *Graben-like Trenches in the East of the Russian Platform*, translated from Sovietskaya Geologico, 1963 by E.R. Hope, Directorate of Scientific Information Services, DRB - Canada, 1965.

Nash, J. Madeleine, *When Life Nearly Died: Researchers link earth's greatest mass extinction to Siberian volcanoes that erupted for a million years*, Time Magazine Vol 146 No 12, September 18, 1995.

National Park Service, *Yukon-Charley Rivers National Preserve, Park Geology*, May 18, 2001.
http://www.aqd.nps.gov/grd/parks/yuch/

Natural Fission Reactors, Proceedings of a Meeting of the Technical Committee, International Atomic Energy Agency, Paris, 19-21 December 1977.

Neal, James T., Editor, *Tectonic Landforms*, New Britannica, Vol 16 p 768, 1987.

Nemiroff, Robert and Jerry Bounell, *Astronomy Picture of the day, South Pole-Aitken Basin*, LHEA at NASA/GSFC, September 6, 1996.

Nininger, H.H., *Meteorite Distribution on the Earth*, Chapter 6 in Barbara M. and Gerard P. Kuiper, Editors, The Moon, Meteorites, and Comets, University of Chicago Press, 1963.

Norman, John, et al, *Astrons-The Earth's Oldest Scars?*, New Scientist, 73:689-692, 1977, reviewed by William R. Corliss in Unknown Earth: A Handbook of Geological Enigmas, The Sourcebook Project, 1980.

Norman, Sleep, *Earth Sciences: The puzzle of the South Pacific - try make sense of islands*, Nature 389, 439-440, 2 October 1997.

Normark, W.R., *Delineation of the main extrusion zone of the East Pacific Rise at lat 21 degrees N*, Geology Vol 4 Issue 11, 1976.

Olsen, P.E., et al, Ascent of Dinosaurs linked to an Iridium anomaly at the Triassic-Jurassic boundary, Science Vol 296.

Olsen, Paul E., *Enhanced: Giant Lava Flows, Mass Extinctions, and mantle plumes, Central Atlantic Magmatic Province* (CAMP) may be world's largest, Science Vol 284 Issue 5414, 2 September 2002.

Ostro, Steven J., et al, *Radar Images of Asteroid 1989 PB*, Science, Vol 248, 22 June 1990.

Palfy, Jozef, et al, *Carbon isotope anomaly and other geochemical changes at the Triassic-Juriassic boundary from a marine section in Hungary*, Geology Vol 29 No 11, 31 October 2001.

Parks, Michael, *30 Dead in quake: jolt of at least 7.0 strikes Soviet Georgia*, Los Angeles Times, April 30, 1991.

Parks, Noreen, *Exploring Loihi: the next Hawaiian island*, Earth, Vol 3 Issue 5, September 1994. (Availabale at EBSCOhost.)

Peltzer, Gilles, Paul Takponnier, and Roland Armijo, *Magnitude of Late Quaternary Left-Lateral Displacements aong the north edge of Tibet*, Science, Vol 249, 8 December 1989.

Penick, James Lal, Jr., *The New Madrid Earthquakes*, Revised Edition, University of Missouri Press, 1981.

Perkins, S., *Extinctions tied to Impact from Space*, Science News, February 24, 2001.
http://ehostvgw3.epnet.com/fulltext.asp?resultSet
Id=R00000005&hitNum=1&booleanTerm=1...

Perkins, Sid, *Presto, Change-O!: Extraterrestrial impacts transform Earth's surface in an instant*, Science News, Vol 161, June 15, 2002.

Perlman, David, Chronicle Science Editor, *Volcanoes may have stolen dinosausr' oxygen supply*, San Francisco Chronicle, October 28, 1993.

Perlman, David, Science Editor, *Big Midwest Quake Expected*, San Francisco Chronicle, February 19, 1990.

Perlman, David, Science Editor, *Experts Issue Dire Statistics*, San Francisco Chronicle, December 9, 1989.

Perlman, David, Science Editor, *Space Object that killed Dinosaurs broke through Earth's Crust*, San Francisco Chronicle, December 18, 2000.

Phillips, Roger J., et al, *Impact Craters on Venus: Initial Analysis from Magellan*, Science, Vol 252, 12 April 1991.

Physiographic Regions of Canada, map 1:5,000,000, Geological Survey of Canada, 1970.

Pittman, Walter C. III, *Plate Tectonics*, Encyclopedia of Science and Technology, McGraw-Hill, 1987.

Polya, George, *How to Solve It*, Princeton University Press, 1971.

Ponte, Lowell, *When Asteroids Shake the Earth*, Reader's Digest (Canadian), August 1989.

Powell, James Lawrence, *Night comes to the Cretaceous, dinosaur Extinction and the Transformation of modern geology*, W.H. Freeman and Company, date unknown.

Probe may have found Impact Ocean, New York Times, republished in San Francisco Chronicle, 10 August 1990.

R., J.D., *Impact Cratering on Earth*, September 1998.
http://gaci/no.agg.e??.ca.crater/paper/cratering_e.html

Rand McNally, *Encyclopedia of Rivers*, 1980.

Raup, David M. and J. John Sepkowski, *Periodicity of Extinctions in the Geologic Past*, Proceedings of the National Academy of Sciences, Vol 81, February 1984.

Raup, David M., *Biogeographic Extinction: A Feasibility Test*, in Silver, 1982.

Raup, David M., The Nemesis Affair, *A Story of the Death of Dinosaurs and the Ways of Science*, Norton, 1986.

Renne, Paul R. and Asish R. Basu, *Rapid eruption of the Siberian Traps flood basalts at the Permo-Triassic boundary*, Science Vol 253, 12 July 1991.

Renne, Paul R., et al, *The Age of Parana flood volcanism, rifting of Godwanaland, and the Jurassic-Cretaceous boundary*, Science Vol 258, 6 November 1992.

Renne, Paul R., *Flood Basalts-bigger and badder*, Science Vol 296, 7 June 2002.

Researchers: Asteroid that killed dinosaurs may have struck Cuba, Palo Alto Times, 20 May 1990.

Rhodes, J.M. and John R. Lockwood, Editors, *Mauna Loa Revealed: Structure, Composition, History, and Hazards*, Geophysical Monograph 92, American Geophysical Union, date unknown.

Rich, Thomas H., Patricia Vickers-Rich, and Roland A. Gangloff, *Polar Dinosaurs*, Science Vol 295, 8 February 2002.

Richards, Mark A., Richard G. Gordon, and Rob D. van der Hilst, *The History and Dynamics of Global Plate Motions*, Preface, On Line Book Catalog, American Geophysical Union, March 15, 2001.

Richards, Mark A., Robert A. Duncan, and Vincent E. Courtillot, *Flood Basalts and Hot-Spot Tracks: Plume Heads and Tails*, Science, Vol 246, October 1989.

Richardson, Robert S., *Getting Acquainted with Comets,* McGraw-Hill, 1967.

Richter, C.F., *Elementary Seismology*, Freedman, 1958, McGraw-Hill Encyclopedia of Science and Technology, Vol 6, p 201, 1977.

Rifting in West Siberia, Internet Geology News Letter No. 41, April 17, 2000.
http://geocities.com/internetgeology/L41.html

Robinson, H.E.C., *Robinson's Pacific Ocean*, Map No.1804, H.E.C. Robinson Pty: Ltd, Melbourne, undated.

Roddy, D.J., R.O. Pepin, R.B. Merrill, Editors, *Impact and Explosion Cratering, Planetary and Terrestrial Implications*, Proceeding of the Symposium on Planetary Cratering Mechanics, Falstaff, Arizona, September 13-17, 1976.

Rood, Robert T., T.M. Bania, and Dana S. Balser, *The Saga of ^{3}He*, Science Vol 295, 1 February 2002.

Rubtsov, V.V., *Three New Books on the Tunguska problem* (in Russian), Research Institute on Anomalous Phenomena, Bulletin, Vol 7 No 1, Kharkov, Ukraine, January-March 2001.

Ruddiman, William F. and John E. Kutzlach, *Plateau uplift and climatic change*, Scientific American, March 1991.

Russian Classification of Hydrocarbon Reserves and Resources, Internet Geology News Letter No. 67, October 16, 2000.
http://geocities.com/internetgeology/L67a.html

Sagan, Carl and Ann Druyan, *Comet*, Random House, 1980.

Sagan, Carl and Richard Turco, *A Path where no Man Thought, Nuclear Winter and the end of the Arms Race*, Random House, 1990.

Sager, William W. and Anthony A.P. Koppers, *Late Cretaceous Polar Wander of the Pacific Plete: Evidence of a rapid true polar wander event*, Science Vol 287 Issue 5452, 21 January 2000.

Sandallioglu, Okan, *What is Petroleum?* Ortadogu Teknik Universtesi Petrol ve Dogalgaz Muhendislig I, Ankara, Turkey, 21 Jully 2001.
<http:onyx.msis.metu.edu.tr/~sandai/petroleum1.html>

Saul, John M., *Known circular structures of large scale and great age on the Earth's surface*, Nature, Vol 271, 26 January 1978.

Saunders, R. Stephen, *The Surface of Venus*, Scientific American, December 1990.

Schaber, Gerald G. and Joseph M. Boyce, Probable Distribution of Large Impact Basins on Venus, Comparison with Mercury and the Moon, in Roddy, 1976.

Schmidt, Robert M. and Keith A. Holsapple, *Estimate of Crater Size for Large-Body Impact: Gravity-Scaling Results*, Special Paper 190, Geological Society of America, 1981. In Silver 1981.

Schneider, David, *Hot-Spotting*, Scientific American Vol 276 Issue 4, April 1997. (Available at EBSCOhost).

Scholz, C.H., *Transform Fault System of California and New Zealand: Similarities in their Tectonic and Sesimic Styles*, Geological Society of London Journal, 133:215-229, 1977.

Schultz, Peter H., *Moon Morphology, Interpretations Based on Lunar Photographs*, University of Texas Press, 1976.

Schumm, Stanley A., Editor, *River Morphology, Benchmark Papers in Geology*, Dowden, Hutchinson & Ross, Inc., 1972.

Seargent, David A., *Comets: Vagabonds of Space*, Doubleday, 1982.

Segall, Paul and Mike Lisowski, *Surface Displacements in the 1906 San Francisco and 1989 Loma Prieta earthquakes*, Science Vol 250, 30 November 1990.

Sepkowski, J. John, Jr., *Mass Extinctions in the Phanerzoic Oceans: A Review,* Special Paper 190, Geological Society of America, 1982.

Seyfert, Carl K. and Leslie A. Sirkin, *Earth History and Plate Tectonics, An Introduction to Historical Geology*, Harper & Row, date unknown.

Seyfert, Carl K., *The Encyclopedia of Structural Geology and Plate Tectonics*, Van Nostrand Reinhold Company, 1987.

Shakleton, R.M., J.F. Dewey, and B.F. Windly, Editors, Tectonic *Evolution of the Himalayas and Tibet*, The Royal Society, London, 1988.

Sharpton, V.L. and P.D. Ward, Eds., *Global Catastrophies in Earth history: An interdisciplinary conference on impact, volcanism, and mass mortality*, Geol. Soc. Am. Special Paper, No. 247, (1990).

Sharpton, Virgil L., *Chicxulub Impact crater provides clues to Earth's history*, Eos, Vol 76, December 26, 1995.

Sharpton, Virgil L., Kevin Burke, Antonio Camargo-Zanoguero, et al, *Chicxulub Multiring Impact Basin: Size and other characteristics derived from Gravity Analysis*, Science Vol 261, 17 September 1993.

Shephard, Francis P., *Submarine Geology*, Second Edition, with Chapters by D.L. Inmen and E.D. Goldberg, Harper & Row, 1962.

Short, Nicholas M., Sr. and Robert W. Blair, *Geomorphology from Space*, National Aeronautics and Space Agency, 1986.

Short, Nicolas M., Paul D. Lowman, Jr., Stanley C. Freden and William A. Finch, Jr., *Mission to Earth: Landsat Views the World*, National Aeronautics and Space Administration, 1976.

Simkin, Tom, et al, *This Dynamic Planet, World Map of Volcanoes, Earthquakes, and Plate Tectonics*, Smithsonian Institution, U.S. Geological Survey, 1989.

Simkin, Tom, et al, *Volcanoes of the World. Regional Directory, Gazetteer, and Chronology during the last 10,000 Years*, Hutchin Ross Publications, 1981.

Sinnett, Roger W., *An Asteroid Whizzes Past Earth*, Sky & Telescope, July 1989.

Sitchin, Zecharia, *"Nemesis"—A New Idea as old as the Bible, Catastrophism and Ancient History*, A Journal of Interdisciplinary Study, Vol VII, Part 1, January 1985.

Sleep, Norman, *The Puzzle of the South Pacific,* Nature Vol 389 Issue 6650, October 2, 1997. (Available at EBSCOhost).

Smith, David G., Editor, *The Cambridge Encyclopedia of Earth Sciences*, Crown Publishers Inc, Cambridge University Press, 1982.

Solomon, Sean C., et al, *Venus Tectonics: Inititial Analysis from Magellan*, Science, 12 April 1991.

Space Age Portrait of a Continent, Atlas of North America, National Geographic Society, 1985.

Spencer, Charles W., *Describing Petroleum Reservoirs of the Future*, U.S. Geological Survey, Energy Resource Surveys Program, USGS Fact Sheet FS-020-97.

Spray, John G., Simon P. Kelley, and David B. Rowley, *Evidence for a late Triassic multiple impact event on Earth*, Nature, 392, Macmillan Publisher Ltd, December 17, 2001.
http://www.nature.com/cgi-taf/DynaPage.taf?file=/nature/journal/v392/n6672/abs/392171a0...

Stager, Curt and Chris Johns, *Africa's Great Rift, Is Africa Breaking Apart?*, National Geographic, Vol 177 No 5, May 1990.

Stanley, George D., Jr., Review of Goddwana and Tethys, Science, Vol 244, 30 June 1989.

Stanley, Steven M., *Extinctions—or, which way did they go?*, Astronomy, Special Issue, Vol 1 No 11, January 1991.

Stearns, Harold T., *Geology of the State of Hawaii*, Second Edition, Pacific Book Publishers, 1985.

Steckler, Michael and Gomma Omar, *Earth Scientists Discover How Continent Broke*, Science, 24 November 1995.

Steward, John, *Antarctica: An Encyclopedia* (2 vol), reviewed by Beth A. Clewis, Library Journal, January 1991.

Stone, Richard, *Caspian Ecology teeters on the brink*, Science Vol 295, 18 January 2002.

Stoneley, Jack with A.T. Lawton, *Cauldron of Hell: TUNGUSKA*, Simon and Schuster, 1977.

Stuart, Joseph Scott, *A Near-Earth asteroid Population estimate from the LINEAR Survey*, Science Vol 294, 23 November 2001.

Sugimura, A. and S. Uteda, *Island Arcs, Japan and Its Environs*, Developments in Geotectonics 3, Elsevier Scientific, date unknown.

Sullivan, Walter, *Continents in Motion, the New Earth Debate*, McGraaw-Hill, 1974.

Suppe, Frederick, Editor, *The Structure of Scientific Revolutions*, University of Illinois Press, 1974.

Suslov, S.P., *Physical Geography of Asiatic Russia*, translation by Noad D. Gershevsky, W.H. Freeman and Company, 1961.

Szep, Robert A., *When big things strike the Earth...The enigmatic Sudbury Star Wound*. http://www.paleoguy.com/sudburyinfo.htm

Tagle, R., *Concentration Patterns of Platinum Group elements and gold in the impact melt rocks*, Eos Trans. 81(48), Fall Meeting Supplement, Abstract P11A-07, 2000.

Talent, J.A., *The Case of the Peripatetic Fossils*, Natjure, 338, 613, 1989.

Tarling, D.H., *Gondwanaland and the Evolution of the Indian Ocean*, in M.G. Audley-Charles and A. Hallam, Editors, Gondwana and Tethys, Oxford University Press, 1988.

Taylor, Thomas N. and Edith L. Taylor, *Antarctic Paleobiology—Its Role in the Reconstruction of Gondwana*, from a workshop, Columbus, Ohio, June 1988.

The Geological Evolution of Tibet, Report of the 1975 Royal Society-Academia Sinica Geotrave of the Qinghai-Xizang Plateau, London, 1988.

The New Madrid Fault, 15 June 2002. http://www21.semo.edu/ces/CES2.HTML

The Ten Largest earthquakes of the 20th Century, Time.com, January 1, 2002. http://www.infoplease.com/ipa/A076403.html

Thomson, Kerr C., *Seismology*, McGraw-Hill Encyclopedia of Science and Technology, 1977.

Torrence, Mark, *Tectonic Motion on North America,* Ratheon Information Technology, 14 April 2000. http://cddisa.gsfc.nasa.gov/926/noamtect.html

Trawer, W. Peter, Editor, *Discovering Alvarez, Selected Works of Luis W. Alvarez with Commentary by his Students and Colleagues*, University of Chicago Press, 1987.

Trends in Water Resources Management, Great Manmade River Project, July 19, 2002. http://www.fao.org/waicent/faoinfo/agricult/aquastat/LIBYA.HTM

Turco, R.P., O.B. Toon, T.P. Ackerman, J.B. Pollack, and C. Saga, *Climate and Smoke: An Appraisal of Nuclear Winter*, Science, Vol 246 12 January 1990.

Turco, R.P., O.B. Toon, T.P. Ackerman, J.B. Pollack, and Carl Sagan, *Nuclear Winter: Global Consequences of Multiple Nuclear Explosions*, Science Vol 222 No 4630, 23 December 1993.

Udo, Reuben K., *Geographical Regions of Nigeria*, University of California, 1970.

United States Satellite View: Advanced very high resolution image, U.S. Geological Survey, 1990.

Unklesbay, A.G., *Midwest Earthquakes*, Earth Science, Winter 1987.

Urey, C. Harold, *Lunar Atlas*, Space Sciences Laboratory, North American Aviation, Inc., 1964.

Vajda, Vivi, J. Ian Raine, and Christopher J. Hollis, *Indication of Global Deforestation at the Cretaceous-Tertiary boundary by New Zealand Fern spike*, Science Vol 294, 25 November 2001. Buffler, Richard T., Gail Christeson, and Yosio Nakamura, *Structure of the Chicxulub KT Impact Crater, Yucatan, Gulf of Mexico*, 6/22/01.

http://www.ig.utexas.edu/research/projects/chix/chix.html

Van Andel, Tjeer H., *Plate Tectonics*, in Yearbook of Science and the Future, Encyclopedia Britannica, 1988.

Velikovsky, Immanuel, *My Challenge to Conventional Views in Science*, Kronos, A Journal of Interdisciplinary Synthesis, Velikovsky and Establishment Science, Vol III No 2, Winter 1977.

Vidale, John E., *Peeling back the layers in Earth's mantle*, Science Vol 294, 12 October 2001.

Volcano Hazards Program, *Synopsis of Historical Eruptions in the United States*, U.S. Geological Survey, October 1, 1998.
http://volcanoes.usgs.gov/volcanoes/Historical.html

Waldropo M. Mitchell, *After the Fall: although the Dust was bad, the Chemical Fallout from the Cretaceous-Teritary impact was worse—much worse*, Science, Vol 239, 26 February 1988.

Wang, Kun, *Glassy Microspherules (microtektites) from an Upper Devonian Limestone*, Science Vol 256, 12 June 1992.

Ward, Peter D., David R. Montgomery, and Roger Smith, *Altered River Morphology in South Africa Related to the Permian-Triassic Extinction*, Science, 8 September 2000.

Warrall, D.M. and S. Snelson, *Evolution of the northern Gulf of Mexico, with emphasis on Cenozoic growth faulting and the role of salt*, Chapter 7, The Geology of North America, Geological Society of America, 1989.

Wells, Stephen G., et al, Cosmogenic ^{3}He surface-exposure dating of stone pavements: Implications for landscape evolution in deserts, Geology Vol 23 No 7, October 31, 2001.

West, Richard, *Update on SL9/Jupiter Collision*, Jet Propulsion Laboratory, December 15, 1994.
http://www.jpl.nasa.gov/sl9/news49.html

Wetherhill, George W., *Apollo Objects*, Scientific American, Vol 240, March, 1979.

Wetherill, George W., *Occurrence of great Impacts during growth of the Terrestrial Planets*, Science Vol 228, 17 May 1985.

White, Robert S., *Ancient Floods of Fire*, Natural History, April 1991.

Wiens, Douglas A. and Nathaniel O. Snider, *Repeating deep earthquakes: Evidence for Fault Activation at Great Depth*, Science Vol 293, 24 August 2001.

Wilchert, U., et al, *Oxygen Isotopes and the Moon-Forming Giant Impact*, Science, Vol 294, 12 October 2001.

Wilkeing, Laurel L., Editor, *Comets, with 48 collaborating authors*, The University of Arizona Press, 1982.

Williams, Darren M. and James F. Kasting, *Earth stabilized by Moon*, Icarus, September 1997. Reviewed in News Notes, Thank The Moon you're here, Sky & Telescope, March 1998.

Williams, Darren M., James F. Kasting, and Lawrence A. Frakes, *Low-latitude glaciation and rapid changes in the Earth's obliquity explained by obliquity-oblateness feedback* (Chaotic obliquity from about 60° to 88° up to 500 My).

Williams, G.E., *History of the Earth's Obliquity (stabilized at ~ 26° about 430 My)*. Earth Science Reviews 34, 1993.

Williams, J.R., *The Energy of Things*, Air Force Special Weapons Center (ARDC), Kirkland Air Force Base, New Mexico, 1963.

Wilson, J. Tuzo, *Continental Drift*, Scientific American, Vol 208 No 4, April 1963.

Winterer, E.L., Donald M. Hussong, and Robert W. Decker, Editors, *The Eastern Pacific Ocean and Hawaii*, Geological Society of America, 1989.

World's Largest Crater?, News Notes, Sky & Telescope, April 1989.

Wortel, M.J.R. and S.A.P.L. Cloetingh, *On the Dynamis of Convergent Plate Boundaries and Stress in the Lithosphere*, in Wezel, F.C., Editor, The Origin of Arcs, Developments in Geotectonics International Conference, University of Urbino, September 22-25, Elsevier, 1986.

Wu, C., *Buckyballs Can Come from outer space*, Science News, April 8, 2001.

Xie, Shang-Ping, W. Timothy Liu, Quiyu Liu, and Masmi Nonake, *Far-reaching effects of the Hawaiian Islands on the Pacific Ocean atmosphere system*, Science Vol 292, 15 June 2001.

Yulsman, Tom, *The Seafloor laid bare; Top-secret data recently declassified by the Navy has enabled scientists to view the seafloor almost as if the oceans had been drained completely of water*, Earth Vol 5 Issue 3, June 1996. (Available at EBSCO).

Zephyr Services, *BEARINGS.EXE* (software), 1900 Murray Avenue, Pittsburgh PA 15217, 1988.

Zeylik, B.S. and E.Y. Segtmuratove, *Giant Meteorite Impact Structure in Central Kazakhastan and its Magma and Ore-controlling Significance*, Doklady Akadmic Nauk, U.S.S.R., Earth Sciences Sections, Vol 218, 1975.

Zich, Arthur, *Botswana (discove y of diamonds)*, National Geographic Vol 178 No 6, December 1990.

Zirbes, M., *Earthquakes with 1,000 or more deaths from 1900*, U.S.G.S. National Earthquake Information Center, November 30, 2000
http://neic.usgs.gov/neis/eqlists/eqmajr.html

The author resisted growth of this bibliography. After consulting a few thousand sources, the list was boiled down to about 20 per cent. The breadth of subject matter required ample clues to the readers where more information could be found. Each citation, therefore, is a doorway into libraries and web sites bursting with details. A lengthy bibliography appeared to be justified because it is the only one assembled on megacraters.

List Of Figures.

Untitle sketches in Catalog Of Megacraters following

List Of Tables.

About the Author

Mr. McCampbell earned a degree in Engineering Physics at the University of California, Berkeley. After graduate studies, his professional specialties were focused upon the effects of nuclear weapons and design of nuclear reactors. He was an early user of large computers for solving technical and economic problems. Later management assignments included a wide variety of projects. Examples are engineering designs for the Space Environment Simulator at the L.B. Johnson Manned Spacecraft Center at Houston and the Fast Flux Test Facility (plutonium fueled reactor) for the Atomic Energy Commission at Hanford, Washington. Other key responsibilities supported the Bay Area Rapid Transit System, Master Plans for Air Force research centers, economic development of third-world countries, and Environmental Protection for the Alaskan Pipeline.